The Patellofemoral Joint

AF386712

Jason L. Koh • Ryosuke Kuroda
João Espregueira-Mendes
Alberto Gobbi
Editors

The Patellofemoral Joint

A Case-Based Approach

Editors
Jason L. Koh
NorthShore Orthopaedic & Spine Institute
NorthShore University HealthSystem
Evanston, IL
USA

João Espregueira-Mendes
Clínica Espregueira - FIFA Medical Centre of Excellence
Porto, Portugal

Ryosuke Kuroda
Orthopaedic Surgery
Kobe University Hospital
Kobe, Japan

Alberto Gobbi
Bioresearch Foundation, Orthopaedic Arthroscopic Surgery Intl
Milano, Italy

ISBN 978-3-030-81547-9 ISBN 978-3-030-81545-5 (eBook)
https://doi.org/10.1007/978-3-030-81545-5

© ISAKOS 2022
This work is subject to copyright. All rights are reserved by the Publisher, whether the whole or part of the material is concerned, specifically the rights of translation, reprinting, reuse of illustrations, recitation, broadcasting, reproduction on microfilms or in any other physical way, and transmission or information storage and retrieval, electronic adaptation, computer software, or by similar or dissimilar methodology now known or hereafter developed.
The use of general descriptive names, registered names, trademarks, service marks, etc. in this publication does not imply, even in the absence of a specific statement, that such names are exempt from the relevant protective laws and regulations and therefore free for general use.
The publisher, the authors and the editors are safe to assume that the advice and information in this book are believed to be true and accurate at the date of publication. Neither the publisher nor the authors or the editors give a warranty, expressed or implied, with respect to the material contained herein or for any errors or omissions that may have been made. The publisher remains neutral with regard to jurisdictional claims in published maps and institutional affiliations.

This Springer imprint is published by the registered company Springer Nature Switzerland AG
The registered company address is: Gewerbestrasse 11, 6330 Cham, Switzerland

Foreword

It is truly a great pleasure to write this Foreword to a timely and exciting book produced by ISAKOS and organized by Drs. Koh, Kuroda, Espregueira-Mendes, and Gobbi, some of the best and smartest knee surgeons in the world. I am proud to say also that three of the editors are members of the Patellofemoral Foundation Board of Directors!

Patellofemoral patients often stimulate a lot of discussion in the clinical office and require considerable understanding of anatomy and complex biomechanical function related to this joint in order to determine optimal treatment. Nothing is better than a case-based approach to learning in this complicated world of the patellofemoral joint.

This timely book is being published when there is an explosion of interest and new information regarding patellofemoral joint anatomy, function, and treatment. Decision-making is even more complicated now as we better understand the intricacies of this important little joint that can cause so much disability. We are steadily improving what we can do for patellofemoral patients as this knowledge is evolving rapidly.

The Patellofemoral Joint: A Case-Based Approach will challenge your knowledge and thinking about the patellofemoral joint. After reading most of the book, I said to Jason L. Koh "your book is a lot more fun to read than a traditional textbook." You will also see that there are few simple answers and a healthy dialog with differences of opinion regarding treatment. What often becomes apparent with regard to patellofemoral problems is how often different treatments can be effective for the same problem. As you read, I encourage you to decide for yourself what you would do, given your knowledge added to the opinions and knowledge of the authors.

I commend and thank the editors and contributors as well as ISAKOS for this important book.

John P. Fulkerson
Yale University
New Haven, CT, USA

Contents

Part I Patellofemoral Anatomy, Mechanics and Evaluation

1 **Patellofemoral Biomechanics** 3
Benjamin C. Mayo, Farid Amirouche, and Jason L. Koh

2 **Patellofemoral Anatomy, Mechanics, and Evaluation: Patient and Family History in the Evaluation of Patellofemoral Patients** 21
John Fritch, Jason L. Koh, and Shital N. Parikh

3 **Examination of the Patients with Patellofemoral Symptoms** . . . 27
Claudia Arias Calderón, Renato Andrade, Ricardo Bastos, Cristina Valente, Antonio Maestro, Rolando Suárez Peña, and João Espregueira-Mendes

4 **Imaging Evaluation in the Patient with Patellofemoral Symptoms** 39
Allison Mayfield and Jason L. Koh

5 **Principles of Prevention and Rehabilitation for the Patellofemoral Joint** 45
Leonard Tiger Onsen and Jason L. Koh

6 **Newest Surgical Treatments for Patellofemoral Osteochondral Lesions** 57
Ignacio Dallo and Alberto Gobbi

Part II Patellofemoral Instability: Case-Based Evaluation and Treatment

7 **Patellofemoral Instability in the Pediatric Patient with Open Physes: A 11-Year-Old Girl with Trochlear Dysplasia** 69
Alexandra H. Aitchison, Daniel W. Green, Jack Andrish, Marie Askenberger, Ryosuke Kuroda, and Geraldo Schuck de Freitas

8 **A 12-Year-Old Girl with Recurrent Patellar Dislocation and Multiple Risk Factors Including Genu Valgum** 89
Shital N. Parikh, Jacob R. Carl, Andrew Pennock, Javier Masquijo, and Franck Chotel

9 Patellofemoral Instability in the Young Male Soccer Player with First-Time Dislocation, Closed Physes, Normal Trochlea: 17-Year-Old 105
Petri Sillanpää, Julian Feller, and Laurie A. Hiemstra

10 The Patellofemoral Joint: A Case-Based Approach. 111
Ashraf Abdelkafy, John P. Fulkerson, and Ryosuke Kuroda

11 Patellofemoral Case Study: Patella Instability, High TT-TG, Patella Alta + Trochlear Dysplasia in a Skeletally Mature Female 117
Jason L. Koh, Andrew Cosgarea, Phillippe Neyret, and Elizabeth A. Arendt

12 Patellofemoral Instability in a Young Patient with a Chondral Defect, Patella Alta and a Lateralized Tuberosity 125
Jack Farr, Jason L. Koh, Christian Lattermann, Julian Feller, and Andrew Gudeman

Part III Patellofemoral Pain. Chondrosis, and Arthritis: Case-Based Evaluation and Treatment

13 Patellofemoral Instability in the Adult Female with Recurrent Instability, Chondrosis of Patella and Trochlea. 137
Kanto Nagai, Ryosuke Kuroda, Stefano Zaffagnini, and Mauro Núñez

14 Patellofemoral Pain, Chondrosis, and Arthritis in the Young to Middle-Aged Patient: A 32-Year-Old Woman with Lateral Patella and Trochlear Chondrosis. 149
Sabrina M. Strickland, Francesca De Caro, and Robert A. Magnussen

15 Patellofemoral Pain, Chondrosis, and Arthritis: A 18-Year-Old with Anterior Knee Pain and Normal Articular Cartilage 157
Sheanna Maine, Stacey Compton, and Ebenezer Francis Arthur

16 A 24-Year-Old Female with Anterior Knee Pain, Normal Looking Cartilage, and Mild Malalignment. 161
Alexandra H. Aitchison, John P. Fulkerson, Marc Tompkins, and Daniel W. Green

17 Patellofemoral Pain, Chondrosis, and Arthritis: A 23-Year-Old with Patellofemoral Pain and Maltorsion of the Lower Limbs: The Place of Torsional Osteotomies. 165
Magaly Iñiguez C, Phillippe Neyret, Sheanna Maine, and Shital N. Parikh

18 Anterior Knee Pain in a Patient with a Central Trochlea Defect: A 32-Year-Old Man with a Central Trochlear Defect 177
Jason L. Koh, Jack Farr, Yukiyoshi Toritsuka, Norimasa Nakamura, Alberto Gobbi, and Ignacio Dallo

19 A 35-Year-Old Woman with Painful Early Lateral Patellofemoral Degenerative Arthrosis 187
Robert A. Magnussen and John P. Fulkerson

20 Patellofemoral Chondrosis and Instability in the Middle Aged Patient 191
Seth L. Sherman, Taylor Ray, Adam Money, Stefano Zaffagnini, Mauro Núñez, and Julian Feller

21 Patellofemoral Pain and Arthritis in Latter Middle-Aged Patient, with Marginal Osteophytes and General Patellofemoral Chondrosis 199
Stefano Zaffagnini, Giacomo Dal Fabbro, Margherita Serra, and Elizabeth A. Arendt

Part IV Traumatic Injuries to the Patellofemoral Joint: Case-Based Evaluation and Treatment

22 Stellate Patella Fracture Case 207
Mauro Núñez, Rajeev Garapati, Stefano Zaffagnini, Khalid Alkhelaifi, and Ashraf Abdelkafy

23 Quadriceps Tendon Tear: Evaluation and Management in a 54-Year-Old Man 223
Jason L. Koh, Roshan Wade, and Chaitanya Waghchoure

24 Chronic Patella Tendon Tear in a 24-Year-Old Man: Revision Procedure 229
Jason L. Koh, Sabrina M. Strickland, and Petri Sillanpää

Part V Tendinopathies of the Patellofemoral Joint: Case-Based Evaluation and Treatment

25 Tendinopathies of the Patellofemoral Joint: A Case-Based Approach: Osgood–Schlatter's Disease in a 12-Year-Old 237
Juan Pablo Martinez-Cano, Sheanna Maine, and Marc Tompkins

26 Patellar Tendinopathy in a 21-Year-Old Long Jumper 245
Timothy L. Miller, Michael Baria, Nicola Maffulli, Seth L. Sherman, and Adam Money

27 Patellar Tendinitis at Osgood–Schlatter's Lesion of a 32 Year Old 255
Orlando D. Sabbag, Miho J. Tanaka, Adam Money, and Seth L. Sherman

Part VI Advances and the Future Treatment of Patellofemoral Disorders

28 Advances in Patellofemoral Disorders 263
Justin T. Smith, Betina B. Hinckel, Miho J. Tanaka, Elizabeth A. Arendt, Renato Andrade, and João Espregueira-Mendes

Introduction

On behalf of myself, co-editors Drs. Ryosuke Kuroda, João Espregueira-Mendes, and Alberto Gobbi, the contributing authors, and the International Society of Arthroscopy, Knee surgery, and Orthopaedic Sports Medicine (ISAKOS), with great pleasure we invite you to enjoy our book **The Patellofemoral Joint: A Case-Based Approach**.

There have been many recent advances in our understanding of the patellofemoral joint and its treatment. This practical case-based book is intended to assist clinicians in the evaluation and treatment of patellofemoral patients by reviewing critical elements for assessment, followed by clinical scenarios where different experts provide their recommendations for treatment. Uniquely, this ISAKOS text utilizes the international experience and perspective of leading experts and researchers from around the world in the treatment of these often complex situations.

In the case-based chapters, the first authors present a clinical case for discussion, with relevant history, physical examination, and imaging findings, followed by a proposed plan of treatment. Similar to a panel discussion, the other authors then provide their unique perspectives and identify areas of similarity and differences. At times, you will find that the assessment and management of the condition is very different depending on the author!

As editors and authors, it has been a pleasure to create this work, and we hope you enjoy the dialog and discussion and find this approach interesting and informative. We are all continuously learning from one another, and hope that this is a small contribution to helping each of us take better care of patients with patellofemoral conditions.

Many thanks,

Jason L. Koh, Ryosuke Kuroda, João Espregueira-Mendes, and Alberto Gobbi

November 2021

Jason L. Koh

Northwestern University McCormick School of Engineering
Evanston, IL, USA

University of Chicago, Pritzker School of Medicine, Chicago, IL, USA

Orthopedic and Spine Institute, Skokie, IL, USA

NorthShore University HealthSystem, Evanston, IL, USA

Mark R. Neaman Family Chair of Orthopedic Surgery, Evanston, IL, USA

Part I

Patellofemoral Anatomy, Mechanics and Evaluation

Patellofemoral Biomechanics

1

Benjamin C. Mayo, Farid Amirouche,
and Jason L. Koh

1.1 Introduction

The knee is a complex joint with dynamic articulations between the femur and tibia as well as the femur and patella. Due to the incidence of patellofemoral pain and related conditions, the patellofemoral joint has been studied extensively. The patellofemoral joint kinematics are affected by both static and dynamic restraints, composed of osseous and soft tissue structures. The normal biomechanics of this joint allow for the forces of the quadriceps muscle to be transferred efficiently to the distal tibia to allow for controlled knee motion. However, these forces can be altered by different morphologic abnormalities, which may result in pain, instability, and dysfunction.

B. C. Mayo
Department of Orthopaedics, University of Illinois at Chicago—College of Medicine, Chicago, IL, USA

F. Amirouche (✉)
Department of Orthopaedics, University of Illinois at Chicago—College of Medicine, Chicago, IL, USA

NorthShore Orthopaedic Institute, NorthShore University Health System, Evanston, IL, USA

Orthopaedic Biomechanics Research Laboratory, Department of Orthopaedic Surgery, Orthopaedic and Spine Institute, Northshore University HealthSystem, Evanston, IL, USA
e-mail: famirouche@northshore.org

J. L. Koh
NorthShore Orthopaedic Institute, NorthShore University Health System, Evanston, IL, USA

Patellofemoral chondral lesions are seen in as many as 60% of patients who undergo knee arthroscopy [1]. However, there are many different pathologies associated with anterior knee pain including arthritis, instability, load transfer, and focal chondral damage. These pathologies can be traumatic in nature, or worsen from overuse. There are several theories as to how these issues cause pain including high joint contact stresses, malalignment, quad muscle weakness, and delayed VMO activity [2]. Risk factors for patellofemoral pain have been attributed to quadriceps weakness and volumetric change viewed as a decrease in size [3–5]. Additionally, delayed VMO activation has been identified as a risk for lateral patellar maltracking [6–8]. This is important as it has been suggested that PFJ pain may be precursor to OA [9].

It is important to have a thorough understanding of the anatomy and biomechanics of the patellofemoral joint to be able to appropriately diagnose and treat the disorder. This chapter will provide an overview of the normal biomechanics and kinematics of the patellofemoral joint, as well as how certain conditions alter the forces and contact stress through the joint.

1.2 Epidemiology

Patellofemoral disorders are extremely prevalent in the general population, with an incidence of 5.9 cases per 100,000, and up to

© ISAKOS 2022
J. L. Koh et al. (eds.), *The Patellofemoral Joint*, https://doi.org/10.1007/978-3-030-81545-5_1

29/100,000 in those 10–17 years old. It has been reported that patellofemoral injuries account for nearly 25% of all knee injuries [10–12], and as high as 30% in those 13–19 years old [13, 14]. In the general population, approximately 10% of all orthopedic complaints are patellofemoral in nature, and as high as 40% of complaints in athletes [15]. Women are especially susceptible to patellofemoral disorders, and experience issues at a 2–3 times higher rate than men [12, 16, 17].

1.3 Anatomy

The patellofemoral joint consists of several different articulating surfaces which engage at different points of knee motion. The interaction of these surfaces is controlled by both the bony geometry and several soft tissue supporting structures that can provide static and dynamic restraints.

1.4 Osseous Anatomy

1.4.1 Patella

The patella is a sesamoid bone, with attachments to the quadriceps tendon superiorly and patella tendon inferiorly. It has normal dimensions of 3.8–5.3 cm in length, 4.0–5.5 cm in width, and 1.9–2.6 cm thickness [18, 19]. The anterior surface of the patella is convex, while the two main surfaces on the posterior surface, the medial and lateral facet, are concave. The posterior surface has 7 facets in total. The medial and lateral facets are the two largest articulations, with a small medial facet called the odd facet also contributing at deep flexion. The facets of the patella are variable from person to person, with variances in the bony and cartilaginous medial ridge. The Wiberg classification of patella is based on the location of the median ridge: Type I has the median ridge more central, progressing to Type IV where the ridge is extremely medial, thus creating a significant larger lateral facet compared to medial (Fig. 1.1). There has been noted to be a higher incidence of patella instability in Type 3 patella [20].

1.4.2 Trochlea

The anterior aspect of the distal femur forms the trochlea, which is the articulating surface for the patella. The trochlea consists of lateral and medial walls, or facets constituting a groove. The lateral facet of the trochlea is larger, extends more proximally, and has a shallower angle to match the lateral femoral condyle, while the medial facet is smaller and shorter. The area between the two facets is the trochlear groove. When this anatomy is altered either with a shallow groove or more distal groove, it increases the risk for patellar instability and dislocations when the quadriceps pulls laterally. Variations in the depth of the trochlea are most often found proximally, thus making early flexion the most likely time for patellar instability. When the trochlea does not have a normal concave shape, there is trochlear dysplasia. This can predispose the patient to instability as the normal static restraint is lessened.

The Dejour classification of trochlear dysplasia classifies the severity of the dysplasia using trochlear groove position on a lateral knee X-ray (Fig. 1.2). In Type A, there is a shallow trochlea with crossover sign, when the floor

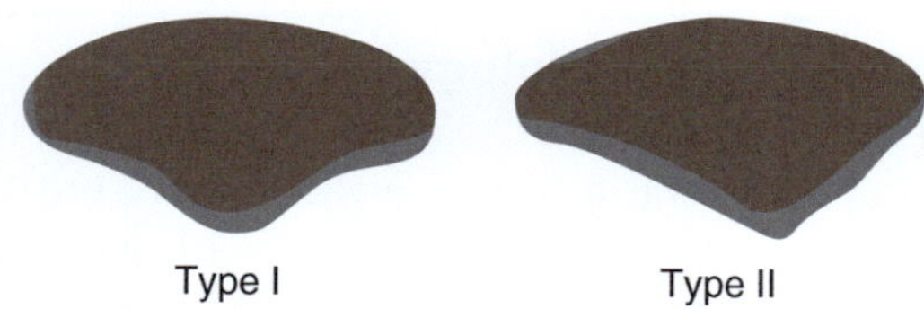

Fig. 1.1 Wiberg classification of patella morphology

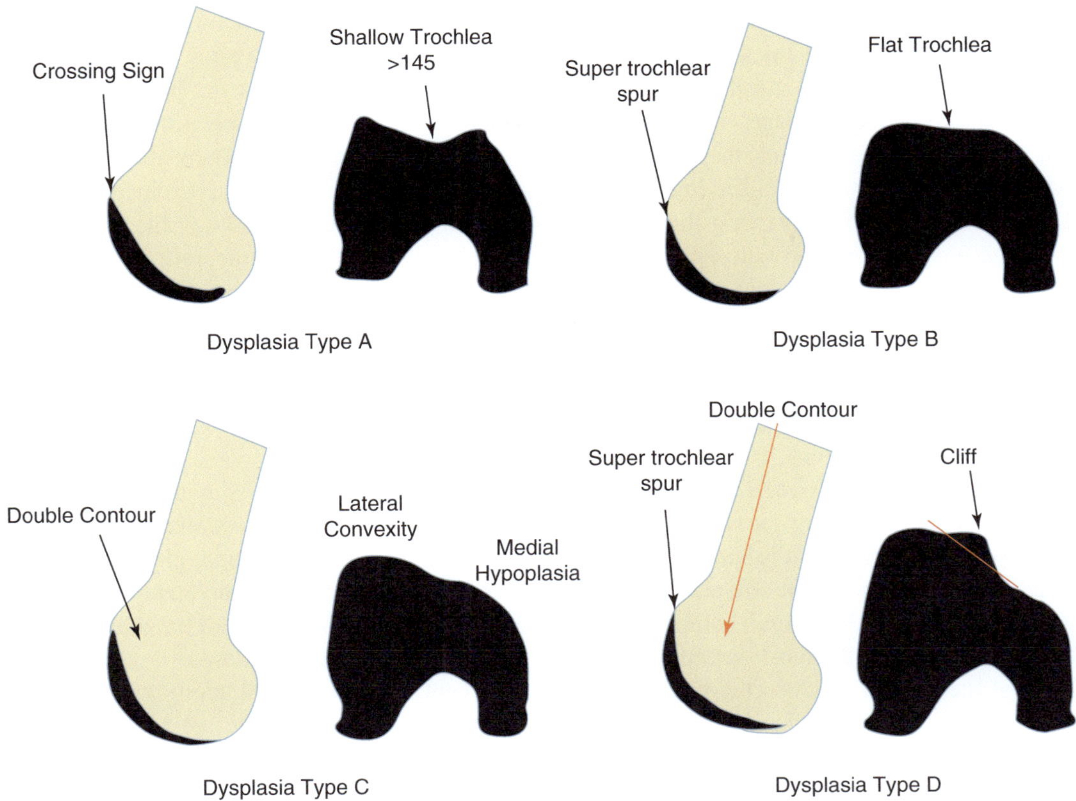

Fig. 1.2 Dejour classification of trochlear dysplasia

of the trochlea crosses over the highest point of the medial femoral condyle [21]. In Type B there is a shallow or flat trochlea with a supratrochlear spur. Type C has a positive crossing sign and a double contour sign. While Type D is an extension of type C, with a supratrochlear spur.

The trochlea can also be evaluated on advanced imaging such as CT or MRI. The trochlea can also be measured by the sulcus angle, which is a measurement between the angles of the medial and lateral walls of the trochlea. The normal angle is 138°, with an angle greater than a 150 representing a dysplastic trochlear groove (Fig. 1.3).

The congruence angle is created by drawing a line from the apex of the trochlear groove through the peak of the ridge of the patella, compared to a line that bisects the angle of the trochlear groove. A value of −6 is normal, while an angle greater than 16° is abnormal (Fig. 1.3).

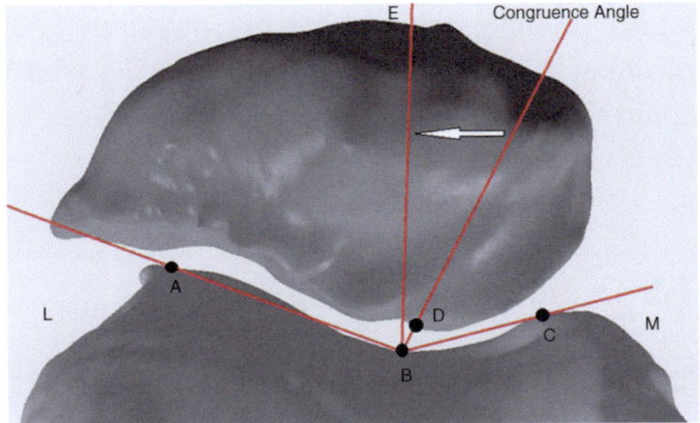

Fig. 1.3 Image demonstrating the measurement of the sulcus angle (angle between lines BA, the lateral wall, and BC, the medial wall), as well as the congruence angle (angle between lines BD, through the ridge of the patella, and BE, bisecting the sulcus angle)

1.4.3 Cartilage

There is only cartilage on the superior two-thirds of patella, which is the surface that articulates with the trochlea. The cartilage on the patella is significantly thicker than cartilage in other areas of the body. However, the thickness is variable in different surfaces of the patella. The average thickness of patellar cartilage is 4.1 ± 1.3 mm, but this increases near 7 mm in the central region, which allows it to withstand the highest forces [22, 23]. Because of the varying thickness, the cartilage is not congruent with underlying subchondral bone. These variations may cause some medial or lateral shift when viewed on X-ray. The peak of the cartilage ridge is often seen slightly lateral to the median bone ridge peak. The cartilaginous median ridge is completely median in 40% of patients, whereas it veers medially as moving more distal in the rest [23]. The cartilaginous ridge can be very prominent in some patients, though in others it is barely identifiable. The thickest portion of the articular cartilage is around 54% from the lateral border and 55% from the superior pole. In osteoarthritic patients, most of the cartilage loss in the patellofemoral joint occurs on the lateral facet of the patella, and to a lesser extent the lateral trochlea [24]. Trochlear groove cartilage is typically between 2 and 3 mm [25], and like the patella, is thicker in the groove of the facets. When compared to other cartilage in the body, patellar cartilage has a higher compressive aggregate modulus [26] meaning it is more permeable and pliable, which could contribute to the higher incidence of lesions [27].

1.5 Soft Tissue Anatomy

1.5.1 Quadriceps/Extensor Mechanism

The extensor mechanism consists of the quadriceps muscle group, quadriceps tendon, patella, patellar tendon, and tibial tubercle. The quadriceps muscle group is formed by the vastus medialis, vastus intermedius, vastus lateralis, and rectus femoris which converge in a single tendon into the superior aspect of the patella. The quadriceps tendon averages 5–8 cm in length. The patellar tendon originates from the inferior pole of the patella and inserts on the tibial tubercle, slightly lateral to midline. Studies have found the length of the tendon to be between 3.5 and 5.5 cm [18]. The vastus lateralis muscle provides a lateral pull on the patella, which is counteracted by the force from the vastus medialis obliquus (VMO). The VMO inserts onto the medial retinaculum and superomedial patella, as distal to the proximal 50% of the patella. The VMO is the only dynamic restraint to lateral translation, while the other soft tissue restraints are static. The vector of the VMO contraction force is between 50 and 65° away from the long axis of the femur [28]. When the VMO is deficient in strength or timing, or orientation of pull, medial pull is overcome by the pull of the vastus lateralis. This is particularly evident from 0 to 15 degrees of flexion. These deficiencies can be partially corrected with proper physical therapy, but in some instances poor anatomy such as direction of the fibers will not provide sufficient medial stabilization despite conservative treatment. Farahmand et al. report that the forces of each quadriceps muscle are proportional to cross-sectional area [29]. The largest contribution is from the vastus lateralis (40%), while the smallest contribution is from VMO (25%).

1.5.2 Medial Restraints

The proximal medial restraints are the MPFL, and medial quadriceps tendon femoral ligament (MQTFL), and the distal restraints which are the medial patellotibial ligament (MPTL) and medial patellomeniscal ligament (MPML). The MPFL is the predominant passive soft tissue restraint on the medial side of the knee and is a flat fan shaped ligament that provides static restraint to lateral translation of the patella. It originates on the femur between the medial epicondyle and adductor tubercle, and inserts onto the proximal 50% of the medial edge of the patella. Cadaveric studies have identified the origin just proximal and posterior to medial epicondyle and distal to adductor tubercle. Placella et al. [30] reported that the femoral origin of the MPFL is at the adductor tubercle in 29.6% of patients, on the medial femoral condyle in 17.8%, and between them in 44% of knees. The anterior insertion is a broad area with an average size of 26.0 mm

(14.0–52.0 mm) [31–33], with a smaller femoral attachment of 12.7 mm (6.0–28.8 mm). The median length is 56.9 mm (46.0–75.0), and has been reported to be anywhere between 3 and 30 mm in width [18, 34], while the thickness is 0.44 ± 0.19 mm [35].

The MPFL has demonstrated to be tightest in the first portion of knee flexion, ensuring the patella enters the trochlear groove without drifting laterally. The MPFL is primary passive restraint to lateral patellofemoral translation from 0 to 30 degrees of flexion where it provides up to 60% of the restraint [34, 36, 37]. Laxity in the MPFL can be due to traumatic tears, or due to congenital laxity. Laxity in the MPFL is thought to be the cause of recurrent dislocation in the setting of other risk factors [38].

It is important to understand the behavior of the native MPFL to allow isometric movement if a reconstruction is required. Victor et al. [39] reported that the MPFL is isometric from 0 to 40 degrees of flexion, and experiences shortening at a rate of 0.5 mm/10 degrees flexion after that for a total of 4 mm of shortening at 120 flexion. Additionally, the superior and inferior portions behave differently. The cephalad portion has the most isometric behavior and is most taut in extension, while the caudal portion is most taut at 30 flexion [39, 40].

The proximal fibers of the MPFL have been named the MQTFL due to their distinct attachment on quadriceps tendon. The MPTL and MLMP are secondary stabilizers to lateral translation during motion. Though not as significant as the MPFL proper, they contribute 26% of the resistance to lateral translation in extension, to 46% of the resistance at 90 of flexion [41]. They also resist lateral patellar tilt of 72% at 90 flexion. When the MPTL is ruptured in isolation, an additional 8.6 mm of translation can be seen [32].

The ultimate load to failure of the MPFL ranges between 72 N and 208 N. While the ultimate failure elongation ranges from 8.4 mm to 26 mm, with stiffness from 8.0 N/mm to 42.5 N/mm. The MPFL and MPTL have similar ultimate loads to failure (178 ± 46 M vs 147 ± 80 N), but both significantly greater than MPML (105 ± 62 N). Sectioning the MPFL reduced force required to displace the patella laterally 1 cm by 14–22% [42].

1.5.3 Lateral Restraints

The lateral soft tissues restraints to the patella are in multiple layers. The superficial layer is oblique lateral retinaculum, while the deep layer has oblique and transverse fibers of the patellotibial and epicondylopatellar bands [43]. The lateral patellofemoral ligament (LPFL) femoral insertion is 19.7 mm anterior to the posterior end of lateral condyle, and 16.5 mm proximal to the distal end of lateral condyle [44]. The strongest lateral patellar restraint is the ITB-patella with 582 N load to failure, whereas LPFL is 172 N [45]. Lateral retinacular tightness may lead to abnormal patellar tilt and increased forces on the lateral trochlea and patellar facet. Patellar tilt is when medial side of patellar is raised, which is often seen with a tight lateral retinaculum or dysplastic VMO. This can be assessed on CT or MRI axial by drawing a line through the medial and lateral aspect of the patella compared to the posterior condyles. The normal patellar tilt is 2°, whereas a value greater than 5° is considered abnormal [22].

Ishibashi et al. examined lateral patellar retinaculum tension in patellar instability by assessing tension at varying degrees of flexion before and after anteromedial tibial tubercle transfer. They noted increased lateral retinaculum tension after anteromedialization at 0 and 30 degrees of flexion, but this was not statistically significant after 30 degrees (60, 90, and 120 degrees) [46] However, if the lateral retinaculum is released, it could worsen lateral instability, or lead to medial patella instability. Performing a lateral retinacular release after MPFL reconstruction reduces the force required to translate laterally by 7–11% [47].

1.5.4 Neurovascular

The primary blood supply to the patella is from the arterial plexus that surrounds the patella [48]. The femoral nerve controls the extensor mechanism.

1.6 Biomechanics/Kinematics

From a biomechanical perspective, the patella is a sesamoid bone which acts as a lever. This creates a mechanical advantage by transmitting forces from the quadriceps muscles to the attachment on the proximal tibia to extend the knee joint at a larger moment from its axis of rotation. It is a Type 3 lever in which small amounts of quadriceps contraction allow for large displacement of the tibia. As the knee flexes and extends, the position of the patella relative to the trochlea is changing, in what is known as a rolling fulcrum. Studies have demonstrated that the increase in moment arm from the center of the knee to the patella improves the efficiency of the quadriceps by nearly 50% [49, 50]. With so many structures contributing to the stability and motion of the patella, the biomechanics are complex. When there is an imbalance between the dynamic and static stabilizers there can be significant alterations in the patellofemoral biomechanics.

The patella articulates with the distal femur trochlear groove. When the knee is fully extended, the patella typically sits superior to the groove, and as it bends, it engages and is pulled medially. The quadriceps pulls the patella in a superolateral direction. The difference in force vector direction from the quadriceps and the patellar tendon create what is called the Quadriceps (Q) Angle. The Q angle is variable from person to person, but the normal Q angle is around 15 ± 5° (Fig. 1.4). Females often have

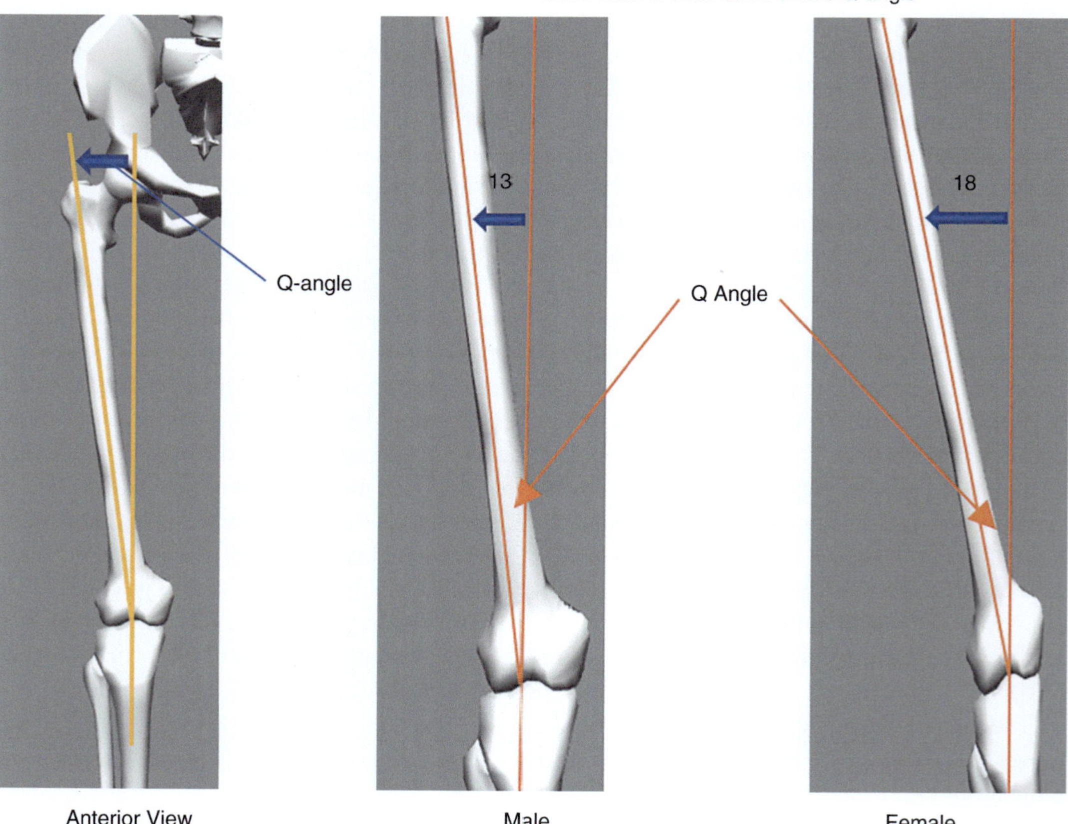

Fig. 1.4 Q angle is measured by a line from the anterosuperior iliac spine to the patella (the pull vector of the quadriceps muscle) and a line from the patella to the tibial tubercle (the pull vector from the patellar tendon). The difference in Q angle between males and females is demonstrated by the change in knee alignment

higher Q angle [51–55], which may be mostly attributable to taller height of males [52]. However measuring the Q angle can be difficult depending on patient positioning. The normal Q angle in males ranges from 8 to 16° when supine and 11–20° while standing, in females it is between 15 and 19° supine, and 15–23° standing [55–57]. Though it has been thought abnormal Q angles can contribute to pain and dysfunction or put patients more at risk for injury, some studies have failed to find a significant correlation [58, 59].

1.7 Kinematics

1.7.1 Tracking

As a sesamoid bone with multiple soft tissue attachments, the patella has motion in all planes including flexion and extension, spin in the coronal plane, and tilt in the axial plane. There is also translation in all planes. The patella has approximately 7 cm of translation from full extension to flexion. When the knee is fully extended the patella sits slightly lateral, and any force through the quadriceps is primarily in a proximal direction with very little force being directed posteriorly. As the knee begins to flex past 20–30°, the median ridge of the patella engages in the trochlear groove and is slightly medialized. After 30°, patellar stability is largely determined by trochlea anatomy [60–63].

Through normal motion in flexion and extension, different anatomical structures are the primary stabilizers. In the early portion of flexion, the medial tissues are the primary restraints such as the MPFL and VMO. As it moves through flexion, the lateral stability is controlled by the bony restraints, and the tension of the patellar and quadriceps tendons pulls the rotation in the sagittal plane to maintain contact perpendicular to the patellofemoral joint reactive force.

During knee flexion, the tibia internally rotates, effectively reducing the Q angle and reducing the lateral pull on the patella. Huberti et al. assessed force on the tendons at different flexion angle and showed the highest force ratio at 30 degrees of flexion [64]. This results in the patella being at highest risk for dislocation in first 30 degrees before it engages the trochlea fully and has a high amount of lateral pull [65]. While a majority of the resistance to lateral translation is due to the medial sided structures, the lateral retinaculum contributes 10% restraint of lateral displacement as well [66].

When the patella engages and translates medially into trochlea, the patella tilts due to the tension of the medial retinaculum as it slides along the trochlear groove [67]. Another major stabilizer of the patella is the VMO, which has more influence over the patellar spin and tilt, and less over the translation [68, 69]. This is due to the orientation of the fibers of the VMO which are still mostly vertically directed, thus more effective and limiting translation with more knee flexion as their force vector becomes more medially oriented. Delays in VMO activation or decrease in force relative to the vastus lateralis have been associated with increased patellar tilt [70] as well as increased lateral patellar shift [71].

During knee flexion, the patella also flexes (distal patella rotates backwards) due to the shape of the distal femur and pull from the patella tendon. However, the patella flexes slower than knee at a rate of 60–70% of the knee flexion [72–76]. This difference is more evident at >100 flexion [77]. The patella also has the ability to rotate around its longitudinal axis, known as patellar tilt. This is typically seen as a movement of lateral border of patella toward femur relative to medial border. The amount of rotation ranges from 1 to 15° while flexing from 0 to 90° [77].

1.7.2 Abnormal Tracking

Any combination of abnormal anatomic features can result in altered patellofemoral forces that lead to pain, instability, chondral damage, or a combination of these issues. When the Q angle is increased to greater than 20°, there is an increased lateral displacement force as well as lateral patellar contact pressure [51]. The mean Q angle has been reported to be slightly higher in women, placing them at higher risk for abnormal patellar tracking. This may be attributable to height differences as mentioned, but also wider hips in women result in a larger valgus alignment at the knee. An increase in Q angle has demonstrated to shift contact pressures to only the lateral facet in nearly half of patients, while others had increased pressures on both facets. Both groups experienced an increase in pressures by 45% at 20 degrees of flexion. Similarly, a decrease in Q angle can increase forces on just the medial or on both facets [65, 78, 79].

On the other end of the patella, changing the tibial tubercle position can alter the facet loading. The tibial tuberosity-trochlear groove (TT-TG) distance is another method to assess alignment of the patellofemoral joint. On axial CT or MRI can measure distance between tibial tubercle and trochlear groove. The normal range is from 10 to 13 mm. When the distance is greater than 15 mm, there is an increased risk of patellar instability [80]. The further lateral the tubercle is relative to the groove, the higher lateral directed force vector produced by the patellar tendon on the patella.

It is also important to consider anatomic variations away from the knee, as the entire kinetic chain can contribute to abnormal patellofemoral tracking. Patellar tracking can also be affected by core weakness or tightness, torsion of femoral neck and shaft, and tibia [81–83]. Internal rotation of the femur causes a lateral displacement of the patella relative to the femur, and has been shown to increase patellar tilt on MRI [84]. Similarly, Kaiser et al. analyzed femoral torsion (both internal and external rotation) and demonstrated that increased internal torsion increased lateral patellar shift from 0 to 30 degrees of flexion. Additionally, there was a noted increase in patellar tilt at both 10 and 20° of internal rotation. Lastly, they noted that if the MPFL was transected, less internal rotation of the femur was needed to identify patellar instability. Functional internal rotation can occur from weakness of the abductors and external rotators. Weakness in these muscles has been correlated with patellofemoral pain [85, 86] and shown to improve after strengthening [87–91]. However, this has not been proven to be a direct cause of patellofemoral pain, and may instead be a symptom of the pain. Women have been noted to have less strength in abduction and external rotation, as well as increased Q angle, dynamic knee valgus, and hip internal rotation angle, all of which potentially contribute to maltracking of the patella [92, 93]. Also of note, patients with an increased femoral anteversion have a higher rate of dysplastic trochlea, further increasing the risk of patellar instability and maltracking [94]. Further distal in the kinetic chain, excessive foot pronation prevents the tibia from external rotation. This can lead to a compensatory increase in femur internal rotation against the fixed patella [95] and potentially place them at increased risk for patellofemoral pain [96].

1.7.3 Sagittal Plane

Patella that sit too high or low to normal can also cause pain or instability. There are numerous ways to measure the height of the patella on radiographs including Insall-Salvati [97], Blackburne-Peel index [98], and the Caton-Deschamps index [99]. A patella that sits too far proximal, known as patella alta, has demonstrated to be a significant contributor to recurrent patella dislocation. With patella alta, there must be more knee flexion before the patella engages in the groove, which results in a relative lack of constraint in early flexion which can result in chondromalacia as well as increased instability rates [51, 100–102]. Additionally, this delays the quadriceps tendon from engaging with the trochlea at deep flexion which leads to an increased amount of force through the patella. Also of note, patella alta can alter the isometry point of the MPFL. Belkin et al. demonstrated that when the Caton-Deschamps index is above normal, the MPFL isometry point can be up to 5–10 mm proximal to the standard Schottle's point [103].

Conversely, a patella that sits too far distal, known as patella baja, is frequently associated with arthropathy from trauma. This can increase joint forces, as well as limit motion and result in a higher rate of arthritis [104]; however, there is a higher contact area seen which leads to lower stresses, as well as sharing by the quadriceps tendon which engages earlier [105–107].

1.8 Mechanics

1.8.1 Contact Area

The patellofemoral contact area changes throughout motion, with only part of the patella articulating at any given time (Fig. 1.5). In early flexion, the more distal aspect articulates with the proximal trochlea [108, 109] and as it flexes more, the more proximal portion contacts with more distal aspects of the trochlea. The total contact area increases from 0 to 60 degrees of flexion [110–112]. Once the knee goes past 60° however, there is debate in the literature as to whether the area increases or decreases [64, 108, 113–115]. Past 90 degrees of flexion, the vertical cartilage of the median ridge is the predominant contact area. Others have shown the maximum contact area is between 80 and 90 degrees of flexion, which is when the joint reactive forces are near peak, thus helping to minimize the stress from increasing too sharply [116, 117].

With patella alta, there is decreased contact area during flexion, which consequently leads to an increase in stress on the articular cartilage [118–120]. This also leads to the quadriceps tendon engaging the trochlear later in flexion, decreasing the amount of stress shared by the tendon and placing more focus onto the patella. Although some disorders can alter contact area, not all have been reported to have a negative effect. One study of patients with mild osteoarthritis of the knee in varus alignment showed no difference in patellar tracking, kinematics, or contact areas when compared to health knees [121].

1.8.2 Contact Stresses

The patellofemoral contact forces are equal and opposite to the forces from the combined patella and quadriceps force vectors (Fig. 1.6). When the knee is in full extension, the forces from the quadriceps and patellar tendons are nearly parallel to the patellofemoral articulation, resulting in

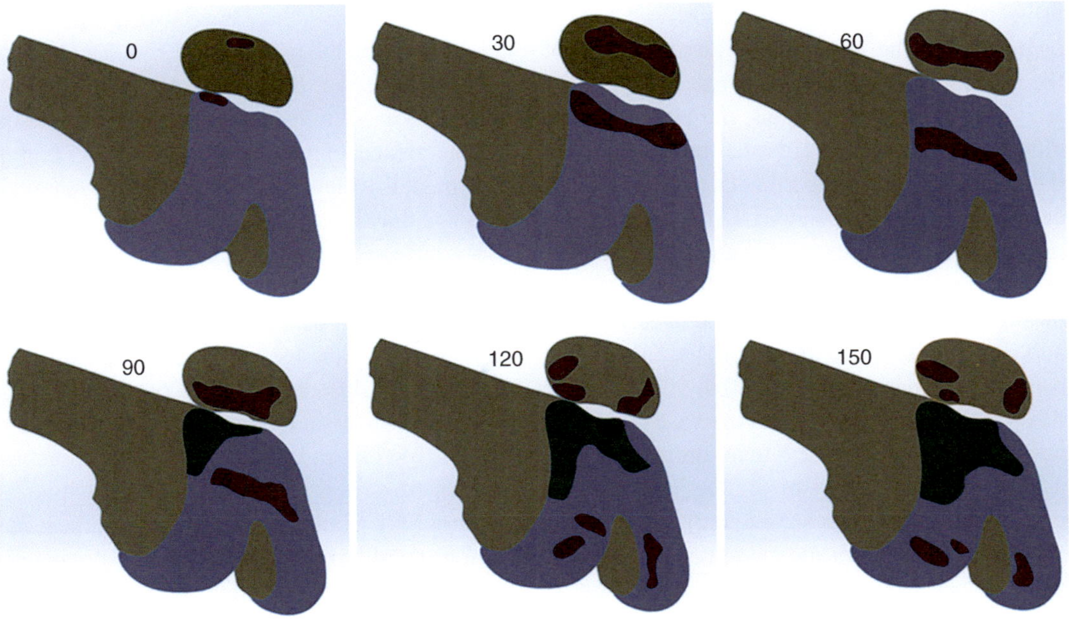

Fig. 1.5 Figure depicting the contact areas of the patella and trochlear and various degrees of flexion. The red areas illustrate the contact between the patella and femur, while the green area is the contact area of the quadriceps tendon

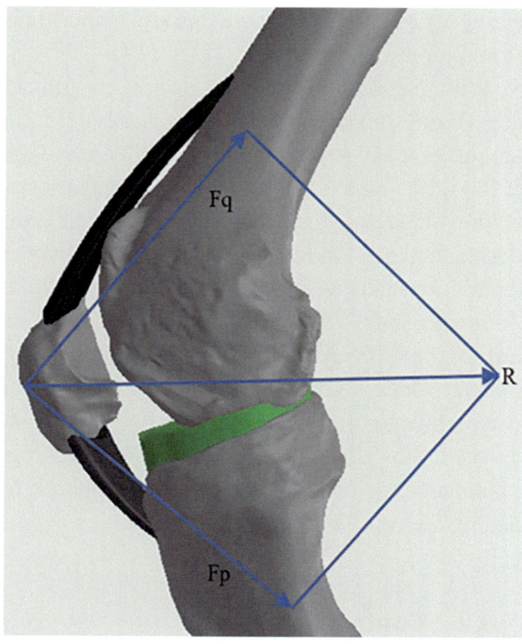

Fig. 1.6 Image demonstrating the resultant joint reactive force (R) due to the forces by the quadriceps (Fq) and patellar tendon (Fp)

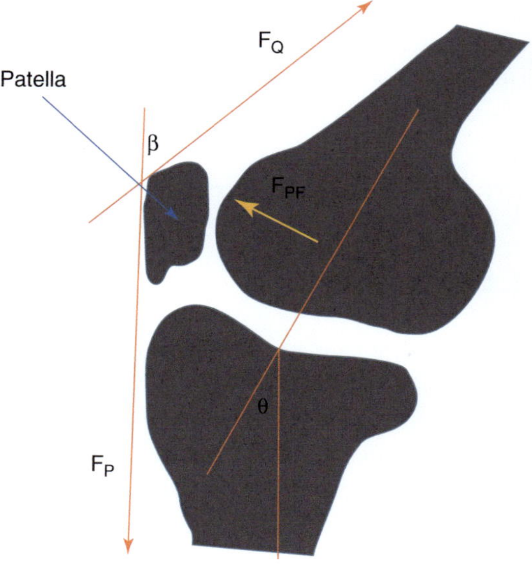

Fig. 1.7 Image demonstrating the change in direction of the resultant joint reactive force (Fpf) as the knee flexes (Angle theta), which leads to an increase in angle beta

little to no joint reactive force. As the knee flexes, the forces are directed more posteriorly, thus increasing the joint reactive force (Fig. 1.7). Past 90 degrees of flexion, the quadriceps tendon engages with the trochlea and absorbs some of the reactive forces between the patella and trochlea [22, 49]. As this happens, the patella rotates along its long axis towards the medial side and the ridge that divides the medial and odd facets engages the femoral condyle. Even though compressive load increases with increased knee flexion, the patellofemoral joint reactive force levels due to the increasing surface contact area. Huberti and Hayes determined that the highest amount of pressure is experienced from 60 to 90 degrees of flexion, but with a concordant increase in contact area and cartilage thickness to minimize the overall stress [79]. A study of female patients with patellofemoral pain has significantly higher peak, and average minimum and maximum strain in the patella when compared to those without pain. It was also noted that cartilage thickness was negatively associated with peak minimum and maximum patella strain [122].

A significant amount of research has been completed to understand the patellofemoral joint mechanics [50, 79, 123–125]. Through normal daily activities, the patella encounters nearly 10 times the force of body weight, and as much as 20 times body weight with sport movements. In a well-tracking knee, there is relatively equal contact pressure on the medial and lateral facets from 20 to 90 degrees of flexion [50, 120]. However, in those with patella maltracking contact forces on the lateral facet of the patella are 4–6 times higher than medial which is where there was maximal cartilage thickness [126].

As different pathologies alter normal patellofemoral tracking, the joint stresses can be altered as well. Pal et al. demonstrated that patellofemoral cartilage stresses were most sensitive to changes with different amounts of vastus medialis forces when compared to other quadriceps muscle forces, and those with patellofemoral tracking disorders were the most sensitive to changes [70]. Similarly, quadriceps tightness has demonstrated to increase the force of the patella against trochlea [2]. In patients with tracking disorders, one surgical management is to move the tibial tubercle in an effort to alter the Q angle and reduce contact stresses. The largest decrease in contact stresses is seen when tibial tubercle is moved anteriorly as opposed to medially [24]. Cadaveric studies have reported that knees with trochlear dysplasia experience increased internal rotation, lateral tilt, lateral translation, contact pressures, and decreased contact area and stability when compared to non-dysplastic knees [127].

Stresses in the patellofemoral joint vary depending on amount of flexion, contact area, and forces applied, but also whether the leg is operating in an open or closed chain. In closed chain exercise, there are increased stresses when the knee is in greater amounts of flexion [64]. However, in open chain exercises where the foot is not planted, the quadriceps forces increase and contact areas decrease when extending from 0 to 90°. However, the force vector of the quadriceps is redirected away from the patellofemoral surface with more extension, limiting the amount of stress transferred to the joint [128].

The effect of different activities on the patellofemoral joint forces is well studied. Atkins et al. analyzed the patellofemoral joint reactive forces with different amounts of forward lean during stair ascent. They reported that patellofemoral joint stresses decreased with trunk flexion, and increased in extension. With forward lean the center of gravity is shifted anteriorly, reducing the length of the moment arm and thus reducing the patellofemoral joint reactive force.

In a study of patients with patella alta performing fast walking have less contact area and higher patellofemoral stress than normal, but there is no overall change in joint reaction forces [118].

Similarly to stair climbing, running on a decline results in higher patellofemoral joint stresses due to the increased moment arm from the trunk extension, while there is no significant difference in forces between flat and incline running [129]. Different running techniques can also alter patellofemoral stresses. Forefoot landing instead of hindfoot as well as faster step rate decreases overall joint stresses [130].

Different positions of squatting and kneeling can place extremely high stresses through the patellofemoral joint. Squatting creates loading conditions that have higher varus and resultant moments than kneeling. Of the kneeling postures, one knee kneeling has the highest force magnitudes of kneeling postures [131]. When squatting, there are higher patellofemoral joint stresses, reaction forces, and quadriceps force when the knees extend anteriorly past the toes than when they stay behind the toes [132].

1.8.3 Modeling Techniques of Patellofemoral Joint

Finite element (FE) analysis has been at the forefront of orthopedic research and has become an integral part of fundamental joint mechanics. The knee patella femoral joint is a complex joint with different articulate surfaces and muscles forces working in coordination to provide the articulation needed and strength to perform the task at hand. A number of pathologies discussed earlier in our kinematics and biomechanics of the knee patella can be fully described by FEA models. These models can give insight into patella tracking, stress conditions and the effects of anatomical changes that create instability at the knee in addition musculoskeletal diseases if they can be adequately described as physiological boundary conditions, can be implemented and an analysis of contact surfaces can be measured along with the stress–strain patterns that can result from the assumptions entered into the model (Fig. 1.8).

A 3D FE knee model can be reconstructed from CT scan images where a detailed CAD model is first generated through some sophisticated reconstruction of surfaces, solid modeling, and smoothing techniques to create renderings mimicking the actual anatomy of the patient (Fig. 1.9). Geometries of joint substructures are developed, a mesh is completed, then the material assignments are assigned along with the boundary conditions to run the analysis (Fig. 1.10).

We usually rely on gait analysis and experimental data to supply the forces values to the FE

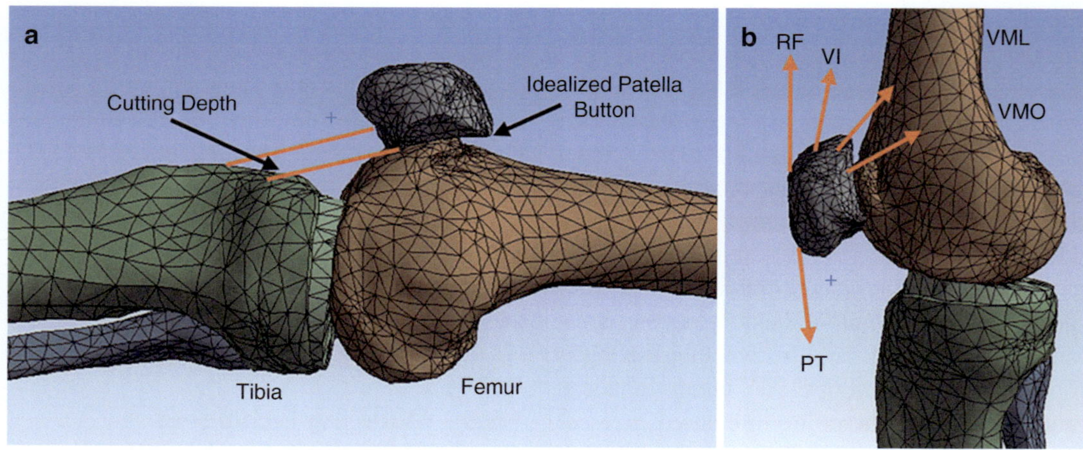

Fig. 1.8 Finite element model of the knee in full extension with an idealized patella subject to Tibia and quadriceps forces. A fully meshed model in extension (**a**) and with a sagittal view (**b**)

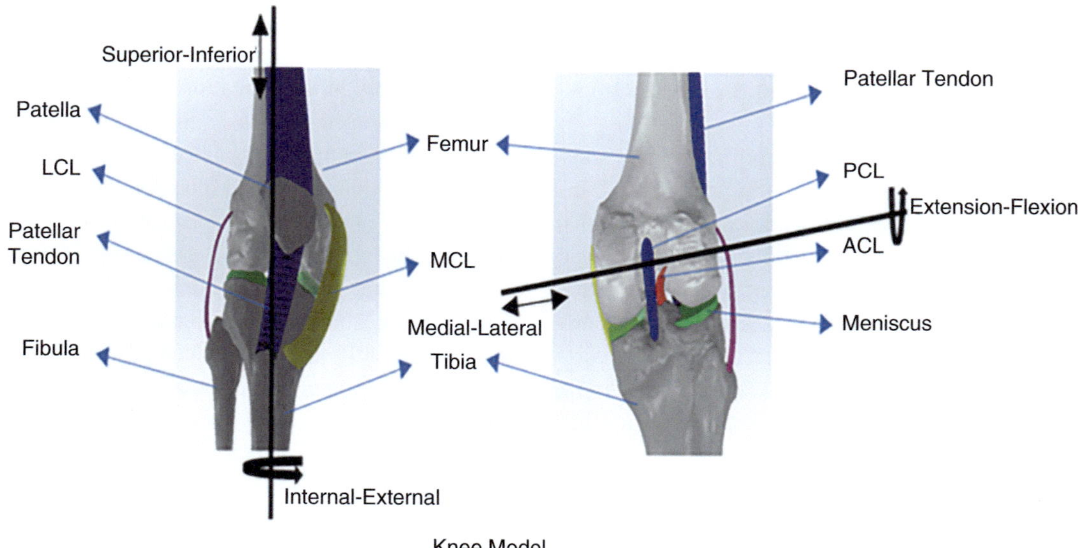

Fig. 1.9 3D knee model built using CT scans and mimics with full ligaments (ACL, PCL), meniscus and patella tendon and quad tendons. Both the internal-external rotation and extension-flexion are shown on both models

model to simulate certain conditions. This usually include measuring forces and moments for one complete gait cycle, and applying these extracted values as a loading condition into the FE model of the knee joint and the simulations are then conducted. The contact conditions including the contact force, total contact area, and the stress conditions in both the medial and the lateral compartments and the tibial will be estimated. Different scenarios can be explored and compared for each case considered. It is important to note that validation of these FE models requires expertise, well designed experiments and proper input conditions to evaluate situations in terms of patellofemoral contact force, contact area, and contact pressure so one can clinical provide an adequate assessment of what cause anterior knee pain. Multiphysics models are currently being developed to include better

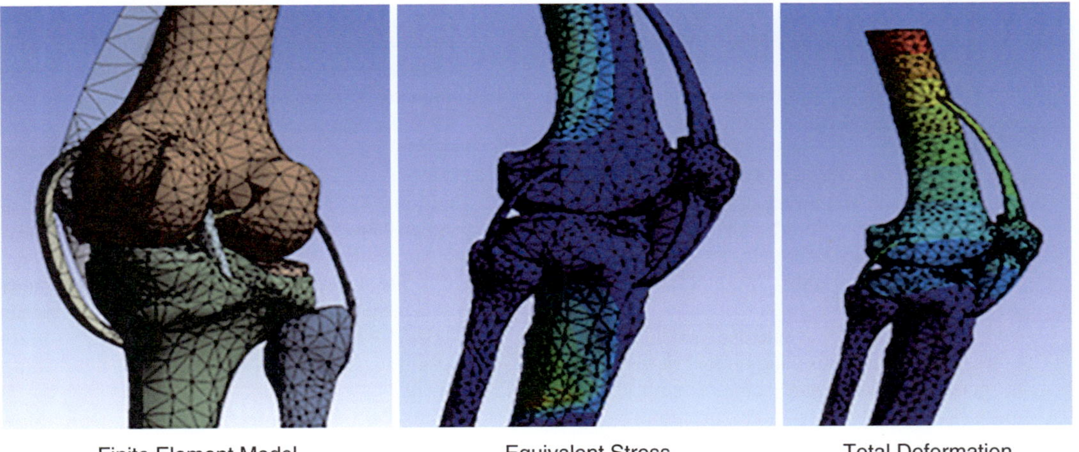

Fig. 1.10 A complete FE mesh of knee model, equivalent stress and total deformation are shown successively

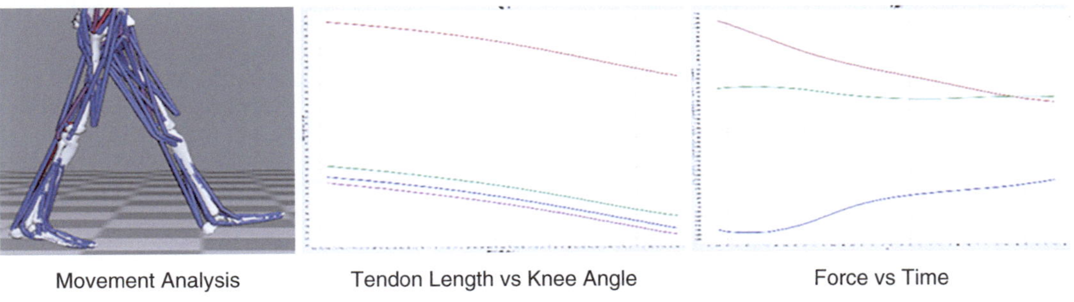

Fig. 1.11 3D gait model to evaluate the kinematics and forces at the knee

understanding of the connective tissues, different bone identification procedures to adequality determine bone quality and enhance the outcome measures from these models (Fig. 1.11).

1.9 Summary

The patellofemoral joint has multiple articulations between the facets of the patella and femur. During knee flexion and extension, the direction and magnitude of the forces on the patella change, resulting in varying joint reactive forces. Additionally, the contact area increases as the knee flexes from zero degrees. This helps mitigate the joint stresses due to increasing joint reactive forces with increased knee flexion. It is important to consider the normal alignment of the knee, as well as the entire kinetic chain when evaluating the patellofemoral joint. Alterations in hip, knee, leg, and foot mechanics can result in increased or abnormal forces in the patellofemoral joint resulting in pain or dysfunction.

References

1. Widuchowski W, Widuchowski J, Trzaska T. Articular cartilage defects: study of 25,124 knee arthroscopies. Knee. 2007;14(3):177–82.
2. Witvrouw E, Lysens R, Bellemans J, Cambier D, Vanderstraeten G. Intrinsic risk factors for the development of anterior knee pain in an athletic population. A two-year prospective study. Am J Sports Med. 2000;28(4):480–9.
3. Callaghan MJ, Oldham JA. Quadriceps atrophy: to what extent does it exist in patellofemoral pain syndrome? Br J Sports Med. 2004;38(3):295–9.
4. Dvir Z, Shklar A, Halperin N, Robinson D, Weissman I, Ben-Shoshan I. Concentric and eccentric torque

variations of the quadriceps femoris in patellofemoral pain syndrome. Clin Biomech (Bristol, Avon). 1990;5(2):68–72.
5. Kaya D, Citaker S, Kerimoglu U, et al. Women with patellofemoral pain syndrome have quadriceps femoris volume and strength deficiency. Knee Surg Sports Traumatol Arthrosc. 2011;19(2):242–7.
6. Chester R, Smith TO, Sweeting D, Dixon J, Wood S, Song F. The relative timing of VMO and VL in the aetiology of anterior knee pain: a systematic review and meta-analysis. BMC Musculoskelet Disord. 2008;9:64.
7. Voight ML, Wieder DL. Comparative reflex response times of vastus medialis obliquus and vastus lateralis in normal subjects and subjects with extensor mechanism dysfunction. An electromyographic study. Am J Sports Med. 1991;19(2):131–7.
8. Witvrouw E, Sneyers C, Lysens R, Victor J, Bellemans J. Reflex response times of vastus medialis oblique and vastus lateralis in normal subjects and in subjects with patellofemoral pain syndrome. J Orthop Sports Phys Ther. 1996;24(3):160–5.
9. Wilson DR, McWalter EJ, Johnston JD. The measurement of joint mechanics and their role in osteoarthritis genesis and progression. Rheum Dis Clin North Am. 2008;34(3):605–22.
10. Baquie P, Brukner P. Injuries presenting to an Australian sports medicine centre: a 12-month study. Clin J Sport Med. 1997;7(1):28–31.
11. Lankhorst NE, Bierma-Zeinstra SM, van Middelkoop M. Factors associated with patellofemoral pain syndrome: a systematic review. Br J Sports Med. 2013;47(4):193–206.
12. Taunton JE, Ryan MB, Clement DB, McKenzie DC, Lloyd-Smith DR, Zumbo BD. A retrospective case-control analysis of 2002 running injuries. Br J Sports Med. 2002;36(2):95–101.
13. Blond L, Hansen L. Patellofemoral pain syndrome in athletes: a 5.7-year retrospective follow-up study of 250 athletes. Acta Orthop Belg. 1998;64(4):393–400.
14. Kannus P, Aho H, Jarvinen M, Niittymaki S. Computerized recording of visits to an outpatient sports clinic. Am J Sports Med. 1987;15(1):79–85.
15. Thijs Y, Van Tiggelen D, Roosen P, De Clercq D, Witvrouw E. A prospective study on gait-related intrinsic risk factors for patellofemoral pain. Clin J Sport Med. 2007;17(6):437–45.
16. DeHaven KE, Lintner DM. Athletic injuries: comparison by age, sport, and gender. Am J Sports Med. 1986;14(3):218–24.
17. Stracciolini A, Casciano R, Levey Friedman H, Stein CJ, Meehan WP 3rd, Micheli LJ. Pediatric sports injuries: a comparison of males versus females. Am J Sports Med. 2014;42(4):965–72.
18. Reider B, Marshall JL, Koslin B, Ring B, Girgis FG. The anterior aspect of the knee joint. J Bone Joint Surg Am. 1981;63(3):351–6.
19. Tria AJ Jr, Palumbo RC, Alicea JA. Conservative care for patellofemoral pain. Orthop Clin North Am. 1992;23(4):545–54.
20. Panni AS, Cerciello S, Maffulli N, Di Cesare M, Servien E, Neyret P. Patellar shape can be a predisposing factor in patellar instability. Knee Surg Sports Traumatol Arthrosc. 2011;19(4):663–70.
21. Dejour D, Saggin P. The sulcus deepening trochleoplasty—the Lyon's procedure. Int Orthop. 2010;34(2):311–6.
22. Grelsamer RP, Proctor CS, Bazos AN. Evaluation of patellar shape in the sagittal plane. A clinical analysis. Am J Sports Med. 1994;22(1):61–6.
23. Kwak SD, Colman WW, Ateshian GA, Grelsamer RP, Henry JH, Mow VC. Anatomy of the human patellofemoral joint articular cartilage: surface curvature analysis. J Orthop Res. 1997;15(3):468–72.
24. Cohen ZA, Henry JH, McCarthy DM, Mow VC, Ateshian GA. Computer simulations of patellofemoral joint surgery. Patient-specific models for tuberosity transfer. Am J Sports Med. 2003;31(1):87–98.
25. Strauss EJ, Galos DK. The evaluation and management of cartilage lesions affecting the patellofemoral joint. Curr Rev Musculoskelet Med. 2013;6(2):141–9.
26. Froimson MI, Ratcliffe A, Gardner TR, Mow VC. Differences in patellofemoral joint cartilage material properties and their significance to the etiology of cartilage surface fibrillation. Osteoarthritis Cartilage. 1997;5(6):377–86.
27. Ateshian GA, Kwak SD, Soslowsky LJ, Mow VC. A stereophotogrammetric method for determining in situ contact areas in diarthrodial joints, and a comparison with other methods. J Biomech. 1994;27(1):111–24.
28. Raimondo RA, Ahmad CS, Blankevoort L, April EW, Grelsamer RP, Henry JH. Patellar stabilization: a quantitative evaluation of the vastus medialis obliquus muscle. Orthopedics. 1998;21(7):791–5.
29. Farahmand F, Senavongse W, Amis AA. Quantitative study of the quadriceps muscles and trochlear groove geometry related to instability of the patellofemoral joint. J Orthop Res. 1998;16(1):136–43.
30. Placella G, Tei M, Sebastiani E, et al. Anatomy of the medial patello-femoral ligament: a systematic review of the last 20 years literature. Musculoskelet Surg. 2015;99(2):93–103.
31. Aragao JA, Reis FP, de Vasconcelos DP, Feitosa VL, Nunes MA. Metric measurements and attachment levels of the medial patellofemoral ligament: an anatomical study in cadavers. Clinics (Sao Paulo). 2008;63(4):541–4.
32. Hinckel BB, Gobbi RG, Demange MK, et al. Medial patellofemoral ligament, medial patellotibial ligament, and medial patellomeniscal ligament: anatomic, histologic, radiographic, and biomechanical study. Arthroscopy. 2017;33(10):1862–73.
33. Steensen RN, Dopirak RM, McDonald WG 3rd. The anatomy and isometry of the medial patellofemoral ligament: implications for reconstruction. Am J Sports Med. 2004;32(6):1509–13.

34. Amis AA, Firer P, Mountney J, Senavongse W, Thomas NP. Anatomy and biomechanics of the medial patellofemoral ligament. Knee. 2003;10(3):215–20.
35. Nomura E, Inoue M, Osada N. Anatomical analysis of the medial patellofemoral ligament of the knee, especially the femoral attachment. Knee Surg Sports Traumatol Arthrosc. 2005;13(7):510–5.
36. Conlan T, Garth WP Jr, Lemons JE. Evaluation of the medial soft-tissue restraints of the extensor mechanism of the knee. J Bone Joint Surg Am. 1993;75(5):682–93.
37. Hautamaa PV, Fithian DC, Kaufman KR, Daniel DM, Pohlmeyer AM. Medial soft tissue restraints in lateral patellar instability and repair. Clin Orthop Relat Res. 1998;(349):174–182.
38. Yeung M, Leblanc MC, Ayeni OR, et al. Indications for medial patellofemoral ligament reconstruction: a systematic review. J Knee Surg. 2016;29(7):543–54.
39. Victor J, Wong P, Witvrouw E, Sloten JV, Bellemans J. How isometric are the medial patellofemoral, superficial medial collateral, and lateral collateral ligaments of the knee? Am J Sports Med. 2009;37(10):2028–36.
40. Stephen J, Ephgrave C, Ball S, Church S. Current concepts in the management of patellofemoral pain—the role of alignment. Knee. 2020;27(2):280–6.
41. Philippot R, Boyer B, Testa R, Farizon F, Moyen B. The role of the medial ligamentous structures on patellar tracking during knee flexion. Knee Surg Sports Traumatol Arthrosc. 2012;20(2):331–6.
42. Gu W, Pandy M. Direct validation of human knee-joint contact mechanics derived from subject-specific finite-element models of the tibiofemoral and patellofemoral joints. J Biomech Eng. 2020;142(7):071001.
43. Fulkerson JP, Gossling HR. Anatomy of the knee joint lateral retinaculum. Clin Orthop Relat Res. 1980;(153):183–188.
44. Capkin S, Zeybek G, Ergur I, Kosay C, Kiray A. An anatomic study of the lateral patellofemoral ligament. Acta Orthop Traumatol Turc. 2017;51(1):73–6.
45. Merican AM, Sanghavi S, Iranpour F, Amis AA. The structural properties of the lateral retinaculum and capsular complex of the knee. J Biomech. 2009;42(14):2323–9.
46. Ishibashi Y, Okamura Y, Otsuka H, Tsuda E, Toh S. Lateral patellar retinaculum tension in patellar instability. Clin Orthop Relat Res. 2002;(397):362–369.
47. Bedi H, Marzo J. The biomechanics of medial patellofemoral ligament repair followed by lateral retinacular release. Am J Sports Med. 2010;38(7):1462–7.
48. Scapinelli R. Blood supply of the human patella. Its relation to ischaemic necrosis after fracture. J Bone Joint Surg Br. 1967;49(3):563–70.
49. Hungerford DS, Barry M. Biomechanics of the patellofemoral joint. Clin Orthop Relat Res. 1979;(144):9–15.
50. Schindler OS, Scott WN. Basic kinematics and biomechanics of the patello-femoral joint. Part 1: the native patella. Acta Orthop Belg. 2011;77(4):421–31.
51. Aglietti P, Insall JN, Cerulli G. Patellar pain and incongruence. I: measurements of incongruence. Clin Orthop Relat Res. 1983;(176):217–224.
52. Grelsamer RP, Dubey A, Weinstein CH. Men and women have similar Q angles: a clinical and trigonometric evaluation. J Bone Joint Surg Br. 2005;87(11):1498–501.
53. Hsu RW, Himeno S, Coventry MB, Chao EY. Normal axial alignment of the lower extremity and load-bearing distribution at the knee. Clin Orthop Relat Res. 1990;(255):215–227.
54. Skalley TC, Terry GC, Teitge RA. The quantitative measurement of normal passive medial and lateral patellar motion limits. Am J Sports Med. 1993;21(5):728–32.
55. Woodland LH, Francis RS. Parameters and comparisons of the quadriceps angle of college-aged men and women in the supine and standing positions. Am J Sports Med. 1992;20(2):208–11.
56. Johnson LL, van Dyk GE, Green JR 3rd, et al. Clinical assessment of asymptomatic knees: comparison of men and women. Arthroscopy. 1998;14(4):347–59.
57. Sojbjerg JO, Lauritzen J, Hvid I, Boe S. Arthroscopic determination of patellofemoral malalignment. Clin Orthop Relat Res. 1987;(215):243–247.
58. Post WR. Clinical evaluation of patients with patellofemoral disorders. Arthroscopy. 1999;15(8):841–51.
59. Post WR. Anterior knee pain: diagnosis and treatment. J Am Acad Orthop Surg. 2005;13(8):534–43.
60. Fox AJ, Wanivenhaus F, Rodeo SA. The basic science of the patella: structure, composition, and function. J Knee Surg. 2012;25(2):127–41.
61. Sanchis-Alfonso V. How to deal with chronic patellar instability: what does the literature tell us? Sports Health. 2016;8(1):86–90.
62. Staubli HU, Durrenmatt U, Porcellini B, Rauschning W. Anatomy and surface geometry of the patellofemoral joint in the axial plane. J Bone Joint Surg Br. 1999;81(3):452–8.
63. Tecklenburg K, Dejour D, Hoser C, Fink C. Bony and cartilaginous anatomy of the patellofemoral joint. Knee Surg Sports Traumatol Arthrosc. 2006;14(3):235–40.
64. Huberti HH, Hayes WC, Stone JL, Shybut GT. Force ratios in the quadriceps tendon and ligamentum patellae. J Orthop Res. 1984;2(1):49–54.
65. Elias JJ, Cech JA, Weinstein DM, Cosgrea AJ. Reducing the lateral force acting on the patella does not consistently decrease patellofemoral pressures. Am J Sports Med. 2004;32(5):1202–8.
66. Desio SM, Burks RT, Bachus KN. Soft tissue restraints to lateral patellar translation in the human knee. Am J Sports Med. 1998;26(1):59–65.
67. Feller JA, Amis AA, Andrish JT, Arendt EA, Erasmus PJ, Powers CM. Surgical biomechanics of the patellofemoral joint. Arthroscopy. 2007;23(5):542–53.

68. Lorenz A, Muller O, Kohler P, Wunschel M, Wulker N, Leichtle UG. The influence of asymmetric quadriceps loading on patellar tracking—an in vitro study. Knee. 2012;19(6):818–22.
69. Wilson NA, Sheehan FT. Dynamic in vivo 3-dimensional moment arms of the individual quadriceps components. J Biomech. 2009;42(12):1891–7.
70. Pal S, Besier TF, Gold GE, Fredericson M, Delp SL, Beaupre GS. Patellofemoral cartilage stresses are most sensitive to variations in vastus medialis muscle forces. Comput Methods Biomech Biomed Engin. 2019;22(2):206–16.
71. Sakai N, Luo ZP, Rand JA, An KN. The influence of weakness in the vastus medialis oblique muscle on the patellofemoral joint: an in vitro biomechanical study. Clin Biomech (Bristol, Avon). 2000;15(5):335–9.
72. Amis AA, Senavongse W, Bull AM. Patellofemoral kinematics during knee flexion-extension: an in vitro study. J Orthop Res. 2006;24(12):2201–11.
73. Cheung RT, Mok NW, Chung PY, Ng GY. Non-invasive measurement of the patellofemoral movements during knee extension-flexion: a validation study. Knee. 2013;20(3):213–7.
74. Nha KW, Papannagari R, Gill TJ, et al. In vivo patellar tracking: clinical motions and patellofemoral indices. J Orthop Res. 2008;26(8):1067–74.
75. Suzuki T, Hosseini A, Li JS, Gill TJ, Li G. In vivo patellar tracking and patellofemoral cartilage contacts during dynamic stair ascending. J Biomech. 2012;45(14):2432–7.
76. Yao J, Yang B, Niu W, et al. In vivo measurements of patellar tracking and finite helical axis using a static magnetic resonance based methodology. Med Eng Phys. 2014;36(12):1611–7.
77. Yu Z, Yao J, Wang X, et al. Research methods and Progress of patellofemoral joint kinematics: a review. J Healthc Eng. 2019;2019:9159267.
78. Cohen ZA, Mow VC, Henry JH, Levine WN, Ateshian GA. Templates of the cartilage layers of the patellofemoral joint and their use in the assessment of osteoarthritic cartilage damage. Osteoarthritis Cartilage. 2003;11(8):569–79.
79. Huberti HH, Hayes WC. Patellofemoral contact pressures. The influence of q-angle and tendofemoral contact. J Bone Joint Surg Am. 1984;66(5):715–24.
80. Balcarek P, Jung K, Frosch KH, Sturmer KM. Value of the tibial tuberosity-trochlear groove distance in patellar instability in the young athlete. Am J Sports Med. 2011;39(8):1756–61.
81. Arendt EA. Core strengthening. Instr Course Lect. 2007;56:379–84.
82. Pollard CD, Sigward SM, Powers CM. Gender differences in hip joint kinematics and kinetics during side-step cutting maneuver. Clin J Sport Med. 2007;17(1):38–42.
83. Willson JD, Dougherty CP, Ireland ML, Davis IM. Core stability and its relationship to lower extremity function and injury. J Am Acad Orthop Surg. 2005;13(5):316–25.
84. Souza RB, Draper CE, Fredericson M, Powers CM. Femur rotation and patellofemoral joint kinematics: a weight-bearing magnetic resonance imaging analysis. J Orthop Sports Phys Ther. 2010;40(5):277–85.
85. Bolgla LA, Malone TR, Umberger BR, Uhl TL. Hip strength and hip and knee kinematics during stair descent in females with and without patellofemoral pain syndrome. J Orthop Sports Phys Ther. 2008;38(1):12–8.
86. Piva SR, Goodnite EA, Childs JD. Strength around the hip and flexibility of soft tissues in individuals with and without patellofemoral pain syndrome. J Orthop Sports Phys Ther. 2005;35(12):793–801.
87. Dolak KL, Silkman C, Medina McKeon J, Hosey RG, Lattermann C, Uhl TL. Hip strengthening prior to functional exercises reduces pain sooner than quadriceps strengthening in females with patellofemoral pain syndrome: a randomized clinical trial. J Orthop Sports Phys Ther. 2011;41(8):560–70.
88. Fukuda TY, Melo WP, Zaffalon BM, et al. Hip posterolateral musculature strengthening in sedentary women with patellofemoral pain syndrome: a randomized controlled clinical trial with 1-year follow-up. J Orthop Sports Phys Ther. 2012;42(10):823–30.
89. Fukuda TY, Rossetto FM, Magalhaes E, Bryk FF, Lucareli PR, de Almeida Aparecida Carvalho N. Short-term effects of hip abductors and lateral rotators strengthening in females with patellofemoral pain syndrome: a randomized controlled clinical trial. J Orthop Sports Phys Ther. 2010;40(11):736–42.
90. Khayambashi K, Mohammadkhani Z, Ghaznavi K, Lyle MA, Powers CM. The effects of isolated hip abductor and external rotator muscle strengthening on pain, health status, and hip strength in females with patellofemoral pain: a randomized controlled trial. J Orthop Sports Phys Ther. 2012;42(1):22–9.
91. Nakagawa TH, Muniz TB, Baldon Rde M, Dias Maciel C, de Menezes Reiff RB, Serrao FV. The effect of additional strengthening of hip abductor and lateral rotator muscles in patellofemoral pain syndrome: a randomized controlled pilot study. Clin Rehabil. 2008;22(12):1051–60.
92. Boling MC, Padua DA, Marshall SW, Guskiewicz K, Pyne S, Beutler A. A prospective investigation of biomechanical risk factors for patellofemoral pain syndrome: the joint undertaking to monitor and prevent ACL injury (JUMP-ACL) cohort. Am J Sports Med. 2009;37(11):2108–16.
93. Leetun DT, Ireland ML, Willson JD, Ballantyne BT, Davis IM. Core stability measures as risk factors for lower extremity injury in athletes. Med Sci Sports Exerc. 2004;36(6):926–34.
94. Liebensteiner MC, Ressler J, Seitlinger G, Djurdjevic T, El Attal R, Ferlic PW. High femoral anteversion is related to femoral trochlea dysplasia. Arthroscopy. 2016;32(11):2295–9.

95. Tiberio D. The effect of excessive subtalar joint pronation on patellofemoral mechanics: a theoretical model. J Orthop Sports Phys Ther. 1987;9(4):160–5.
96. Neal BS, Griffiths IB, Dowling GJ, et al. Foot posture as a risk factor for lower limb overuse injury: a systematic review and meta-analysis. J Foot Ankle Res. 2014;7(1):55.
97. Insall J, Salvati E. Patella position in the normal knee joint. Radiology. 1971;101(1):101–4.
98. Blackburne JS, Peel TE. A new method of measuring patellar height. J Bone Joint Surg Br. 1977;59(2):241–2.
99. Caton J, Deschamps G, Chambat P, Lerat JL, Dejour H. [Patella infera. Apropos of 128 cases]. Rev Chir Orthop Reparatrice Appar Mot. 1982;68(5):317–325.
100. Geenen E, Molenaers G, Martens M. Patella alta in patellofemoral instability. Acta Orthop Belg. 1989;55(3):387–93.
101. Neyret P, Robinson AH, Le Coultre B, Lapra C, Chambat P. Patellar tendon length—the factor in patellar instability? Knee. 2002;9(1):3–6.
102. Simmons E, Jr., Cameron JC. Patella alta and recurrent dislocation of the patella. Clin Orthop Relat Res. 1992;(274):265–269.
103. Belkin NS, Meyers KN, Redler LH, Maher S, Nguyen JT, Shubin Stein BE. Medial patellofemoral ligament isometry in the setting of patella alta. Arthroscopy. 2020;36(12):3031–6.
104. Lancourt JE, Cristini JA. Patella alta and patella infera. Their etiological role in patellar dislocation, chondromalacia, and apophysitis of the tibial tubercle. J Bone Joint Surg Am. 1975;57(8):1112–5.
105. Bertollo N, Pelletier MH, Walsh WR. Relationship between patellar tendon shortening and in vitro kinematics in the ovine stifle joint. Proc Inst Mech Eng H. 2013;227(4):438–47.
106. Meyer SA, Brown TD, Pedersen DR, Albright JP. Retropatellar contact stress in simulated patella infera. Am J Knee Surg. 1997;10(3):129–38.
107. Upadhyay N, Vollans SR, Seedhom BB, Soames RW. Effect of patellar tendon shortening on tracking of the patella. Am J Sports Med. 2005;33(10):1565–74.
108. Goodfellow J, Hungerford DS, Zindel M. Patellofemoral joint mechanics and pathology. 1. Functional anatomy of the patello-femoral joint. J Bone Joint Surg Br. 1976;58(3):287–90.
109. White BJ, Sherman OH. Patellofemoral instability. Bull NYU Hosp Jt Dis. 2009;67(1):22–9.
110. Ahmed AM, Burke DL. In-vitro measurement of static pressure distribution in synovial joints—part I: tibial surface of the knee. J Biomech Eng. 1983;105(3):216–25.
111. Ahmed AM, Burke DL, Hyder A. Force analysis of the patellar mechanism. J Orthop Res. 1987;5(1):69–85.
112. Retaillaud JL, Darmana R, Devallet P, Mansat M, Morucci JP. [Experimental biomechanical study of the advancement of tibial tuberosity]. Rev Chir Orthop Reparatrice Appar Mot. 1989;75(8):513–523.
113. Ahmed AM, Burke DL, Yu A. In-vitro measurement of static pressure distribution in synovial joints—part II: retropatellar surface. J Biomech Eng. 1983;105(3):226–36.
114. Hehne HJ. Biomechanics of the patellofemoral joint and its clinical relevance. Clin Orthop Relat Res. 1990;(258):73–85.
115. Matthews LS, Sonstegard DA, Henke JA. Load bearing characteristics of the patello-femoral joint. Acta Orthop Scand. 1977;48(5):511–6.
116. Bellemans J. Biomechanics of anterior knee pain. Knee. 2003;10(2):123–6.
117. Salsich GB, Ward SR, Terk MR, Powers CM. In vivo assessment of patellofemoral joint contact area in individuals who are pain free. Clin Orthop Relat Res. 2003;(417):277–284.
118. Ward SR, Powers CM. The influence of patella alta on patellofemoral joint stress during normal and fast walking. Clin Biomech (Bristol, Avon). 2004;19(10):1040–7.
119. Ward SR, Terk MR, Powers CM. Influence of patella alta on knee extensor mechanics. J Biomech. 2005;38(12):2415–22.
120. Ward SR, Terk MR, Powers CM. Patella alta: association with patellofemoral alignment and changes in contact area during weight-bearing. J Bone Joint Surg Am. 2007;89(8):1749–55.
121. Hinterwimmer S, von Eisenhart-Rothe R, Siebert M, Welsch F, Vogl T, Graichen H. Patella kinematics and patello-femoral contact areas in patients with genu varum and mild osteoarthritis. Clin Biomech (Bristol, Avon). 2004;19(7):704–10.
122. Ho KY, Keyak JH, Powers CM. Comparison of patella bone strain between females with and without patellofemoral pain: a finite element analysis study. J Biomech. 2014;47(1):230–6.
123. Kuroda R, Kambic H, Valdevit A, Andrish JT. Articular cartilage contact pressure after tibial tuberosity transfer. A cadaveric study. Am J Sports Med. 2001;29(4):403–9.
124. Lee TQ, Sandusky MD, Adeli A, McMahon PJ. Effects of simulated vastus medialis strength variation on patellofemoral joint biomechanics in human cadaver knees. J Rehabil Res Dev. 2002;39(3):429–38.
125. Schindler OS. Basic kinematics and biomechanics of the patellofemoral joint part 2: the patella in total knee arthroplasty. Acta Orthop Belg. 2012;78(1):11–29.
126. Akbarshahi M, Fernandez JW, Schache AG, Pandy MG. Subject-specific evaluation of patellofemoral joint biomechanics during functional activity. Med Eng Phys. 2014;36(9):1122–33.
127. Van Haver A, De Roo K, De Beule M, et al. The effect of trochlear dysplasia on patellofemoral biomechanics: a cadaveric study with simulated trochlear deformities. Am J Sports Med. 2015;43(6):1354–61.

128. Cohen ZA, Roglic H, Grelsamer RP, et al. Patellofemoral stresses during open and closed kinetic chain exercises. An analysis using computer simulation. Am J Sports Med. 2001;29(4):480–7.
129. Ho KY, French T, Klein B, Lee Y. Patellofemoral joint stress during incline and decline running. Phys Ther Sport. 2018;34:136–40.
130. Dos Santos AF, Nakagawa TH, Serrao FV, Ferber R. Patellofemoral joint stress measured across three different running techniques. Gait Posture. 2019;68:37–43.
131. Pollard JP, Porter WL, Redfern MS. Forces and moments on the knee during kneeling and squatting. J Appl Biomech. 2011;27(3):233–41.
132. Kernozek TW, Gheidi N, Zellmer M, Hove J, Heinert BL, Torry MR. Effects of anterior knee displacement during squatting on patellofemoral joint stress. J Sport Rehabil. 2018;27(3):237–43.

2 Patellofemoral Anatomy, Mechanics, and Evaluation: Patient and Family History in the Evaluation of Patellofemoral Patients

John Fritch, Jason L. Koh, and Shital N. Parikh

2.1 Event History

A patient's first patellar dislocation is a memorable and often traumatic event. Despite this, often the patient's description of the event will not clearly lend itself to a patellar dislocation. Usually patients will not present with a frank dislocation as they often spontaneously reduce. Patients often describe a painful, swollen, and guarded knee after an acute event [1, 2]. The clinician should ask if the kneecap appeared out of place or if it required a reduction maneuver. Particular attention should be paid to pain location and provocation. Pain at the medial epicondyle, medial patellar facet, and lateral femoral condyle should provide clues to a dislocation event. Similarly, an effusion and locking of the knee should raise suspicion for a possible associated osteochondral injury. Careful questioning should help discern between sensations of patellar subluxation, dislocation, and the knee giving way. The clinician must also obtain a thorough history about the exact activity the patient was performing at the time of injury, such as cutting, pivoting, and twisting.

J. Fritch
Department of Orthopaedic Surgery, Pritzker School of Medicine, University of Chicago, Chicago, IL, USA

J. L. Koh
Orthopaedic and Spine Institute, NorthShore University HealthSystem, Skokie, IL, USA

Pritzker School of Medicine, University of Chicago, Chicago, IL, USA

Northwestern University McCormick School of Engineering, Evanston, IL, USA

S. N. Parikh (✉)
Orthopaedic Sports Medicine, Cincinnati Children's Hospital, Cincinnati, OH, USA

University of Cincinnati School of Medicine, Cincinnati, OH, USA
e-mail: Shital.Parikh@cchmc.org

2.2 Previous Instability

The biggest predictor of recurrent instability is any previous dislocation or subluxation. Several recent studies estimate the rate of recurrent instability is between 17 and 30% [3, 4]. This rate varies widely across studies most likely due to the multifactorial nature of patellar instability and design flaws within some of the historical studies. Despite this, it is widely accepted that those patients sustaining more than one prior dislocation are at much higher risk for subsequent dislocation events. In a large epidemiological study, Fithian et al. found that 49% of patients with two prior dislocations sustained another in the same knee [4].

Not surprisingly, patients with a history of patellar instability were at higher risk for instability in the contralateral knee. In one study, risk of contralateral instability was six times higher for patients with recurrent instability in the index knee [4]. Conversely, the odds of recurrent insta-

© ISAKOS 2022
J. L. Koh et al. (eds.), *The Patellofemoral Joint*, https://doi.org/10.1007/978-3-030-81545-5_2

bility was about three times higher in patients with a history of a contralateral dislocation [5]. Clearly, those patients with bilateral instability are at higher risk for recurrence in either knee.

2.3 Age

Young age has been consistently identified as a risk factor for recurrent patellar instability. The association is not completely understood, but is most likely due to high risk individuals sustaining injury early and frequently in life. Peak incidence of patellar instability is the second decade of life. Median age for first-time dislocation is 16 years and recurrent dislocation is 21 years [4]. One study found when instability started before age 16, the odds ratio of recurrence was 11.2 [6]. Recurrent instability rates of 52–60% have been reported in patients under the age of 15 compared to 26–33% in patients 15–18 years old [7, 8]. When considering bone age or maturity, skeletally immature patients have more than twice the risk of recurrent instability than skeletally mature patients, 43.3% and 21.6%, respectively [5, 9]. Although young age is a risk factor for repeat dislocation, increasing age is protective. In one study, for each year increase in age after first dislocation the risk of recurrence decreased by 8% with no recurrence past age 40 [9].

2.4 Gender

Historically, it has been accepted that females between 10 and 17 years old are at highest risk for first time and recurrent patellar instability [4]. The same study reports that the risk of recurrent instability is three times greater in females than males [4]. This gender discrepancy was thought to be due to increased incidence of malalignment and joint laxity in females. More recent studies suggest no difference in patellar instability between males and females [5, 6, 10]. A systematic review by Stefancin et al. reports that the overall incidence of patellar dislocation between males and females is nearly equal: 47% versus 53% [3]. It is possible that previous studies captured a disproportionate patient population; however, this has not been proven.

2.5 Activity

A critical aspect of the patient's history is the activity they were engaged in at the time of patellar dislocation. Injuries can occur from either direct or indirect trauma to the knee; however, non-contact injuries occur more commonly [11]. Multiple studies have cited 50–60% of patellar dislocations occur during sports activities [4, 12, 13]. This is most likely due to sports activities placing the knee in a vulnerable position for patellar instability. Specifically, a position of knee valgus and internal rotation on a planted foot produces a lateral vector on the patella which may result in injury [2]. Sports that have been implicated as higher risk in patellar instability include gymnastics, football, wrestling, basketball, and dancing [4, 12, 14].

Additionally, patients that initially dislocated during a sport activity were at higher risk for recurrence (HR 1.97) presumably due to return to high risk activity [9].

Importantly, there appears to be two subsets of activities associated with patellar instability. Patellar dislocations in the setting of significant trauma such as during contact sports or a direct blow to the knee usually demonstrate normal patellofemoral anatomy and have a lower risk of recurrent instability. Conversely, patellar instability after low risk activities are likely due to the presence of underlying anatomical risk factors or joint hyperlaxity and are at a higher risk for recurrence [15, 16].

2.6 Personal History

Obtaining a complete history is important as 9–15% of patients with a patellar dislocation will have a positive family history [11, 17]. Additionally, patients with factors associated with developmental dysplasia of the hip at the time of birth or delivery by cesarean section had higher odds of contralateral instability [4].

Connective tissue disorders such as Ehlers-Danlos Syndrome or generalized ligamentous hyperlaxity are important to identify as those patients are at higher risk of recurrent instability. Any underlying disorders such as cerebral palsy and Down syndrome should be identified as they may have chronic patellar dislocations and require different treatment algorithms.

2.7 Previous Treatment History

2.7.1 Conservative Management

Although most first-time patellar dislocations are treated conservatively, it is important to understand the natural history of non-operative management and recurrence of instability. Out of patients treated conservatively for a first-time dislocation, one third return to activity without consequences, one third have another dislocation requiring surgical stabilization, and one third do not have another dislocation but continue to have symptoms and are unable to return to previous level of activity [18]. Most conservative treatment involves physical therapy and varying degrees of immobilization or bracing. In a long term study of patients with a primary patellar dislocation, those treated in a patellar brace had greater than three times the risk of repeat dislocation compared to those treated in a posterior splint, however those treated in a splint or cast had the highest rate of stiffness [19]. In general, rehabilitation protocols now focus on early protected mobilization to minimize stiffness while preventing repeat instability.

When examining the outcomes of first-time patellar dislocations treated conservatively versus operatively, it is still unclear which treatment is superior. A prospective randomized study compared 62 patients treated either operatively or conservatively and found no difference in outcome, function, instability, or activity [20]. Similarly, Buchner et al. showed no difference between patients treated with early operative stabilization and those treated conservatively for a first-time dislocation with respect to repeat dislocation, activity level, functional and subjective outcomes [7]. Multiple randomized control trials of operatively treated primary dislocations failed to improve recurrent dislocation rate compared to conservative management [21–25]. A study by Arnbjornsson et al. followed patients with bilateral patellar dislocations for 14 years. One side was treated operatively and the other conservatively. They found that at long term follow-up the operative extremity had worse arthritis and instability than the nonoperatively treated side [26]. It appears that operative treatment of first-time dislocations carries a more substantial complication profile while offering only similar protection from repeat dislocation as non-operative treatment.

2.7.2 Surgical Treatment

Similar to conservative treatment, it is important to know any operative treatment a patellar instability patient had undergone to better understand their risk for recurrence. Although more recent research has led to treatment algorithms resulting in better stability, historically there has been increased risk of recurrent instability and other complications with certain procedures. One such procedure is an arthroscopic lateral retinacular release, which can lead to lateral patellar mobility and medial instability and is no longer supported as treatment for patellar instability [27, 28]. Similarly, MPFL repair or refixation after primary dislocation does not prevent further instability [20–24].

Over time, the success of various surgical techniques has been tested by other patient factors. For example, results of medial plication techniques had satisfactory results at short-term follow-up, however, in the presence of trochlear dysplasia, early redislocation occurred commonly [25, 29].

More recently it has been widely accepted that MPFL reconstruction is a reliable procedure for patellar instability. Redislocation rates after MPFL reconstruction are low; several different studies have published rates of 0–5% [30–33]. However, MPFL reconstruction should not be considered a universal treatment as other patient factors can affect its success. For example, Hopper et al. reported a 100% rate of redislocation in patients with an isolated MPFL reconstruction in the setting of severe trochlear dysplasia [34]. In cases of rotational malalignment, recent literature suggests the addition of tibial tubercle transfer can help decrease recurrent instability. Allen et al. report combined MPFL reconstruction with tibial tubercle anteromedialization has promising results in patients with recurrent patellar instability, citing a redislocation rate of 3% [35]. Tibial tubercle transfer has its own complication profile including overmedialization which can result in increased medial patellofemoral pressure and pain. One author suggests this can be avoided by reserving medialization for cases of lateral patellofemoral chondrosis only and intraoperatively correcting the tubercle sulcus angle to zero [36].

2.8 Importance and Implications

The accurate identification of demographics and risk factors associated with patellar instability are becoming increasingly important in determining treatment algorithms to improve patient outcomes and reduce disability. There is still much unknown about patellar instability. Meetings of experts in the field such as the AOSSM/PFF and International Patellofemoral Study Group (IPSG) help by publishing consensus statements providing useful insight into the evaluation and treatment of patellar instability [37, 38]. Additionally, multicenter trials such as JUPITER (Justifying Patellar Instability Treatment by Early Results) will be critical to learning more about treatment and prognosis of patellar instability across an increasingly diverse patient population [39].

Take Home Message

- Patellar instability is multifactorial in nature.
- Careful history taking can help identify risk factors for recurrence.
- Patient history including age, gender, family history, and ligamentous laxity plays a role.
- Pay attention to mechanism, history of ipsilateral or contralateral instability, and previous treatments.
- Future multicenter research will help define demographics, prognostic factors, and treatment algorithms.

References

1. Koh JL, Stewart C. Patellar instability. Clin Sports Med. 2014;33(3):461–76.
2. Hinton RY, Sharma KM. Acute and recurrent patellar instability in the young athlete. Orthop Clin North Am. 2003;34(3):385–96.
3. Stefancin JJ, Parker RD. First-time traumatic patellar dislocation: a systematic review. Clin Orthop Relat Res. 2007;455:93–101.
4. Fithian DC, Paxton EW, Stone ML, Silva P, Davis DK, Elias DA, et al. Epidemiology and natural history of acute patellar dislocation. Am J Sports Med. 2004;32(5):1114–21.
5. Jaquith BP, Parikh SN. Predictors of recurrent patellar instability in children and adolescents after first-time dislocation. J Pediatr Orthop. 2017;37(7):484–90.
6. Balcarek P, Oberthur S, Hopfensitz S, Frosch S, Walde TA, Wachowski MM, et al. Which patellae are likely to redislocate? Knee Surg Sports Traumatol Arthrosc. 2014;22(10):2308–14.
7. Buchner M, Baudendistel B, Sabo D, Schmitt H. Acute traumatic primary patellar dislocation—long-term results comparing conservative and surgical treatment. Clin J Sport Med. 2005;15(2):62–6.
8. Cash JD, Hughston JC. Treatment of acute patellar dislocation. Am J Sports Med. 1988;16(3):244–9.
9. Lewallen L, McIntosh A, Dahm D. First-time patellofemoral dislocation: risk factors for recurrent instability. J Knee Surg. 2015;28(4):303–9.
10. Lewallen LW, McIntosh AL, Dahm DL. Predictors of recurrent instability after acute patellofemoral dislocation in pediatric and adolescent patients. Am J Sports Med. 2013;41(3):575–81.
11. Atkin DM, Fithian DC, Marangi KS, Stone ML, Dobson BE, Mendelsohn C. Characteristics of patients with primary acute lateral patellar dislocation and their recovery within the first 6 months of injury. Am J Sports Med. 2000;28(4):472–9.
12. Waterman BR, Belmont PJ Jr, Owens BD. Patellar dislocation in the United States: role of sex, age, race, and athletic participation. J Knee Surg. 2012;25(1):51–7.
13. Sillanpaa P, Mattila VM, Iivonen T, Visuri T, Pihlajamaki H. Incidence and risk factors of acute traumatic primary patellar dislocation. Med Sci Sports Exerc. 2008;40(4):606–11.
14. Mitchell J, Magnussen RA, Collins CL, Currie DW, Best TM, Comstock RD, et al. Epidemiology of patellofemoral instability injuries among high school

athletes in the United States. Am J Sports Med. 2015;43(7):1676–82.
15. Parikh SN, Lykissas MG, Gkiatas I. Predicting risk of recurrent patellar dislocation. Curr Rev Musculoskelet Med. 2018;11(2):253–60.
16. Beasley LS, Vidal AF. Traumatic patellar dislocation in children and adolescents: treatment update and literature review. Curr Opin Pediatr. 2004;16(1):29–36.
17. Stanitski CL. Patellar instability in the school age athlete. Instr Course Lect. 1998;47:345–50.
18. Magnussen RA, Verlage M, Stock E, Zurek L, Flanigan DC, Tompkins M, et al. Primary patellar dislocations without surgical stabilization or recurrence: how well are these patients really doing? Knee Surg Sports Traumatol Arthrosc. 2017;25(8):2352–6.
19. Maenpaa H, Lehto MUK. Patellar dislocation - the long-term results of nonoperative management in 100 patients. Am J Sports Med. 1997;25(2):213–7.
20. Palmu S, Kallio PE, Donell ST, Helenius I, Nietosvaara Y. Acute patellar dislocation in children and adolescents: a randomized clinical trial. J Bone Joint Surg Am. 2008;90(3):463–70.
21. Christiansen SE, Jakobsen BW, Lund B, Lind M. Isolated repair of the medial patellofemoral ligament in primary dislocation of the patella: a prospective randomized study. Arthroscopy. 2008;24(8):881–7.
22. Nikku R, Nietosvaara Y, Kallio PE, Aalto K, Michelsson JE. Operative versus closed treatment of primary dislocation of the patella—similar 2-year results in 125 randomized patients. Acta Orthop Scand. 1997;68(5):419–23.
23. Nikku R, Nietosvaara Y, Aalto K, Kallio PE. Operative treatment of primary patellar dislocation does not improve medium-term outcome: a 7-year follow-up report and risk analysis of 127 randomized patients. Acta Orthop. 2005;76(5):699–704.
24. Sillanpaa PJ, Maenpaa HM, Mattila VM, Visuri T, Pihlajamaki H. Arthroscopic surgery for primary traumatic patellar dislocation a prospective, nonrandomized study comparing patients treated with and without acute arthroscopic stabilization with a median 7-year follow-up. Am J Sports Med. 2008;36(12):2301–9.
25. Sillanpaa PJ, Mattila VM, Maenpaa H, Kiuru M, Visuri T, Pihlajamaki H. Treatment with and without initial stabilizing surgery for primary traumatic patellar dislocation a prospective randomized study. J Bone Joint Surg Am. 2009;91A(2):263–73.
26. Arnbjornsson A, Egund N, Rydling O, Stockerup R, Ryd L. The natural-history of recurrent dislocation of the patella—long-term results of conservative and operative treatment. J Bone Joint Surg Br. 1992;74(1):140–2.
27. Christoforakis J, Bull AMJ, Strachan RK, Shymkiw R, Senavongse W, Amis AA. Effects of lateral retinacular release on the lateral stability of the patella. Knee Surg Sports Traumatol Arthrosc. 2006;14(3):273–7.
28. Hughston JC, Deese M. Medial subluxation of the patella as a complication of lateral retinacular release. Am J Sports Med. 1988;16(4):383–8.
29. Schoettle PB, Scheffler SU, Schwarck A, Weiler A. Arthroscopic medial retinacular repair after patellar dislocation with and without underlying trochlear dysplasia: a preliminary report. Arthroscopy. 2006;22(11):1192–8.
30. Howells NR, Barnett AJ, Ahearn N, Ansari A, Eldridge JD. Medial patellofemoral ligament reconstruction: a prospective outcome assessment of a large single centre series. J Bone Joint Surg Br. 2012;94B(9):1202–8.
31. Buckens CF, Saris DB. Reconstruction of the medial patellofemoral ligament for treatment of patellofemoral instability: a systematic review. Am J Sports Med. 2010;38(1):181–8.
32. Enderlein D, Nielsen T, Christiansen SE, Faunø P, Lind M. Clinical outcome after reconstruction of the medial patellofemoral ligament in patients with recurrent patella instability. Knee Surg Sports Traumatol Arthrosc. 2014;22(10):2458–64.
33. Parikh SN, Nathan ST, Wall EJ, Eismann EA. Complications of medial patellofemoral ligament reconstruction in young patients. Am J Sports Med. 2013;41(5):1030–8.
34. Hopper GP, Leach WJ, Rooney BP, Walker CR, Blyth MJ. Does degree of trochlear dysplasia and position of femoral tunnel influence outcome after medial patellofemoral ligament reconstruction? Am J Sports Med. 2014;42(3):716–22.
35. Allen MM, Krych AJ, Johnson NR, Mohan R, Stuart MJ, Dahm DL. Combined tibial tubercle osteotomy and medial patellofemoral ligament reconstruction for recurrent lateral patellar instability in patients with multiple anatomic risk factors. Arthroscopy. 2018;34(8):2420–6.e3.
36. Arendt EA. Editorial commentary: reducing the tibial tuberosity-trochlear groove distance in patella stabilization procedure. Too much of a (good) thing? Arthroscopy. 2018;34(8):2427–8.
37. Post WR, Fithian DC. Patellofemoral instability: a consensus statement from the AOSSM/PFF patellofemoral instability workshop. Orthop J Sports Med. 2018;6(1):2325967117750352.
38. Liu JN, Steinhaus ME, Kalbian IL, Post WR, Green DW, Strickland SM, et al. Patellar instability management: a survey of the international patellofemoral study group. Am J Sports Med. 2018;46(13):3299–306.
39. Bishop ME, Brady JM, Ling D, Parikh S, Stein BES. Descriptive epidemiology study of the justifying patellar instability treatment by early results (Jupiter) cohort. Orthop J Sports Med. 2019;7(3 Suppl):2325967119S00046.

Examination of the Patients with Patellofemoral Symptoms

3

Claudia Arias Calderón, Renato Andrade, Ricardo Bastos, Cristina Valente, Antonio Maestro, Rolando Suárez Peña, and João Espregueira-Mendes

Patellofemoral disorders can range from traumatic injuries, instability or those leading to patellofemoral pain syndrome (PFPS). PFPS is one of the most common diagnoses in patients complaining from anterior knee pain. The cause of pain is multifactorial, can be idiopathic with an incidence between 15 and 25%. A manifold of risk factors has been proposed [1, 2], but only a few shows significant association with patellofemoral pain, including quadriceps weakness and hip abduction strength [2]. A detail medical history and a structured physical exam will help to reach a more accurate diagnosis. In this chapter we present step-by-step examination of patients with patellofemoral symptoms.

3.1 Examination

The exam of the patellofemoral joint can be challenging and we have to consider all the contributing factors. Examination has to be accurate and comprehensive to correctly identify the most suitable treatment. However, examination findings are often subtle and poorly reproducible and not always related to the patient symptoms. There

C. A. Calderón · R. S. Peña
Hospital Nacional Edgardo Rebagliati Martins, Lima, Peru

R. Andrade
Clínica do Dragão, Espregueira-Mendes Sports Centre—FIFA Medical Centre of Excellence, Porto, Portugal

Dom Henrique Research Centre, Porto, Portugal

Faculty of Sports, University of Porto, Porto, Portugal
e-mail: randrade@espregueira.com

R. Bastos · C. Valente
Clínica do Dragão, Espregueira-Mendes Sports Centre—FIFA Medical Centre of Excellence, Porto, Portugal

Dom Henrique Research Centre, Porto, Portugal
e-mail: rbastos@espregueira.com; cvalente@saudeatlantica.pt

A. Maestro
Hospital Begona, Gijón, Spain

Real Sporting de Gijón SAD, Gijón, Spain

J. Espregueira-Mendes (✉)
Clínica do Dragão, Espregueira-Mendes Sports Centre—FIFA Medical Centre of Excellence, Porto, Portugal

Dom Henrique Research Centre, Porto, Portugal

ICVS/3B's–PT Government Associate Laboratory, Braga, Portugal

3B's Research Group—Biomaterials, Biodegradables and Biomimetics, University of Minho, Headquarters of the European Institute of Excellence on Tissue Engineering and Regenerative Medicine, Braga, Portugal

School of Medicine, University of Minho, Braga, Portugal
e-mail: espregueira@dhresearchcentre.com

© ISAKOS 2022
J. L. Koh et al. (eds.), *The Patellofemoral Joint*, https://doi.org/10.1007/978-3-030-81545-5_3

is no single definitive test used to diagnose PFPS [3].

The experience of the clinician is essential to understand what is considered normal and what is pathologic. Physical examination is examiner-dependent [4–8] and can be influenced by many factors including the patient level of pain.

3.1.1 Clinical History

It is important to collect a detailed and comprehensive clinical history. Critically listen to the patients and ask the appropriate questions. The mechanism of injury, location, situations aggravate or alleviate symptoms [4, 9–11]. Rule out other symptoms like weight loss, erythema, or other unusual findings [12] find any other relying medical problem. Review the patient's daily living and sports activities as these may be related to the symptoms, and sometimes modifying these activities may help to solve the problem.

Patellofemoral disorders can range from traumatic injuries (patella dislocation or fracture), instability (patellofemoral instability) or those leading to PFPS.

Traumatic injuries are inherently acute and most often painful. Most commonly these include patellar fractures and dislocations. Clinician should suspect if the patient reports a fall with direct trauma during activity and need to be referred for imaging assessment. In these cases, it is not unusual to find osteochondral injuries or bony contusions.

Patients with patellofemoral instability often describe giving away or slipping symptoms, with not necessarily a previous traumatic injury. In these patients, the clinician should look for history of patellar dislocation and anatomic risk factors.

PFPS may be caused by different conditions, including but not exclusive to patella or trochlear chondral or osteochondral lesions and excessive lateral pressure syndrome. Sometimes the source of pain is idiopathic and the clinician should look for other potential predisposing factors. In compressing syndromes, pain aggravates while squatting, during long periods of time of knee flexion or after sitting a long time and then standing. In acute conditions the patient will report pain for both ascending and descending stairs. In more chronic cases the patient will elicit pain only when ascending. The concentric contraction of the quadriceps muscle during ascending stairs does not cause pain; however, eccentric contraction while descending increases compressive load to the joint [11]. Patella or trochlear chondral or osteochondral lesions are a common cause of patellofemoral pain and may derive from previous trauma. As these lesions are located at the patella or trochlea are subjected to contact pressure during activities which will elicit pain. Common signs and symptoms are intermittent inflammatory knee effusions and painful weight-bearing or impact activities (e.g., landing). Common risk factors associated with patellofemoral cartilage lesions include trochlear dysplasia, patella alta, and excessive lateral patellar tilt [13]. In these cases, the clinician should ask for imaging study to confirm or rule out chondral/osteochondral lesions.

3.1.2 Physical Examination

The physical examination should follow a structured and systematized approach. There are many clinical tests and scores available for the examination of the patellofemoral joint [6, 7]. Most of tests are assessed qualitatively (rather than quantitatively) and not supported accuracy and/or validity for the existent methods.

3.1.2.1 Static Evaluation

Effusion, erythema, increased local temperature can suggest infection, acute trauma, or inflammatory arthropathy. Numbness related to previous procedure, scars, or portals can suggest the presence of neuromas [14]. Palpate the distal quadriceps muscle and tendon, also iliotibial band. Examine the knee in extension and in flexion, move the patella to evaluate physiological movements and displacements.

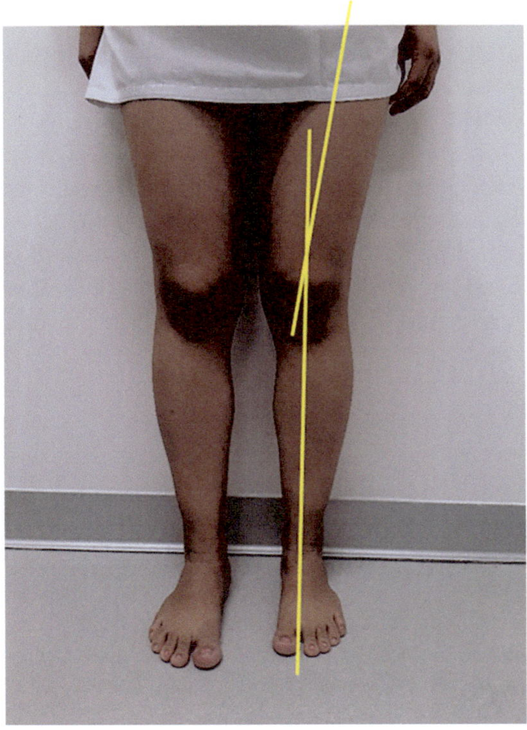

Fig. 3.1 Observe the patient standing and barefoot, check changes in the skin and alignment

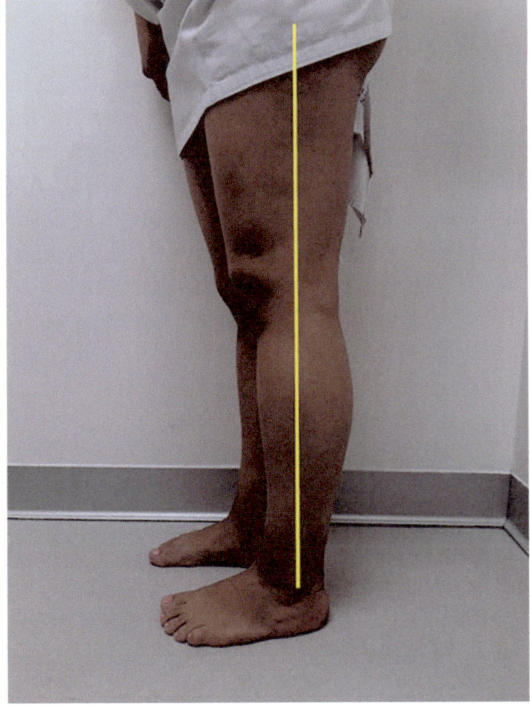

Fig. 3.2 Observe the presence of recurvatum

With the patients standing, examine the three planes, the patient's gait (barefoot and with footwear), posture, and footwear to try to identify potential contributing causes. Observe dystrophic changes, skin color, scaring tissue, scratches or rashes, barefoot gait, and excessive subtalar pronation [15]. Examine limb alignment, both frontal (varus/valgus) and sagittal (flexum/recurvatum). Genu valgus indicates a larger lateral force which leads to maltracking and subluxation, coxa vara also produces genu valgum [11]. Evaluated the patella position to determine its internal or external rotation, the presence or the bayonet sign or excessive external tibial torsion, the presence of tibia varum and the foot to check a midtarsal or subtalar joint position dysfunction (Fig. 3.1) [11].

In the lateral view look for genu recurvatum or flexion contracture (Fig. 3.2). Hyperextension (recurvatum) can indicate hyperlaxity. The Beighton score to test is applied for hypermobility evaluation [16]. An excessive recurvatum is related to Hoffa's syndrome caused by the impingement of the fat pad. Quadriceps weakness can also be present in hyperextension. However, flexion contracture can be caused by trauma, post-surgery, or excessive hamstrings tightness. In the lateral view you can evaluate the camel sign as well with the knee presenting two striking bulges: one represents the tibial tuberosity, the other one the patella.

The transverse plane or anterior view is the best for evaluating femoral rotation. The clinician must differentiate if the rotation is due or not to foot pronation. Check atrophy of the quadriceps can cause strength loss and consequently knee hyperextension. Check for any discrepancy in patella length or height. Patellofemoral instability is more frequent related to patella alta as chondromalacia with patella baja. Palpate the patella during knee flexion and extension, and search for crepitation. Crepitation near full extension suggests distal patellar cartilage lesion, and in flexion proximal patellar cartilage lesion [17].

Fact Box 1

The **static evaluation** should include the assessment of:

- The quadriceps angle (Q angle)
- Limb discrepancy
- Supine position
- Sitting position
- Foot posture

3.1.2.1.1 The Quadriceps Angle (Q Angle)

Is the angle formed by a line from the anterior superior iliac spine to the mid- patella intersecting with a line from the mid patella to the tibial tubercle with the knee in full extension. The average for males is 10–13° and 15–17° for females [11, 18]. The patella has a natural tendency to track laterally during knee flexion-extension movements [19]. An increase angle can indicate abnormal patellar tracking. It is important to interpret this angle with caution and check for hip rotation [18]. Also check for the direction of the patella; a medial orientation is indicative of a “winking” patella due to femoral excessive anteversion, with tibial external rotation and a compensatory hind foot valgus, causing the forces to dislocate the patella laterally. Despite the different techniques to calculate the Q angle, the clinical utility cannot be supported [20]. It is not considered a risk factor to develop patellofemoral pain. The relation between sings is not consistent [21].

3.1.2.1.2 Limb Discrepancy

Assessment of limb discrepancy should be performed in the standing weightbearing position to identify any potential predisposing or compensatory pattern. Differences of leg length produce gait asymmetry and contribute to increase the stress of the joint [22]. Measure from both anterosuperior iliac spines into the corresponding medial malleolus. More than 1.5 cm of difference is pathologic and can be the cause of anterior knee pain [11, 12]. Measurement of limb discrepancy can also be done with the patient supine or using the Weber-Barstow method [11].

3.1.2.1.3 Supine Position

Start by palpating the knee to search for effusion or tenderness. With the knee flexed at 90 degrees, palpate quadriceps and patellar tendons. Tenderness over the tibial tubercle suggests osteochondritis in younger patients. Palpate the fat pad against the femoral condyles and pain is indicative of Hoffa’s syndrome. Pain located in lateral or medial retinaculum may be indicative of previous dislocation (Fig. 3.3). Test the knee range of motion compared to the contralateral limb. Check for any extension deficit in active motion to evaluate any disorder of the extensor mechanism. Evaluate crepitus for chondral damage or impingement of the peripatellar soft tissues as anterior fat pad, plica, or synovial hypertrophy [23].

3.1.2.1.4 Sitting Position

When the patient is sitting a second inspection is made to examine the quadriceps bulk. Check again the tracking of the patella. A laterally tilted patella (grasshopper eye sign) [10] indicates weakness of the vastus medialis obliquus (Fig. 3.4). The patella height can be estimated in the seated position, with the proximal pole of the patella normally found at the same height as the anterior cortex of the distal femur (Fig. 3.5). The tibial tubercle sulcus angle is then determined by drawing a vertical line from the center of the patella tendon to the center of the tibial tubercle. An imagine line is then drawn perpendicular to the femoral epicondyle axis. The angle is determined where these two lines subtend each other. At 90 degrees of flexion, the patella should be centrally located in the femoral sulcus, and the tibial tubercle sulcus angle should be zero [24]. Always examine meniscus and ligaments to rule out associate injuries.

3.1.2.1.5 Foot Posture

Foot posture can be assessed with the patient standing and during gait. Excessive pronation can be cause by flatfoot, flattened medial longitudinal arch or a valgus hindfoot can be assessed. Feiss line and the medial longitudinal arch angle are other methods that have been described to identify a static pronated foot [11].

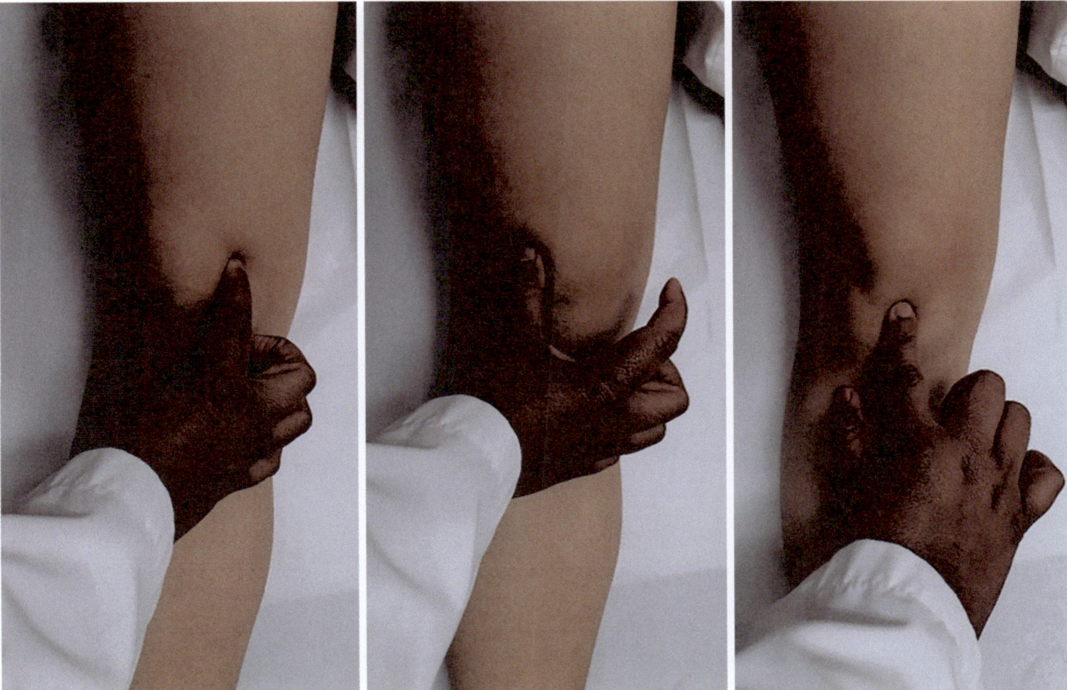

Fig. 3.3 Examine medial, lateral retinaculum, patellar tendon and fat pad

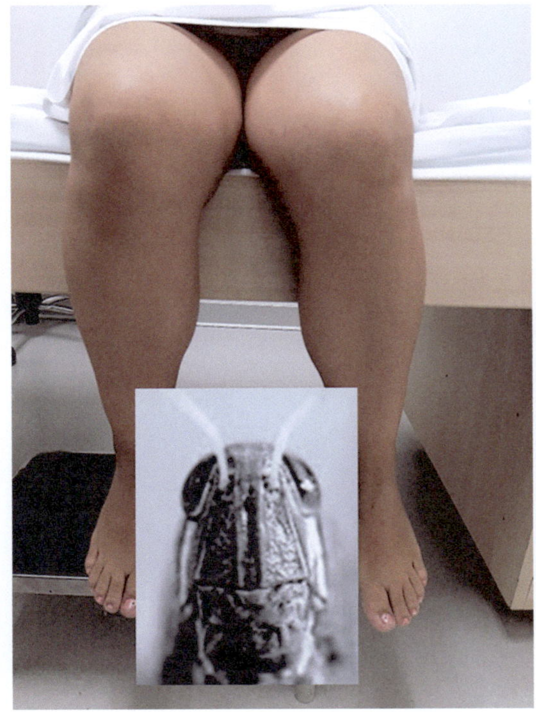

Fig. 3.4 Grasshopper eyes sign

3.1.2.2 Dynamic Evaluation

Ask the patient to walk naturally, and also backwards. Observe the gait and check any limping that can indicate pain, length discrepancy, or motor weakness. If the patient is an athlete, simulate running and jumping conditions.

> **Fact Box 2**
>
> The **dynamic evaluation** should include the assessment of:
>
> - Patellar tracking
> - Step-Down Test
> - Single-Leg Squat
> - Two-leg jump test
> - Single-leg hop test

3.1.2.2.1 Patellar Tracking

Evaluated passive and active knee range of motion. The patient is asked to extend the knee

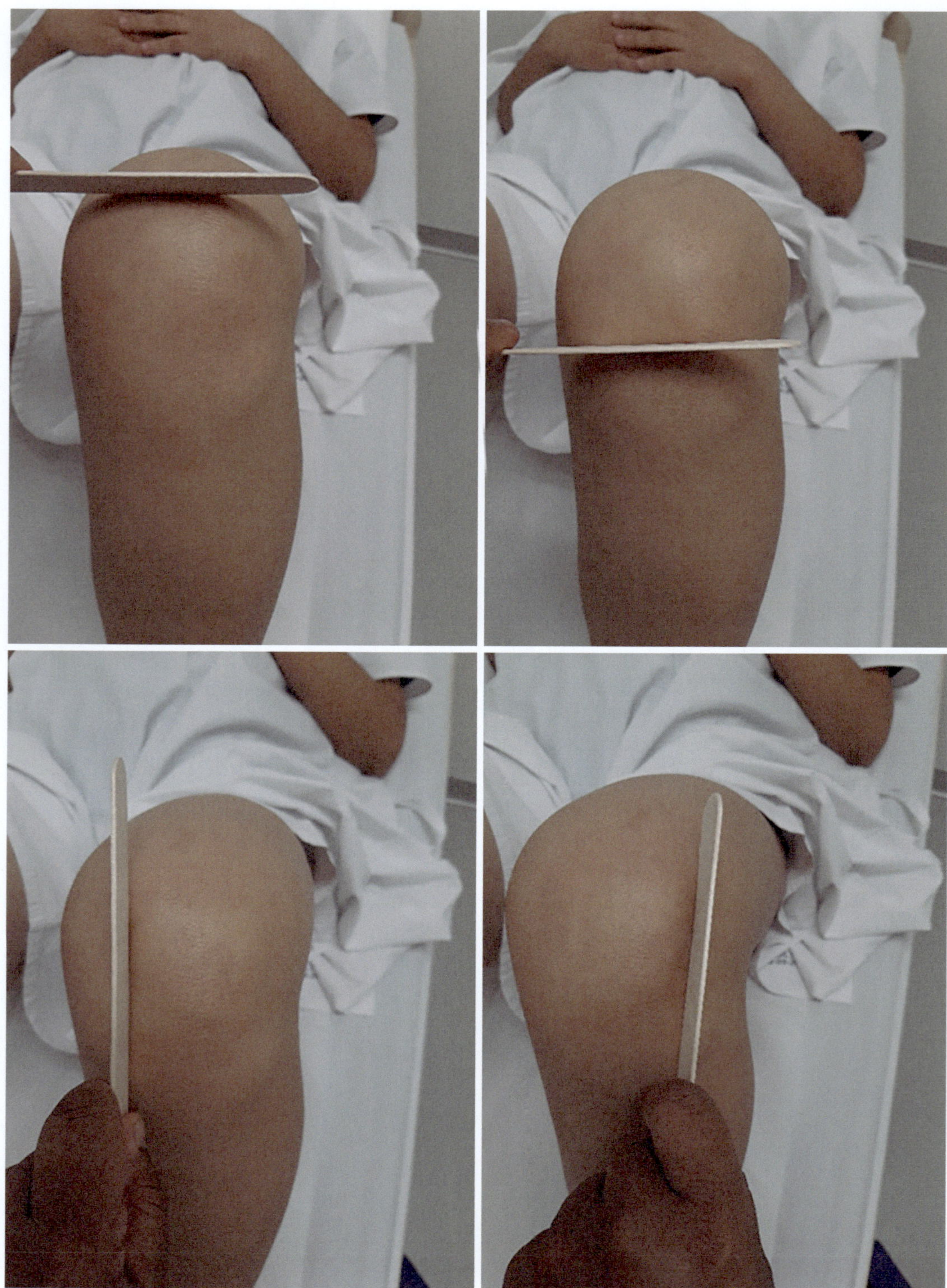

Fig. 3.5 Observe and palpate superior, inferior, medial and lateral border

from 90° to full extension. A normal patella has to move a little medial and then laterally back to the neutral position.

Observe the tracking for the presence of the J-sign. This is a pathological inverted J-path when the patella is in early flexion as the patella begins laterally subluxated and then shift medially to engage the femoral groove [25]. Occurs between 20 and 30 degrees of flexion to full extension which is the position that most subluxation events occur. A lateral J-sign or abrupt lateral deviation near terminal extension during an active quadriceps' contraction may be indicative of a dysfunctional vastus medialis obliquus muscle (Fig. 3.6).

3.1.2.2.2 Step-Down Test

To evaluate hip stability and lower-limb strength. The patient should stand in a box, with arms folded across the chest and squat down on one limb 5 to 10 times. The patient has to maintain the balance at least 1 squat per 2 s. Abnormal test can indicate decreased onset timing of the gluteus medius, decreased hip abduction torque, and decreased lateral trunk strength.

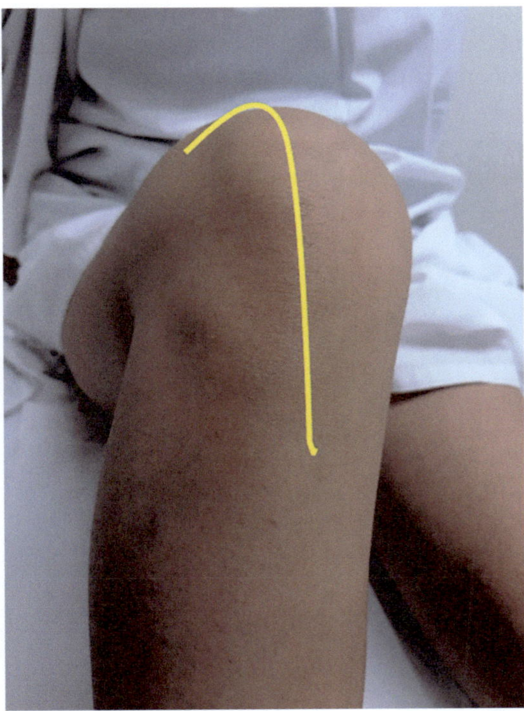

Fig. 3.6 "J" sign

3.1.2.2.3 Single-Leg Squat

The single-leg squat is one of the best tests for patellofemoral pain as 80% of these patients display positive sign [3]. It evaluates the dynamic hip and quadriceps strength [3] and can evidence abnormal movement/postural patterns such as increased ipsilateral trunk lean, contralateral pelvic drop, hip adduction, and knee abduction [26].

3.1.2.2.4 Two-Leg Jump Test

This test prepares the patient for the concentric propulsive push-off motion and the eccentric deceleration landing phase, performed in four gradients (25, 50, 75, and 100%) to prepare for the rest of the tests [11]. Observe any asymmetries during landing phase.

3.1.2.2.5 Single-Leg Hop Test

There are several single-leg hop tests that can be used, including single hop for distance, triple hop for distance, triple crossover hop test for distance, medial side triple hop test, 90° rotation hop test, and 6-m timed hop test. It is important to test both limb three times and check for any bilateral asymmetry.

3.1.2.3 Special Tests

There are many references regarding the evaluation of the best test for the patellofemoral disorders, with inconclusive results, Nunes et al. [3] describe in a meta-analysis that there is no test with good diagnostic accuracy which is able. Also, Smith et al. [7] describe about the inter-observer reliability is poor and intra-observer is moderate, and suggest the standardization of physical exam assessment. In another systematic review, Cook et al. [27] conclude that the tests that demonstrated the best diagnosis accuracy were the active instability test, pain during stair climbing, Clarke's test, pain during prolonged sitting, inferior pole tilt, and pain during squatting. Nevertheless, this requires careful considerations. The majority of the studies demonstrate quality bias and the best test for PFPS is still unknown.

Fact Box 3

The **special tests** can help to reach a more accurate diagnosis and help to rule out some potential sources of pain, and these should comprise:

- Patella glide
- Apprehension or Fairbank test
- Patellar tilt test
- Compression test
- Clarke's test or patellar grind
- The gravity subluxation test
- Flexibility tests

3.1.2.3.1 Patella Glide

In this test, the patella is forced medially and laterally as its medial and lateral dislocations are evaluated. It is performed at full extension and then at 20 degrees of knee flexion to assess the integrity of the medial and lateral restraints. In full extension, the patella is out of the trochlea sulcus and at 20° of knee flexion is engaged within the sulcus. If it is positive at 20 degrees of flexion, it has to be retested at 45 degrees of flexion. Any instability is not normal and can be associated with patella alta.

The patella is divided horizontally into quadrants, moving the patella 50% of its width is two quadrants. A positive test is a movement of three quadrants or more in either direction. Less of a quadrant indicates tightness [18] (Fig. 3.7).

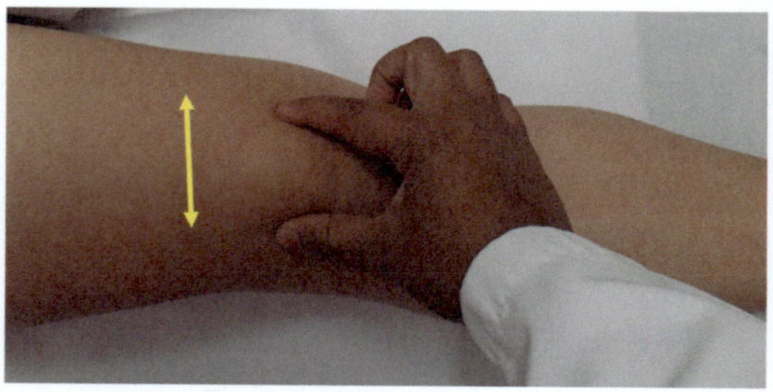

Fig. 3.7 Patellar Glide test

3.1.2.3.2 Apprehension or Fairbank Test

Is a test to detect patellar lateral instability. Performed in two parts, first (provocation) the knee in full extension, apply a lateral force with the thumb of the clinician, then move the knee from full extension to 90 degrees of flexion and then back while keeping the force. The second part (relief) is repeat the first part with the force applied medially [28]. Positive is the oral or facial expression of apprehension when the patient experiences a sense of undeniable dislocation or popping out of the position [28]. This test has a sensitivity of 100%, specificity of 88.4%, predictive positive value 89.2%, and predictive negative value 100% (Fig. 3.8).

3.1.2.3.3 Patellar Tilt Test

This test evaluates the lateral retinaculum tightness, represents a short retinaculum, and is related to an increased lateral tension. It is a negative test if the lateral border cannot be elevated above the medial one [24]. The patella at 30–40 degrees of flexion creates a tightness of the deep retinacular fibers that no tilt appears.

The test is performed with the knee in full extension, while the clinician is on the lateral side of the knee to be tested. The patella is pushed posteriorly on the medial border while pulling anteriorly with the thumbs under the lateral border to assess whether the patella corrects its tilt to at least neutral. If the patient's patella is unable to tilt back

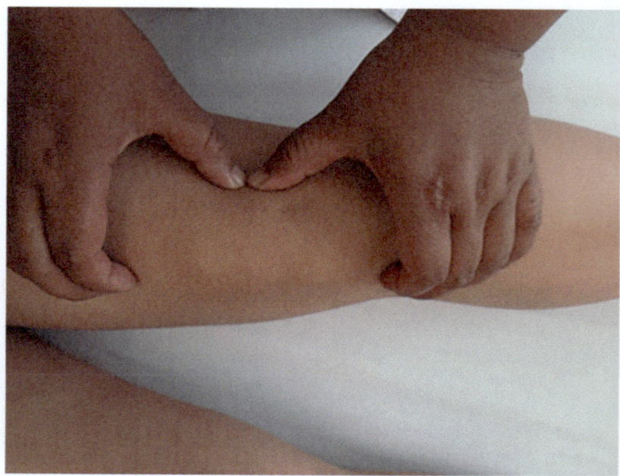

Fig. 3.8 Apprehension or Fairbank test

to neutral, it is indicative of excessive lateral tightness and the potential to have excessive lateral pressure syndrome [11]. Normally, the patella should tilt 15°. Compare both limbs for symmetry.

3.1.2.3.4 Compression Test

To assess arthritis or chondral injuries, the knee in full extension and the patella are directly compressed as the knee is flexed. Positive test is when the compression elicits pain (Fig. 3.9).

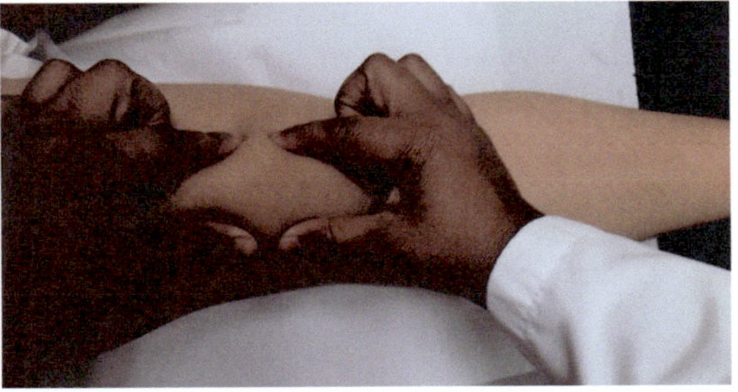

Fig. 3.9 Compression patellar test

3.1.2.3.5 Clarke's Test or Patellar Grind

The patella is compressed against the trochlea while the patient contracts their quadriceps. A positive test is the exacerbation of the pain and is confirmatory for chondromalacia. However, it has low sensitivity and specificity and a high rate of false-positive cases [29].

3.1.2.3.6 The Gravity Subluxation Test

Performed with the patient in side lying with the suspected leg abducted in the air. Then the quadriceps contracts. A positive test is the inability to pull the subluxate patellar lateral in the trochlea groove and indicates laxity of the lateral retinaculum [30].

3.1.2.3.7 Flexibility Tests

Evaluate tightness of quadriceps and hamstrings (Fig. 3.10). Thomas test is used to evaluate the rectus femoris and iliopsoas tightness, Ely's test for quadriceps flexibility, and Ober's test for iliotibial band tightness. Keep on mind to evaluate hip and foot. During the exam the clinician has to determine which restraints are deficient or not.

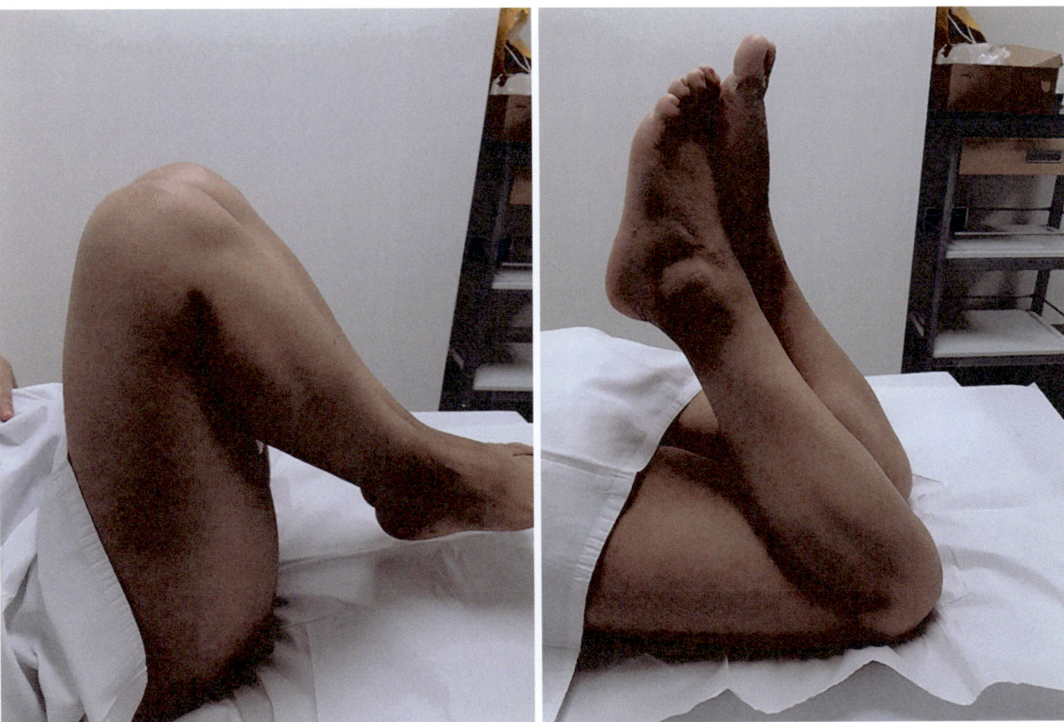

Fig. 3.10 Evaluations of the tightness of quadriceps and hamstrings

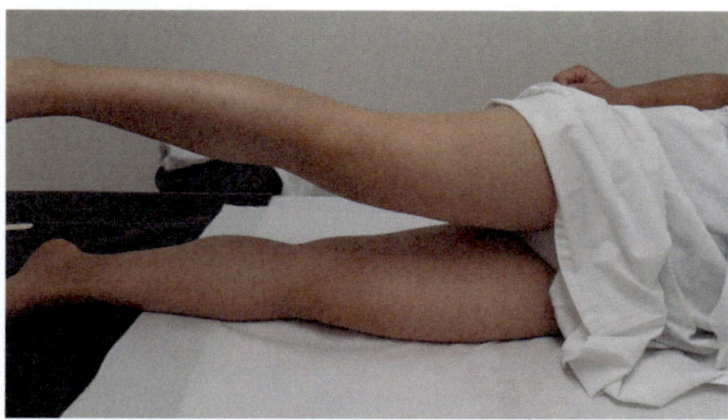

Fig. 3.11 Evaluation of the iliotibial band and hip muscles

The exam has to be done in supine and prone position, in which the examiner can evaluate quadriceps tightness and abnormal rotation of the hip that sometimes could be the reason for the pain. Lying on the side can evaluate iliotibial band (Fig. 3.11).

3.1.2.4 Instrumented Assessment

Several instrumented laxity devices have been proposed for the patellofemoral joint, but showing heterogenous results [31]. In our experience we use the Porto Patella Testing Device (PPTD) to measure the patellofemoral joint laxity combined with magnetic resonance imaging or computed tomography scanning. The PPTD applies lateral-directed stress (at 30°) or a posterior-directed stress (at 70°) to test the lateral translation and external tilt laxity, respectively. The assessment using the PPTD has shown to be valid in assessing the patellofemoral joint laxity and more reliable than manual physical examination [5]. The device is a useful tool to assess patellar movement in patellofemoral conditions including patellofemoral pain [32, 33] and instability [33] .

Take Home Message

- The patellofemoral joint is the most complex area of the knee. It is important to know its anatomy and biomechanics in order to establish an adequate diagnosis.
- Understanding contributing risk factors, not only the local, but also the proximal and distal factors, can enhance the understanding of patellofemoral disorders causes and improve diagnostic.
- It is important to follow a systematic and structured examination and correctly identify the causes of pain.

References

1. Lankhorst NE, Bierma-Zeinstra SM, van Middelkoop M. Risk factors for patellofemoral pain syndrome: a systematic review. J Orthop Sports Phys Ther. 2012;42:81–94.
2. Neal BS, Lack SD, Lankhorst NE, Raye A, Morrissey D, van Middelkoop M. Risk factors for patellofemoral pain: a systematic review and meta-analysis. Br J Sports Med. 2019;53:270.
3. Nunes GS, Stapait EL, Kirsten MH, de Noronha M, Santos GM. Clinical test for diagnosis of patellofemoral pain syndrome: systematic review with meta-analysis. Phys Ther Sport. 2013;14:54–9.
4. Kantaras AT, Selby J, Johnson DL. History and physical examination of the patellofemoral joint with patellar instability. Oper Tech Sports Med. 2001;9:129–33.
5. Leal A, Andrade R, Hinckel BB, Tompkins M, Flores P, Silva F, Espregueira-Mendes J, Arendt E. A new device for patellofemoral instrumented stress-testing provides good reliability and validity. Knee Surg Sports Traumatol Arthrosc. 2020;28:389–97.
6. Leal A, Silva F, Flores P, Pereira H, Espregueira-Mendes J. On the development of advanced methodologies to assist on the diagnosis of human articulations pathologies: a biomechanical approach. 2013.
7. Smith TO, Clark A, Neda S, Arendt EA, Post WR, Grelsamer RP, Dejour D, Almqvist KF, Donell ST. The intra- and inter-observer reliability of the physical examination methods used to assess patients with patellofemoral joint instability. Knee. 2012;19:404–10.
8. Smith TO, Davies L, O'Driscoll ML, Donell ST. An evaluation of the clinical tests and outcome measures used to assess patellar instability. Knee. 2008;15:255–62.
9. Fulkerson JP. Diagnosis and treatment of patients with patellofemoral pain. Am J Sports Med. 2002;30:447–56.
10. Lester JD, Watson JN, Hutchinson MR. Physical examination of the patellofemoral joint. Clin Sports Med. 2014;33:403–12.
11. Manske RC, Davies GJ. Examination of the patellofemoral joint. Int J Sports Phys Ther. 2016;11:831–53.
12. Magee DJ. Orthopedic physical assessment. Elsevier Health Sciences; 2013.
13. Ambra L, Hinckel B, Arendt E, Farr J, Gomoll A. Anatomic risk factors for focal cartilage lesions in the patella and trochlea: a case-control study. Am J Sports Med. 2019;47:036354651985932.

14. Fulkerson JP, Tennant R, Jaivin JS, Grunnet M. Histologic evidence of retinacular nerve injury associated with patellofemoral malalignment. Clin Orthop Relat Res. 1985:196–205.
15. Tiberio D. The effect of excessive subtalar joint pronation on patellofemoral mechanics: a theoretical model. J Orthop Sports Phys Ther. 1987;9:160–5.
16. Juul-Kristensen B, Rogind H, Jensen DV, Remvig L. Inter-examiner reproducibility of tests and criteria for generalized joint hypermobility and benign joint hypermobility syndrome. Rheumatology (Oxford). 2007;46:1835–41.
17. Fulkerson JP. Patellofemoral pain disorders: evaluation and management. J Am Acad Orthop Surg. 1994;2:124–32.
18. Rodríguez-Merchán EC, Liddle AD. Disorders of the patellofemoral joint: diagnosis and management. Springer; 2019.
19. Buuck DA. Disorders of the patellofemoral joint. Philadelphia: Lippincott Williams & Wilkins; 2004.
20. Smith TO, Hunt NJ, Donell ST. The reliability and validity of the Q-angle: a systematic review. Knee Surg Sports Traumatol Arthrosc. 2008;16:1068–79.
21. Livingston LA. The quadriceps angle: a review of the literature. J Orthop Sports Phys Ther. 1998;28:105–9.
22. Golightly YM, Allen KD, Helmick CG, Renner JB, Jordan JM. Symptoms of the knee and hip in individuals with and without limb length inequality. Osteoarthritis Cartilage. 2009;17:596–600.
23. Goodfellow J, Hungerford DS, Woods C. Patellofemoral joint mechanics and pathology. 2. Chondromalacia patellae. J Bone Joint Surg Br. 1976;58:291–9.
24. Kolowich PA, Paulos LE, Rosenberg TD, Farnsworth S. Lateral release of the patella: indications and contraindications. Am J Sports Med. 1990;18:359–65.
25. Sheehan FT, Derasari A, Fine KM, Brindle TJ, Alter KE. Q-angle and J-sign: indicative of maltracking subgroups in patellofemoral pain. Clin Orthop Relat Res. 2010;468:266–75.
26. Nakagawa TH, Moriya ET, Maciel CD, Serrao FV. Trunk, pelvis, hip, and knee kinematics, hip strength, and gluteal muscle activation during a single-leg squat in males and females with and without patellofemoral pain syndrome. J Orthop Sports Phys Ther. 2012;42:491–501.
27. Cook C, Mabry L, Reiman MP, Hegedus EJ. Best tests/clinical findings for screening and diagnosis of patellofemoral pain syndrome: a systematic review. Physiotherapy. 2012;98:93–100.
28. Ahmad CS, McCarthy M, Gomez JA, Shubin Stein BE. The moving patellar apprehension test for lateral patellar instability. Am J Sports Med. 2009;37:791–6.
29. Doberstein ST, Romeyn RL, Reineke DM. The diagnostic value of the Clarke sign in assessing chondromalacia patella. J Athl Train. 2008;43:190–6.
30. Nonweiler DE, DeLee JC. The diagnosis and treatment of medial subluxation of the patella after lateral retinacular release. Am J Sports Med. 1994;22:680–6.
31. Leal A, Andrade R, Flores P, Silva FS, Espregueira-Mendes J, Arendt E. High heterogeneity in in vivo instrumented-assisted patellofemoral joint stress testing: a systematic review. Knee Surg Sports Traumatol Arthrosc. 2019;27:745–57.
32. Leal A, Andrade R, Flores P, Silva FS, Fulkerson J, Neyret P, Arendt E, Espregueira-Mendes J. Unilateral anterior knee pain is associated with increased patellar lateral position after stressed lateral translation. Knee Surg Sports Traumatol Arthrosc. 2020;28:454–62.
33. Leal A, Andrade R, Hinckel B, Tompkins M, Bastos R, Flores P, Samuel F, Espregueira-Mendes J, Arendt E. Patients with different patellofemoral disorders display a distinct ligament stiffness pattern under instrumented stress testing. J ISAKOS. 2020;5:74–9.

Imaging Evaluation in the Patient with Patellofemoral Symptoms

4

Allison Mayfield and Jason L. Koh

4.1 Introduction

In addition to detailed history and physical examination of the patient with patellofemoral instability, formal imaging provides additional diagnostic information. Standardized imaging including radiographs, computed tomography, and magnetic resonance imaging can guide further decision-making and treatment recommendations, as well as assist with intraoperative procedures.

This chapter will review current standard imaging of the patient with patellar pain and instability symptoms, as well as address pertinent imaging findings and their applications regarding treatment decisions.

4.2 Radiographs

A standard series of radiographs include weight-bearing anteroposterior, lateral at 30 degrees of flexion, tunnel views, and a sunrise view of the lower extremity being evaluated. Radiographs of the contralateral extremity may be helpful for comparison although patients may often have bilateral symptoms. Standing long-leg radiographs are not always used, but are valuable in assessing coronal alignment. Standardized technique protocols within the radiology suite assist in recreating reliable, reproducible radiographs.

The anteroposterior and tunnel radiographs may be useful for evaluation tibiofemoral alignment in the coronal plane. Weight-bearing films are important to assess true joint space under physiologic loads. Arthritis of the medial or lateral compartments of the knee may be appreciated, as well as loose bodies or osteochondritis dissecans lesions within the knee. Additional bony morphology including bipartite patella or patellar fractures may be identified.

The axial views of the knee include the sunrise view and the Merchant view. The sunrise view or skyline view requires the patient to be prone with the knee in maximal flexion. This view demonstrates the posterior surface of the patella and the anterior surface of the distal femur, but does not accurately represent the articulation of the patellofemoral joint or tilt. Patients also may find this position difficult to tolerate. The Merchant view is obtained with the patient supine and the knee flexed down over the edge of the table 45°. The X-ray beam is angled 30° downward [1]. (Fig. 4.1). This view is the X-ray of choice when evaluating patellofemoral joint articulation and measuring sulcus angle of the trochlea [2]. Measurement of the sulcus angle is one tool to assist in the evaluation of trochlear dysplasia. This measurement is performed by taking the angle between two lines

A. Mayfield (✉)
Springfield Clinic Orthopaedic Surgery Group, Springfield, IL, USA

J. L. Koh
NorthShore Orthopaedic & Spine Institute, Skokie, IL, USA

© ISAKOS 2022
J. L. Koh et al. (eds.), *The Patellofemoral Joint*, https://doi.org/10.1007/978-3-030-81545-5_4

Fig. 4.1 Radiographic positioning for Merchant view of the patellofemoral joint

from the deepest part of the femoral trochlea extending out to the highest point of the medial and lateral condyles. Values greater than 145° suggest trochlear dysplasia [3]. In one study, 75% of patients with primary patellar dislocation had sulcus angles greater than 145 degrees, compared with only 9% of normal knees [4]. The axial views may demonstrate dislocation, subluxation, or other findings such as ossification at the medial patella or loose bodies in the gutters.

The lateral radiograph of the knee is taken with the joint in 30 degrees of flexion. Care should be taken to have a high-quality image, with superimposed femoral condyles with less than 10% deviation. The use of a foam block to stabilize the knee may be utilized to assist with standard flexion degrees and rotation. Evaluation of the patellar height on lateral radiographs will provide information regarding patellar baja or patellar alta. When the patella is located proximally, the knee requires greater degree of flexion to engage the patella within the trochlea. Therefore, patellar alta is one factor associated with patellar instability [5]. There are many described ways to measure relative patellar height, including the Insall-Salvati [6], Blackburne-Peele [7], Caton-Deschamps [8], and Labelle-Laurin [9] methods (Fig. 4.2). The most commonly used methods are the Caton-Deschamps and the Insall-Salvati measures. The Caton-Deschamps is preferred since in the Insall-Salvati method changes in patella height from a tibial tubercle osteotomy are not measured. The most reproducible method is the Blackburne-Peel ratio [10]. This also has been shown to have the least intra-observer error [11]. This has been described as the ratio of the perpendicular distance from the lower articular margin of the patella to the tibial plateau divided by the length of the articular surface of the patella [7]. A normal ratio is 0.8, with values over 1.0 indicate patella alta, and values under 0.5 indicate patella baja. See Table 4.1.

The lateral radiograph of the knee also lends information regarding the anatomy of the femoral trochlea. Variations from normal trochlear anatomy are considered trochlear dysplasia, which is another risk factor for patellar instability [12]. Trochlear morphology is best evaluated on the lateral radiograph with identification of three anterior lines. The most anterior line is the medial femoral condyle. The middle line is the lateral femoral condyle. The most posterior line is the trochlear floor. In normal knee anatomy, the line of the lateral ridge terminates proximal to the lie of the medial ridge. In the setting of a shallow, dysplastic trochlea the posterior trochlear line will prematurely cross the anterior aspect of the femoral condyles. This has been described by Dejour [12] as the "crossing sign." Dejour et al. reported the presence of the crossing sign in 96% of patients with a true patellar dislocation. In contrast, this sign was found in just 3% of the normal control group [13]. Additional findings of dysplasia on lateral X-ray include supratrochlear spur and a double contour, indicating medial condyle hypoplasia.

Utilizing lateral and axial radiographs, a classification system from trochlear dysplasia has been developed (Fig. 4.3). Type A has a crossing sign (shallow trochlea) on the lateral X-ray and a sulcus angle greater than 145° on the axial. Type B has a supratrochlear spur on the lateral X-ray and a flat or convex trochlea on the axial view. Type C has a crossing sign and double contour on lateral X-ray and medial condyle hypoplasia on axial view. Type D is the most severe dysplasia, with a crossing sign, supratrochlear spur, and

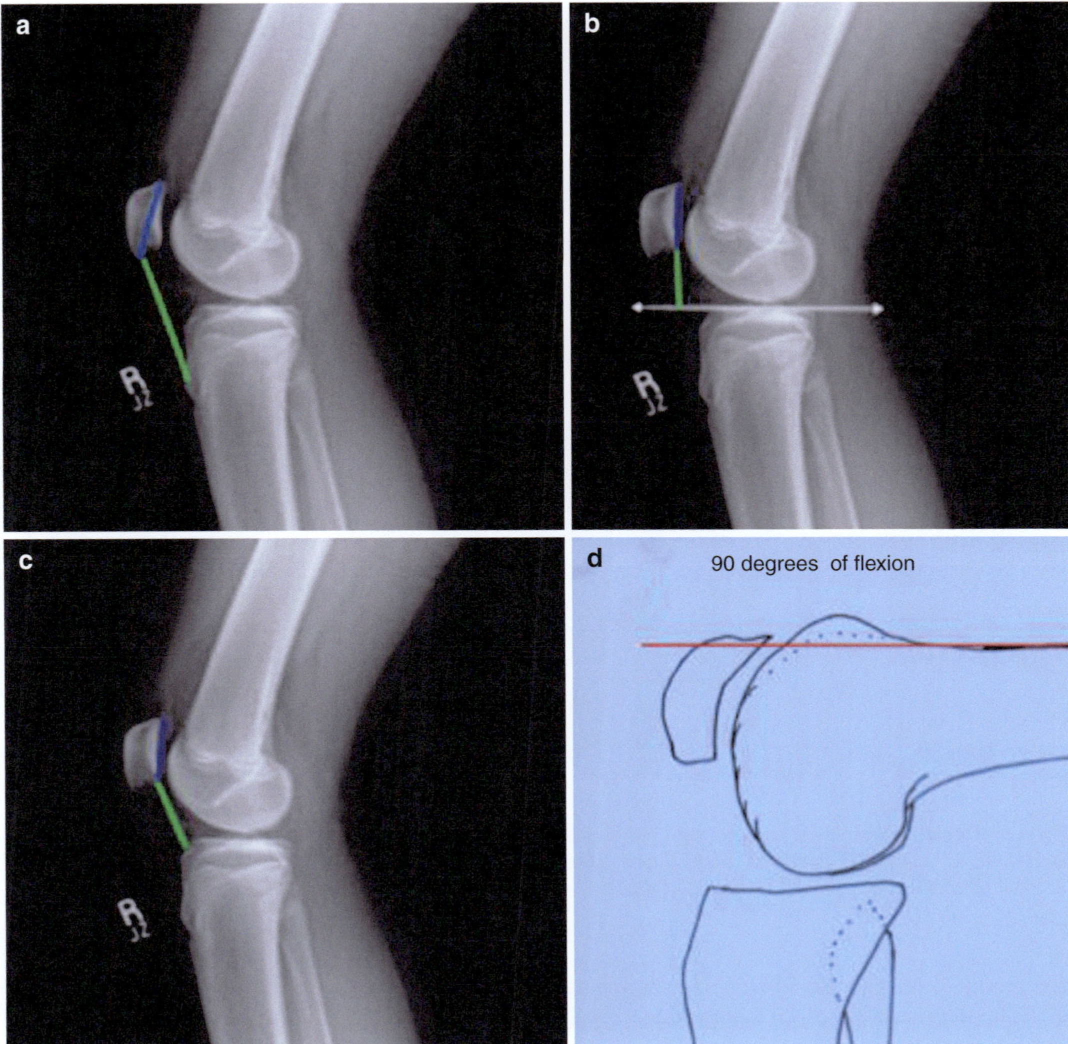

Fig. 4.2 (**a**) The Insall-Salvati ratio: diagonal length of the patella (blue) in relation to the length of the patellar tendon (green). (**b**) The Blackburne-Peel method: Ratio of the articular surface length of the patella (blue) to the height of the lower pole of the articular surface (green) above a line tangential to the tibial plateau (white). (**c**) The Caton-Deschamps ratio: distance between the distal point of the patellar articular surface and the anterosuperior border of the tibia (green) divided by the length of the articular surface of the patella (blue). (**d**) The Labelle-Laurin method: Patella is defined as alta if its most proximal pole lies above the tangent of the ventral cortical line (red) of the femur on a lateral X-ray in 90 degrees of flexion

Table 4.1 Patellar height ratios with respective values for classification

Method	Alta	Normal	Baja
Insall-Salvati	>1.0	0.75–0.99	<0.75
Blackburne-Peel	>1.0	0.8–1.0	<0.8
Caton-Deschamps	>1.2	0.9–1.2	<0.9

double contour on the lateral, with a cliff between trochlear facets on the axial.

Lateral radiograph also allows for evaluation of trochlear depth and trochlear bump. Trochlear depth is measured by the distance between the trochlear floor (most posterior line) and the most anterior trochlear contour, measured along a line 15° from the perpendicular to the tangent of the posterior femoral cortex. Values measuring less than 4 mm were found in 85% of knees with patellar instability compared with 3% of the con-

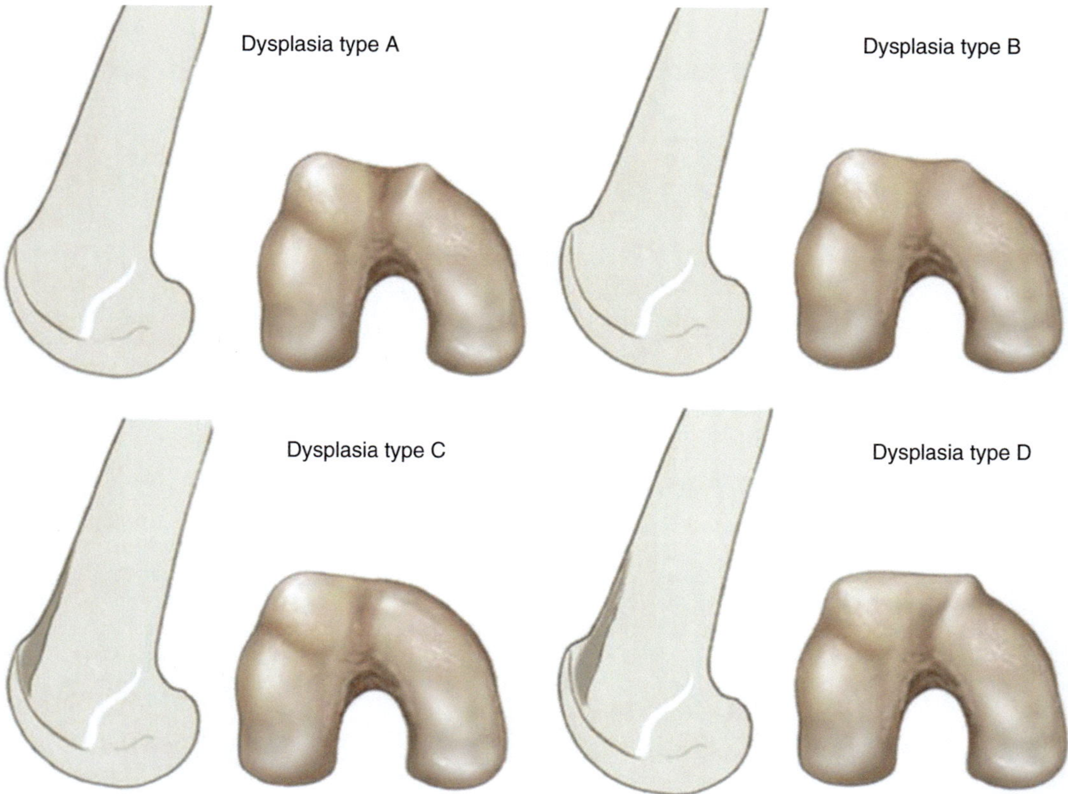

Fig. 4.3 Trochlear Dysplasia. Utilized with permission. Grelsamer, R; Dejour, D; Gould, J. The pathophysiology of Patellofemoral Arthritis. 2008 Clin Orthop N Am 39(2008)269–74 (*Reprinted with permission Dejour, D. Taken from The pathophysiology of Patellofemoral Arthritis. 2008 Clin Orthop N Am 39(2008)269–74*)

trol group13. Magnitude of trochlear bump can be evaluated by using a true lateral X-ray, drawing a straight line tangential with the anterior femoral cortex. The trochlear floor may be anterior, posterior, or flush with this line. When measuring the size of the bump, normal knee values are zero. Positive values indicate anterior positioning of the trochlear floor relative to the anterior femoral cortex, suggesting a shallow trochlea. Values of 3+ mm were found in 66% of patients with patellar instability.

4.3 Computed Tomography

Computed tomography (CT) has the advantage of obtaining a better visualization and representation of the knee and patellofemoral joint in the axial plane. CT imaging provides greater detail regarding osseous anatomy and relationship of the patella to the trochlea. Measurements which may help guide treatment decisions can be taken on CT axial slides, including patellar tilt and the tibial tuberosity-trochlear distance (TT-TG) measurements. Bony loose bodies or osseous defects may also be identified more clearly on a CT examination. Finally, measurements of femoral and tibial version can be performed on CT, which has relevance in determining the presence or absence of pathologic malrotation.

While originally described on axial radiograph [14], patellar tilt is measured on an axial CT series as the angle drawn through the axis of the patella and a line tangential to the posterior femoral condyles. Angles greater than 20° are present in more than 80% of patients with patellar instability [12].

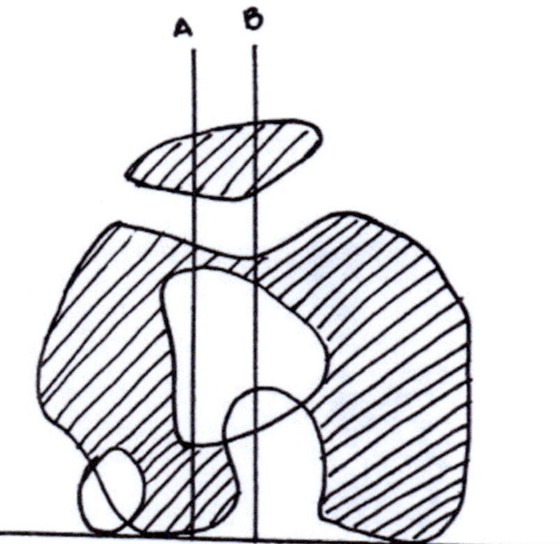

Fig. 4.4 The TT-TG distance: measured distance along a line tangential to the posterior femoral condyles, from a point drawn from the center of the trochlear groove and the center of the tibial tuberosity

When evaluating the relationship of the distal attachment of the extensor mechanism relative to the trochlear groove, the TT-TG distance can be measured on axial CT series. This identifies excessive lateralization, which may predispose to patellar instability. When measuring the TT-TG distance, begin by superimposing axial images of the tuberosity and the trochlear groove. The TT-TG distance is the measured distance along a line tangential to the posterior femoral condyles, from a point drawn from the center of the trochlear groove and the center of the tibial tuberosity [15] (Fig. 4.4). A TT-TG distance greater than 15–20 mm is associated with patellar instability [12].

4.4 Magnetic Resonance Imaging

While CT imaging provides additional detail to osseous anatomy, magnetic resonance imaging (MRI) provides detail in evaluation of soft tissue and articular cartilage. MRI imaging provides evaluation of articular cartilage thickness, morphology of the trochlear cartilage, soft tissue loose bodies, and integrity of soft tissue structures such as the MPFL [16]. Classic characteristic bone bruise patterns in patellar dislocation can be identified, which include impaction at the lateral femoral condyle and medial patellar facet [17].

In the setting of acute patellar dislocation, integrity of the medial retinaculum and MPFL may also be assessed. The location of failure (proximal or distal) may be identified and assist in surgical decision-making. MRI has been found to be 85% sensitive and 70% accurate in identifying injury to the MPFL [18].

Measurements previously described with CT imaging such as TT-TG distance have been shown to be effectively measured on MRI imaging, without the added risk of radiation [15].

4.5 Dynamic Imaging

The advancement in imaging technology has allowed for a more dynamic evaluation of patellofemoral tracking with dynamic computed tomography. While studies evaluating dynamic imaging are increasing, there is much variation in protocols and imaging capture [3]. In one study, results tend to demonstrate high variability in tracking patterns [1]. A group established a grading system of ten categories with high correlation with severity of symptoms, however only one of these was labeled as "normal tracking." A recent review of dynamic patellar tracking concluded that "there may be no normal pattern." [2] Therefore further studies with standardized protocols may provide additional relevant information in the future.

Kinematic MRI has also been utilized in the assessment of patellofemoral tracking and instability. Results and application into clinical practice remain limited due to lack of widespread access to such technology.

References

1. Merchant AC, Mercer RL, Jacobsen RH, et al. Roentgenographic analysis of patellofemoral congruence. J Bone Joint Surg Am. 1974;56(7):1391–6.

2. Merchant AC, Mercer RL, Jacobsen RH, Cool CR. Roentgenographic analysis of patellofemoral congruence. J Bone Jt Surg. 1974;56(7):1391–6.
3. Dejour H, Walch G, Nove-Josserand L, Guier C. Factors of patellar instability: an anatomic radiographic study. Knee Surg Sports Traumatol Arthrosc. 1994;2(1):19–26.
4. Askenberger M, Janarv P, Finnbogason T, Arendt E. Morphology and anatomic patellar instability risk factors in first-time traumatic lateral patellar dislocations: a prospective magnetic resonance imaging study in skeletally immature children. Am J Sports Med. 2016;45(1):50–8.
5. Insall J, Goldberg V, Salvati E. Recurrent dislocation and the high-riding patella. Clin Orthop Relat Res. 1972;88:67–9.
6. Insall J, Salvati E. Patella position in the normal knee joint. Radiology. 1971;101(1):101–4.
7. Blackburne JS, Peel TE. A new method of measuring patellar height. J Bone Joint Surg Br. 1977;59(2):241–2.
8. Caton J, Deschamps G, Chambat P, et al. Patella infera. Apropos of 128 cases. Rev Chir Orthop Reparatrice Appar Mot. 1982;68(5):317–25.
9. Labelle H, Laurin CA. Radiological investigation of normal and abnormal patellae. J Bone Joint Surg Br. 1975;57:530.
10. Berg EE, Mason SL, Lucas MJ. Patellar height ratios. A comparison of four measurement methods. Am J Sports Med. 1996;24(2):218–21.
11. Seil R, Muller B, Georg T, et al. Reliability and interobserver variability in radiological patellar height ratios. Knee Surg Sports Traumatol Arthrosc. 2000;8(4):231–6.
12. Dejour H, Walch G, Nove-Josserand L, et al. Factors of patellar instability: an anatomic radiographic study. Knee Surg Sports Traumatol Arthrosc. 1994;2(1):19–26.
13. Dejour H, Walch G, Nove-Josserand L, Guier C. Factors of patellar instability: an anatomic radiographic study. Knee Surg Sports Traumatol Arthrosc. 1994;2(1):12–26.
14. Grelsamer R, Bazos A, Proctor C. Radiographic analysis of patellar tilt. J Bone Joint Surg Br. 1993;75(5):822–4.
15. Schoettle P, Zanetti M, Seifert B, Pfirmann C, Fucentese S, Romero J. The tibial tuberosity-trochlear groove distance: a comparative study between CT and MRI scanning. Knee. 2006;13(1):26–31.
16. Nomura E, Horiuchi Y, Inoue M. Correlation of MR imaging findings and open exploration of medial patellofemoral ligament injuries in acute patellar dislocations. Knee. 2002;9(2):139–43.
17. Earhart C, Patel DB, White EA, Gottsegen CJ, Forrester DM, Matcuk GR Jr. Transient lateral patellar dislocation: review of imaging findings, patellofemoral anatomy, and treatment options. Emerg Radiol. 2013;20(1):11–23.
18. Sanders TG, Morrison WB, Singleton BA, et al. Medial patellofemoral ligament injury following acute transient dislocation of the patella: MR findings with surgical correlation in 14 patients. J Comput Assist Tomogr. 2001;25(6):957–62.

5 Principles of Prevention and Rehabilitation for the Patellofemoral Joint

Leonard Tiger Onsen and Jason L. Koh

5.1 Introduction

Patellofemoral disorders encompass a wide variety of pathologies resulting in anterior knee pain. The complexity of patellofemoral anatomy and biomechanics creates unique challenges in the diagnosis as well as treatment of these disorders. Anterior knee pain is frequently reported among the most common musculoskeletal complaints in adult and pediatric populations making it likely to be encountered by providers of various specialties [1–5]. Examples of these disorders include but are not limited to patellar tendonitis, patellar instability, chondromalacia, patella maltracking, Sinding-Larsen-Johansson syndrome, and various overuse or muscle imbalance disorders. Prevalence of patellofemoral pain in various populations is commonly found to be near 25%, however others have reported ranges of 15–45% [6–8]. Peak prevalence is frequently found in an active younger population between ages 12 and 22. Despite this, patellofemoral issues can be found in a variety of patients ranging from young active groups to more sedentary elderly populations. Multiple risk factors are associated with patellofemoral pain such as increased activity level, female gender, quadriceps weakness, and hamstring inflexibility [1, 9]. Patellofemoral pain can alter native muscle function and biomechanics, which further convolutes the clinical picture. When evaluating patellofemoral pain, it is essential to consider all contributing factors in order to properly diagnose and treat patients. The purpose of this chapter is to review the rehabilitation principles of the patellofemoral joint to better understand the evaluation and treatment of these conditions.

L. T. Onsen (✉)
Department of Orthopaedic Surgery, College of Medicine, University of Illinois at Chicago, Chicago, IL, USA

J. L. Koh
Orthopaedic & Spine Institute, Department of Orthopaedic Surgery, Northshore University HealthSystem, Skokie, IL, USA

5.2 Anatomy

The patellofemoral joint is an intricate aspect of the anterior knee with significant freedom of movement. Osseous anatomy includes the patella and the trochlear groove of the distal femur which is demonstrated in Fig. 5.1. The patella is the largest sesamoid bone in the body located in the retinaculum layer of the knee and serves to connect the patella and quadriceps tendons of the extensor mechanism. Its articular surface is covered with a thick layer of cartilage and includes two concave facets located medially and laterally separated by a vertical ridge with the lateral facet being larger on average [11]. The genicular and recurrent tibial arteries form a ring around the patella and provide a robust blood supply [12]. The trochlear groove is a concave depression in the anterior distal femur between the medial and

© ISAKOS 2022
J. L. Koh et al. (eds.), *The Patellofemoral Joint*, https://doi.org/10.1007/978-3-030-81545-5_5

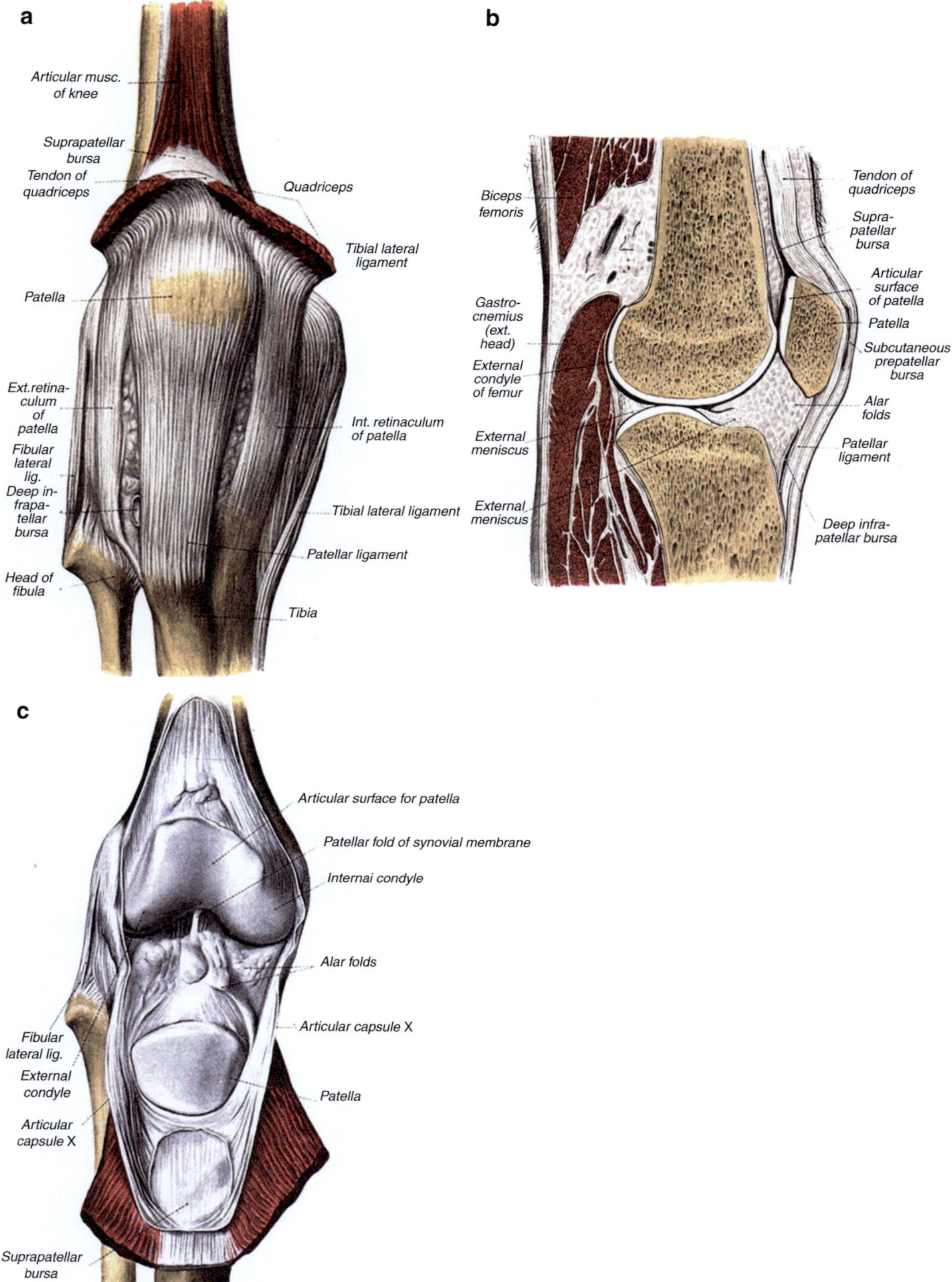

Fig. 5.1 (**a**) AP view of the knee demonstrating the patella within its attachments to the extensor mechanism and retinaculum layer of the knee. (**b**) Lateral cross- section view of the knee again demonstrating the patellofemoral joint and extensor mechanism. (**c**) AP view of the knee with the extensor mechanism reflected to demonstrate the trochlear groove of the distal femur [10]

lateral femoral condyles. Normal groove depth is noted to be greater than 4 mm and shallower grooves have been indicated in patellofemoral pain [13, 14]. The lateral aspect of the trochlear groove is noted to be larger and more prominent as compared to medially, which matches the larger lateral patellar facet.

Pertinent muscular anatomy of the patellofemoral joint involves the quadriceps muscle group made up of the rectus femoris as well as the vastus medialis, lateralis, and intermedius. This group produces force through the extensor mechanism ultimately resulting in knee extension and contributes to dynamic stabilization of the patellofemoral joint. More specifically, the vastus medialis obliquus (VMO) is the primary dynamic restraint to lateral patella translation by providing a proximal and medial force [15]. Intricate ligamentous and soft tissue anatomy also aid in patellofemoral joint stability. Medial structures include the retinaculum, medial patellofemoral ligament (MPFL), and medial meniscopatellar ligament. Figure 5.2 shows a representation of MPFL in relation to other medial knee structures. The MPFL is noted to be the primary restraint to lateral patella translation accounting for 60% of restraint from 0–30 degrees of knee flexion [17]. Lateral structures can be divided into superficial and deep layers. The superficial layer involves the oblique lateral retinaculum, whereas the deep layer involves the transverse retinaculum made up of the epicondylopatellar and patellotibial bands. These structures resist medial translation of the patella and are secondary stabilizers against lateral patellar translation. The relationship and function of these structures will be described in greater detail in the biomechanics section.

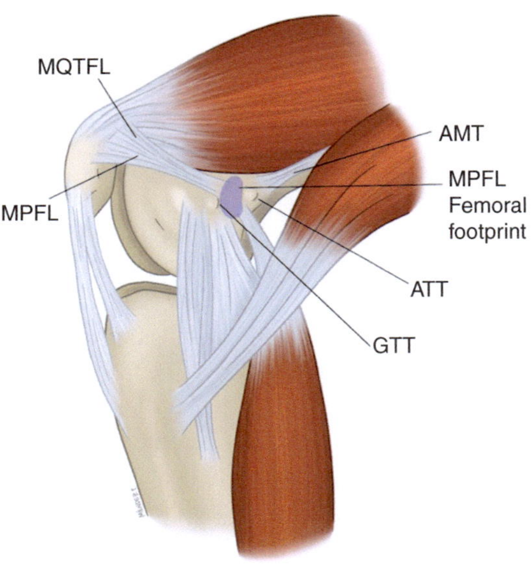

Fig. 5.2 Medial view of the knee demonstrating the MPFL in relation to other structures. *MQTFL* medial quadriceps tendon femoral ligament, *ATT* adductor tendon tubercle, *GTT* gracilis tendon tubercle, *AMT* adductor mangus tendon [16]

5.3 Biomechanics

Biomechanics of the patellofemoral joint are inherently complex with multiple osseous and soft tissue contributors. Forces from the quadriceps are applied to the patellofemoral joint with the ultimate goal of producing knee extension. The patella improves the efficiency of the quadriceps to produce knee extension. More specifically, it increases the distance of the extensor mechanism from the flexion-extension axis and acts as a lever to increase the amount of force produced by the quadriceps for any given amount of contraction [15]. Simplified forces and line of pull on the patella can be described with Q-angle. This is defined as the intersection of a line from the anterior superior iliac spine to the center of patella compared to a line from the center of patella to the tibial tubercle. This is demonstrated in Fig. 5.3. In other words, it can be described as the angle between the quadriceps and patella tendon. Thus, when the quadriceps contracts superior and lateral forces are produced on the patella. The concept can be applied to understand how changes in alignment can impact patellofemoral tracking. Normal Q-angle measures approximately 14° in males and 17° for females [19].

With the knee in full extension at rest, the patella sits at the proximal aspect of the trochlea. In extension the tibia is externally rotated to better lock in with the femur, this further emphasizes Q-angle. Quadriceps contraction in this setting pulls the patella proximal and laterally. With

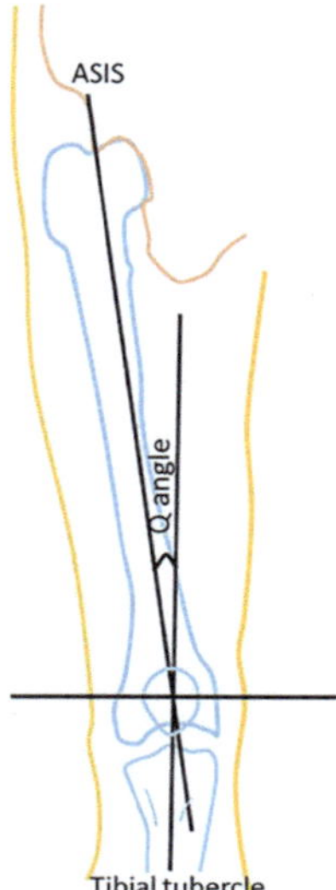

Fig. 5.3 Representation of Q angle measured between the intersection of a line from the anterior superior iliac spine to the center of patella and a line from the center of patella to the tibial tubercle [18]

knee flexion the patella moves distally and medially on the femur towards the trochlea. During flexion the tibia internally rotates to unlock the knee. This aids patellar tracking into the trochlea by reducing Q-angle and lateral forces on the patella. Prior to engaging the trochlea the MPFL serves as the primary restraint to lateral patellar translation accounting for 60% [17, 20]. The patella engages the trochlea at 20–30 degrees of knee flexion resulting in additional stability from through bony contact [21]. With further flexion from 30 to 60°, the patella continues to move distally where it reaches the deeper region of the trochlea increasing total bony contact area and stability. Additionally, with knee flexion, soft tissues such as the retinaculum, quadriceps, and patella tendons exert a posterior force on the patella further increasing contact pressure and stability. This posterior force on the patella is demonstrated in Fig. 5.4. After 90 degrees of flexion the medial patellar facet engages the trochlea.

Various dynamic and static factors can impact patellofemoral tracking and biomechanics. In patella alta greater amounts of knee flexion are required for the patella to engage the trochlea. These individuals rely on ligamentous stability over a great arc of motion prior to the patella reaching the trochlea. Iliotibial (IT) band changes can have a dynamic impact on patellofemoral tracking. This occurs with knee flexion as the IT band exerts a posterior force to the lateral aspect of the patella. IT band tightness can result in altered patellofemoral tracking and pain. More specifically, with IT band tightness there is an increased posterior force which can lead to lateral patella tilt, increase lateral contact forces, and pain. Hamstring tightness can also produce anterior knee pain by increasing the posterior force on the patella as the knee moves from flexion to extension. Hip muscle strength can also have a dynamic impact on patellofemoral biome-

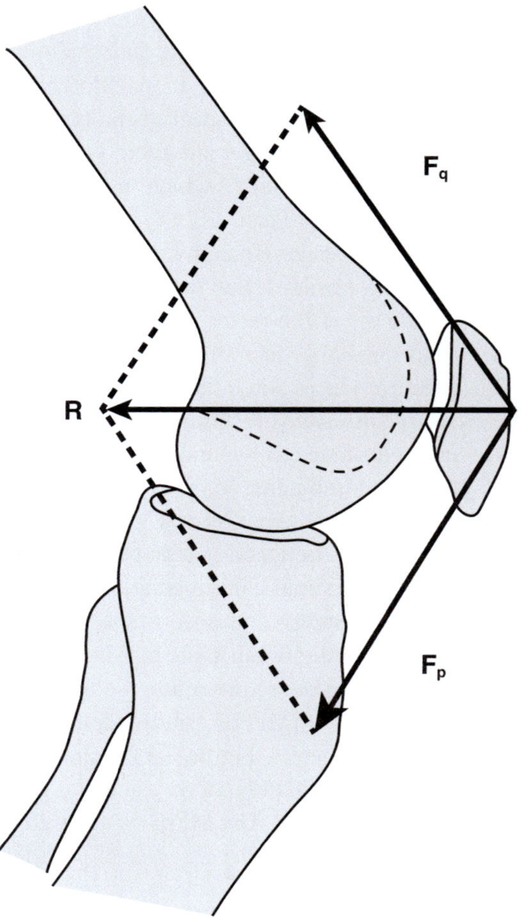

Fig. 5.4 Representation of forces acting on the patella as the knee goes into flexion. Forces from the quadriceps tendon (Fq) and the patella tendon (Fp) ultimately produce a posterior force on the patella [22]

chanics [23]. This is of particular importance in weight bearing where hip muscles influence movement of the femur relative to the patella as demonstrated by Powers et al. [24] Hip abductor and external rotator weakness can result in dynamic knee valgus [25]. When weight bearing, weakness leads to excessive hip adduction and internal rotation. Hip adduction is reported as a major contributor to dynamic knee valgus [25]. Figure 5.5 demonstrates dynamic knee valgus when landing from a jump. With the foot engaged on the ground the knee moves inwards, tibia abducts, and foot pronates to produce dynamic knee valgus [27]. This can in turn causes altered tracking via increased lateral patella pull and contact forces [28]. Moreover, excessive femoral internal rotation secondary to hip external rotator weakness leads to increased lateral patella pull and maltracking. Quadriceps muscles play a significant role in patellofemoral biomechanics. Of the quadriceps, the vastus medialis obliquus is particularly important. These muscles provide a medial force to the patella and contribute to medial patellar rotation to aid in tracking [29]. Additionally, weakness in the VMO can also lead to increased lateral patellar tracking and lateral which can result in increased contact forces and pain. Given the impact of altered biomechanics, it is essential for providers to identify and develop treatment programs to address these alterations.

5.4 Evaluation of the Knee

Evaluation of patellofemoral pain requires a thorough history and physical exam. History begins by identifying the patient's complaint as pain, instability, mechanical symptoms, or a combination of these issues. Further information on the complaint should be elucidated such as inciting events, duration of symptoms and activities or movements that aggravate the issue. It is also important to address a patient's activity level through their job, sports participation, exercise, or other daily activities. In addition, determining any activity level changes either increase or decrease. Other aspects to address include previous history of knee injuries and treatment if any. Medical history should also be noted as many conditions have musculoskeletal effects.

Equally important, a thorough focused physical exam is vital in the evaluation of patellofemoral pain. Any exam begins with inspection of the knee both seated, standing, and supine to look for

Fig. 5.5 (**a**, **b**) An athlete landing from a double leg jump and exhibiting bilateral dynamic knee valgus. (**c**, **d**) An athlete landing from a sling leg jump and demonstrating dynamic knee valgus [26]

any skin changes, swelling, deformities, and general alignment of the limb. Then with the patient supine the knee can be palpated throughout various areas of the patella, retinaculum, tibial tubercle, iliotibial band, and joint line to better pinpoint the location of pain. While in this position the patella can be translated medially and laterally to determine if a stable endpoint exists or if any apprehension occurs. Active and passive range of motion (ROM) can be examined at the knee and hip to look for any quadriceps or hamstrings tightness. Also, during ROM testing patellar tracking can be observed during which a J-sign can be observed in maltracking where lateral patella translation over the anterolateral femur occurs as the knee goes from extension to flexion. Strength can be tested in the major muscle groups crossing both the hip and knee can be tested. Other patella specific exams such as patella grind or compression tests can be performed. Overall ligamentous laxity can be assessed by obtaining a Beighton score. Additionally, core strength can be tested through a single leg squat. In general, these exams should be performed on both the affected and unaffected limbs so one has a control to compare against to find differences. Once a cause is identified treatment can be initiated with physical therapy being a common starting place. General principles of patellofemoral rehab will be described in the next section.

5.5 Rehabilitation Principles

Physical therapy remains a mainstay of treatment for patellofemoral disorders and is often times the first line treatment. General principles of patellofemoral rehab focus on selective muscle strengthening, stretching, proprioception, and functional movement training. These are done with the goal of improving patellofemoral stability, tracking, and pain. Strengthening of lower extremity muscles about the hip and knee is a common area of focus. This is in direct response to knee extensor weakness being common among those with anterior knee pain [30]. Quadriceps weakness can be both a cause and result of patellofemoral pain. Although strengthening of knee extensors has been found to improve patellofemoral pain, programs that include hip strengthening produce greater improvements in pain and function compared to knee alone [31, 32]. Open and closed chain exercises can be used in lower extremity strengthening. Good results have been found with both forms and thus a program including both is recommend [33, 34]. Open chain exercises are those where the foot can move freely with weight or resistance can be added to the distal aspect of the leg. Care must be taken with open chain exercises as they may increase patellofemoral joint contact forces, which can result in pain and hinder progress in this patient population [28]. Examples of open chain exercises include straight leg raises, hamstring curls as well as hip abduction, adduction, extension, and flexion. In closed chain movement, the foot is fixed in place either to the ground or a machine platform. Squats, step downs, lunges, leg press, and biking are close chain exercise options. Examples of open and closed chain exercises are shown in Fig. 5.6. These exercises offer isometric, concentric, and eccentric contractions. Performing these activities with knee internal rotation can allow for further isolation and strengthening of the VMO to better resist lateral patellar translation [35]. Programs are created with combinations of these movements to increase strength, improve tracking, and reduce pain.

Stretching is a key aspect in the rehabilitation of patellofemoral pain. This is commonly done in combination with strengthening and functional movement training rather than on its own. Reduced flexibility of knee flexors, extensors, and the IT band have all been associated with anterior knee pain [36, 37]. Various programs that include stretching with other rehab modalities have been found successful in treating patellofemoral pain [38, 39]. Similar to strengthening, stretching focuses on muscles and soft tissues that cross the knee and hip. Quadriceps stretching can be accomplished with passive knee flexion while standing, prone, or laying on one's side. Hamstrings can be addressed with forward bending at the waist and reaching for one's toes while standing or seated. Modifications such as leg

Fig. 5.6 Various strengthening exercises demonstrated. Open chain exercises (**a**) hip abduction and (**b**) unilateral supine bridge. Closed chain exercises (**c**) squat and (**d**) forward lunge [26]

crossing, leg abduction, and hurdlers stretch can also be done with these movements. The IT band can be stretched with hip adduction in standing or seated position and can be modified by pulling the knee to the chest with adduction. Foam rolling can also be done to help stretch any of the above listed groups. These stretches are done to aid in warm up and cool down but also serve to reduce tension on tight structures can cause patellofemoral pain.

Proprioception in patellofemoral rehab is addressed through neuromuscular training. Reduced core muscle strength can result in altered lower extremity stability and patellofemoral pain [40]. Furthermore, altered balance and posture have been found in these patients [41]. Neuromuscular training focuses on these issues through strength, balance, and proprioception [42]. Exercises discussed in the strengthening section are again used in neuromuscular training, but with modifications. Modifications aim to induce poor form in movements that participants must overcome. These modifications increase participant awareness when poor form occurs and improves their proprioception. This reinforces and strengthens proper form so that it can be repeated in daily life. Greater core muscle activation is induced by these movements to improve lower extremity alignment and ultimately reduce pain. Band squats provide a hip adduction and knee valgus forces that one must overcome while performing a squat. The movement can be further modified to emphasize proprioception by adding an uneven surface. Additional examples include single leg squats on various surfaces and lunges on uneven surfaces. Examples of these movements are shown in Fig. 5.7. Although relatively new compared to traditional strengthening and stretching methods, neuromuscular training has shown positive results. Early studies have found improvements in pain, strength, function, and balance [43]. This is an incredibly important aspect of patellofemoral rehab that can be gradually introduced as deficits in strength and flexibility are improved throughout a program.

Functional movement training aims to address the same issues as neuromuscular training, but is applied to movement an individual may perform in sports or daily activities. This is commonly added later in rehab programs once appropriate

Fig. 5.7 (**a**, **b**) An athlete demonstrating a single leg squat on flat ground and on a raised platform. (**c**) Forward lunge performed with medial resistance band force allowing the athlete to overcome forced knee valgus in the movement and maintain proper form [26]

progress has been made from the initial strengthening and stretching exercises. Individuals are observed performing the various activities where their form and mechanics are evaluated. A program can then be implemented that focuses on improving altered mechanics. Gait is frequently addressed in these programs. Abnormalities such as hip adduction can result in knee valgus which is previously noted and can induce patellofemoral pain [44]. Similar to described earlier strengthening is done to reinforce proper mechanics. Gait retraining can also be done by providing redirection and direct feedback while running [45]. Again these activities emphasize proper form so that it can be repeated in daily life so as reduce factors that can produce patellofemoral pain.

Supportive treatments such as braces and taping can be done in addition to physical therapy. Bracing options focus on providing medial force to reinforce proper patellar tracking. These braces have been found to improve patella control and proprioception [46, 47]. Despite this, studies found mixed results on the clinic impact of bracing [38, 48]. However, one studied showed improved outcomes when combined with therapy [49]. Similar to bracing, the goal of taping is to provide forces to improve tracking. Two styles of taping include McConnell and kinesio taping. In McConnell taping the patella is medially translated then maintained in this position by tape applied to the medial aspect of the knee and patella [50]. Kinesio taping involves a variety of techniques that aim to improve patella tracking by relieving lateral tension, stimulating VMO contraction, and providing medial force [51]. Variations of these taping styles are shown in Fig. 5.8. Despite these goals, studies show mixed results on the impact of these techniques for pain reduction, patella position, proprioception, and VMO activation [52]. Overall supportive therapies on the whole have not been found to significantly impact patellofemoral pain; however, they do provide a low risk adjuvant treatment that can help some patients.

Rehabilitation principles

	Goals	Examples
Strengthening	Improve quadriceps and hip muscle strength	Squats, lunges, straight leg raise, hip abduction, and hip flexion/extension

(continued)

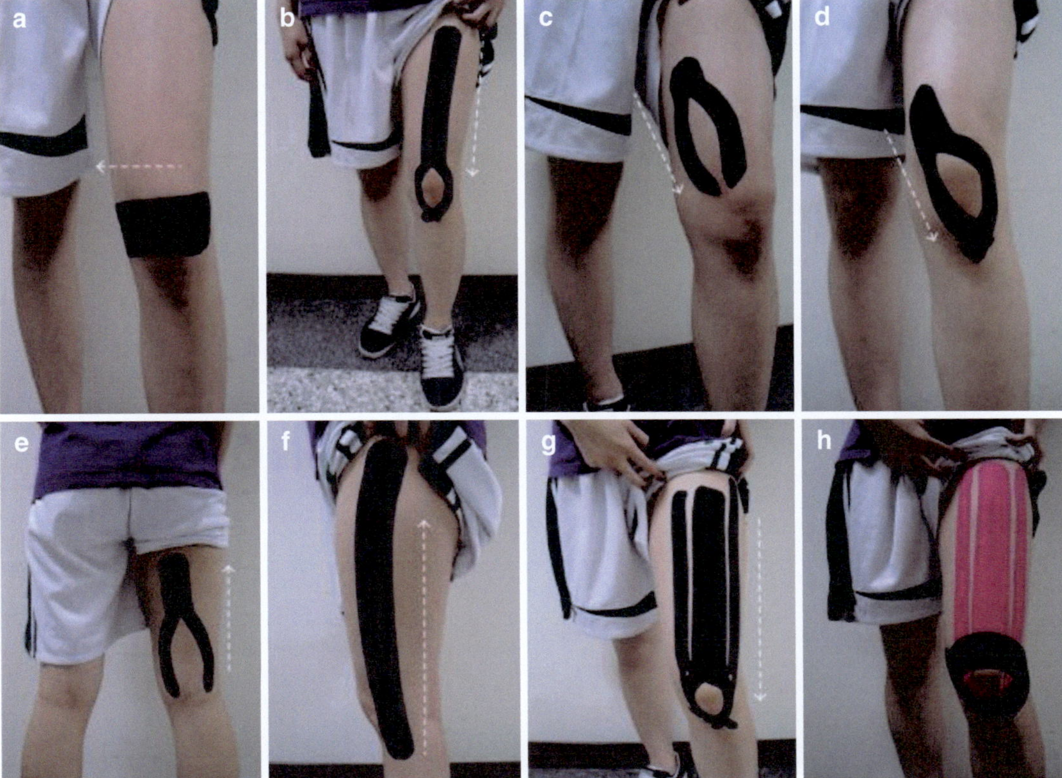

Fig. 5.8 (**a**) McConnell style taping with medial pull of the patella. (**b–d**) Kinesio taping of the patella and quadriceps to improve muscle tension and proprioception of the patella. (**e**) Kinesio taping to improve tension of the hamstrings. (**f**) Kinesio taping to tension the iliotibial band. (**g, h**) Multiple tapings of the quadriceps to increase tension and the patella is wrapped to improve stability [52]

(continued)

	Goals	Examples
Stretching	Improve flexibility of tight structures to improve tracking and reduce joint contact forces	Passive knee flexion, hurdlers stretch, toe touches, knee to check, and hip adduction
Proprioception	Improve core strength, balance, body awareness, and alignment in controlled movements	Single leg squats, band squats, movements on uneven surfaces
Functional movement training	Improve form and mechanics of daily activities or sport specific movements	Gait training
Supportive therapy	Provide a medial force to the patella reinforce proper tracking	Bracing and taping

5.6 Conclusion

Patellofemoral pain is among the most common musculoskeletal complaints. Given this it is important of providers of various specialties to be informed on this issue. A thorough understanding of the anatomy as well as biomechanics of the patellofemoral joint is essential to the evaluation and treatment of these conditions. Physical therapy remains the first line treatment for many patellofemoral disorders. Rehab focuses on lower extremity strengthening, stretching, neuromuscular training, and functional movement training. This aims to improve mechanics and patellar tracking to ultimately reduce pain. Additional supportive treatments including bracing and taping may provide some relief, but their efficacy is largely unproven.

References

1. Boling M, Padua D, Marshall S, Guskiewicz K, Pyne S, Beutler A. Gender differences in the incidence and prevalence of patellofemoral pain syndrome. Scand J Med Sci Sports. 2010;20(5):725–30.
2. Devereaux MD, Lachmann SM. Patello-femoral arthralgia in athletes attending a sports injury clinic. Br J Sports Med. 1984;18(1):18–21.
3. Heino Brechter J, Powers CM. Patellofemoral stress during walking in persons with and without patellofemoral pain. Med Sci Sports Exerc. 2002;34(10):1582–93.
4. Kannus P, Aho H, Järvinen M, Niittymäki S. Computerized recording of visits to an outpatient sports clinic. Am J Sports Med. 1987;15(1):79–85.
5. Wood L, Muller S, Peat G. The epidemiology of patellofemoral disorders in adulthood: a review of routine general practice morbidity recording. Prim Health Care Res Dev. 2011;12(2):157–64.
6. Roush JR, Curtis BR. Prevalence of anterior knee pain in 18-35 year-old females. Int J Sports Phys Ther. 2012;7(4):396–401.
7. Smith BE, Selfe J, Thacker D, Hendrick P, Bateman M, Moffatt F, et al. Incidence and prevalence of patellofemoral pain: a systematic review and meta-analysis. PLoS One. 2018;13(1):e0190892.
8. Callaghan M, Selfe J. Has the incidence or prevalence of patellofemoral pain in the general population in the United Kingdom been properly evaluated? Phys Ther Sport. 2007;8(1):37–43.
9. Dutton RA, Khadavi MJ, Fredericson M. Patellofemoral pain. Phys Med Rehabil Clin N Am. 2016;27(1):31–52.
10. Sobotta J. Atlas and text book of human anatomy. London: W.B. Saunders; 1909. 183 p.
11. Walsh W. Recurrent dislocation of the knee in the adult. In: Delee and Drez's orthopaedic sports medicine. Philadelphia: Saunders; 2003. p. 1710–49.
12. Anatomy. Patellofemoral disorders. p. 19–28.
13. Dejour H, Walch G, Nove-Josserand L, Guier C. Factors of patellar instability: an anatomic radiographic study. Knee Surg Sports Traumatol Arthrosc. 1994;2(1):19–26.
14. Pfirrmann CW, Zanetti M, Romero J, Hodler J. Femoral trochlear dysplasia: MR findings. Radiology. 2000;216(3):858–64.
15. Grelsamer RP, Proctor CS, Bazos AN. Evaluation of patellar shape in the sagittal plane. A clinical analysis. Am J Sports Med. 1994;22(1):61–6.
16. Negrín R, Reyes NO, Iñiguez M, Gaggero N, Sandoval R, Jabes N, et al. Dynamic-anatomical reconstruction of medial patellofemoral ligament in open physis. Arthrosc Tech. 2020;9(7):e1027–32.
17. Amis AA, Firer P, Mountney J, Senavongse W, Thomas NP. Anatomy and biomechanics of the medial patellofemoral ligament. Knee. 2003;10(3):215–20.
18. Khasawneh RR, Allouh MZ, Abu-El-Rub E. Measurement of the quadriceps (Q) angle with respect to various body parameters in young Arab population. PLoS One. 2019;14(6):e0218387.
19. Butler-Manuel PA, Justins D, Heatley FW. Sympathetically mediated anterior knee pain. Scintigraphy and anesthetic blockade in 19 patients. Acta Orthop Scand. 1992;63(1):90–3.
20. Desio SM, Burks RT, Bachus KN. Soft tissue restraints to lateral patellar translation in the human knee. Am J Sports Med. 1998;26(1):59–65.
21. Meira EP, Brumitt J. Influence of the hip on patients with patellofemoral pain syndrome: a systematic review. Sports Health. 2011;3(5):455–65.
22. Loudon JK. Biomechanics and pathomechanics of the patellofemoral joint. Int J Sports Phys Ther. 2016;11(6):820–30.
23. Andrish J. The biomechanics of patellofemoral stability. J Knee Surg. 2004;17(1):35–9.
24. Powers CM, Ward SR, Fredericson M, Guillet M, Shellock FG. Patellofemoral kinematics during weight-bearing and non-weight-bearing knee extension in persons with lateral subluxation of the patella: a preliminary study. J Orthop Sports Phys Ther. 2003;33(11):677–85.
25. Hollman JH, Ginos BE, Kozuchowski J, Vaughn AS, Krause DA, Youdas JW. Relationships between knee valgus, hip-muscle strength, and hip-muscle recruitment during a single-limb step-down. J Sport Rehabil. 2009;18(1):104–17.
26. Ford KR, Nguyen AD, Dischiavi SL, Hegedus EJ, Zuk EF, Taylor JB. An evidence-based review of hip-focused neuromuscular exercise interventions to address dynamic lower extremity valgus. Open Access J Sports Med. 2015;6:291–303.
27. Powers CM. The influence of abnormal hip mechanics on knee injury: a biomechanical perspective. J Orthop Sports Phys Ther. 2010;40(2):42–51.
28. Huberti HH, Hayes WC. Patellofemoral contact pressures. The influence of q-angle and tendofemoral contact. J Bone Joint Surg Am. 1984;66(5):715–24.
29. Wilson NA, Press JM, Koh JL, Hendrix RW, Zhang LQ. In vivo noninvasive evaluation of abnormal patellar tracking during squatting in patients with patellofemoral pain. J Bone Joint Surg Am. 2009;91(3):558–66.
30. Werner S. An evaluation of knee extensor and knee flexor torques and EMGs in patients with patellofemoral pain syndrome in comparison with matched controls. Knee Surg Sports Traumatol Arthrosc. 1995;3(2):89–94.
31. Lack S, Neal B, De Oliveira Silva D, Barton C. How to manage patellofemoral pain—understanding the multifactorial nature and treatment options. Phys Ther Sport. 2018;32:155–66.
32. Alba-Martín P, Gallego-Izquierdo T, Plaza-Manzano G, Romero-Franco N, Núñez-Nagy S, Pecos-Martín D. Effectiveness of therapeutic physical exercise in the treatment of patellofemoral pain syndrome: a systematic review. J Phys Ther Sci. 2015;27(7):2387–90.
33. Witvrouw E, Danneels L, Van Tiggelen D, Willems TM, Cambier D. Open versus closed kinetic chain

exercises in patellofemoral pain: a 5-year prospective randomized study. Am J Sports Med. 2004;32(5):1122–30.

34. Werner S. Anterior knee pain: an update of physical therapy. Knee Surg Sports Traumatol Arthrosc. 2014;22(10):2286–94.
35. Roush MB, Sevier TL, Wilson JK, Jenkinson DM, Helfst RH, Gehlsen GM, et al. Anterior knee pain: a clinical comparison of rehabilitation methods. Clin J Sport Med. 2000;10(1):22–8.
36. Smith AD, Stroud L, McQueen C. Flexibility and anterior knee pain in adolescent elite figure skaters. J Pediatr Orthop. 1991;11(1):77–82.
37. Doucette SA, Goble EM. The effect of exercise on patellar tracking in lateral patellar compression syndrome. Am J Sports Med. 1992;20(4):434–40.
38. Saltychev M, Dutton RA, Laimi K, Beaupré GS, Virolainen P, Fredericson M. Effectiveness of conservative treatment for patellofemoral pain syndrome: a systematic review and meta-analysis. J Rehabil Med. 2018;50(5):393–401.
39. Collins NJ, Barton CJ, van Middelkoop M, Callaghan MJ, Rathleff MS, Vicenzino BT, et al. 2018 consensus statement on exercise therapy and physical interventions (orthoses, taping and manual therapy) to treat patellofemoral pain: recommendations from the 5th international patellofemoral pain research retreat, Gold Coast, Australia, 2017. Br J Sports Med. 2018;52(18):1170–8.
40. Willson JD, Dougherty CP, Ireland ML, Davis IM. Core stability and its relationship to lower extremity function and injury. J Am Acad Orthop Surg. 2005;13(5):316–25.
41. Capin JJ, Snyder-Mackler L. The current management of patients with patellofemoral pain from the physical therapist's perspective. Ann Jt. 2018;3:40.
42. Rabelo ND, Lima B, Reis AC, Bley AS, Yi LC, Fukuda TY, et al. Neuromuscular training and muscle strengthening in patients with patellofemoral pain syndrome: a protocol of randomized controlled trial. BMC Musculoskelet Disord. 2014;15:157.
43. Motealleh A, Mohamadi M, Moghadam MB, Nejati N, Arjang N, Ebrahimi N. Effects of core neuromuscular training on pain, balance, and functional performance in women with patellofemoral pain syndrome: a clinical trial. J Chiropr Med. 2019;18(1):9–18.
44. Willy RW, Manal KT, Witvrouw EE, Davis IS. Are mechanics different between male and female runners with patellofemoral pain? Med Sci Sports Exerc. 2012;44(11):2165–71.
45. Agresta C, Brown A. Gait retraining for injured and healthy runners using augmented feedback: a systematic literature review. J Orthop Sports Phys Ther. 2015;45(8):576–84.
46. Thijs Y, Vingerhoets G, Pattyn E, Rombaut L, Witvrouw E. Does bracing influence brain activity during knee movement: an fMRI study. Knee Surg Sports Traumatol Arthrosc. 2010;18(8):1145–9.
47. Worrell T, Ingersoll CD, Bockrath-Pugliese K, Minis P. Effect of patellar taping and bracing on patellar position as determined by MRI in patients with patellofemoral pain. J Athl Train. 1998;33(1):16–20.
48. Lun VM, Wiley JP, Meeuwisse WH, Yanagawa TL. Effectiveness of patellar bracing for treatment of patellofemoral pain syndrome. Clin J Sport Med. 2005;15(4):235–40.
49. Petersen W, Ellermann A, Rembitzki IV, Scheffler S, Herbort M, Brüggemann GP, et al. Evaluating the potential synergistic benefit of a realignment brace on patients receiving exercise therapy for patellofemoral pain syndrome: a randomized clinical trial. Arch Orthop Trauma Surg. 2016;136(7):975–82.
50. Aminaka N, Gribble PA. Patellar taping, patellofemoral pain syndrome, lower extremity kinematics, and dynamic postural control. J Athl Train. 2008;43(1):21–8.
51. Akbaş E, Atay AO, Yüksel I. The effects of additional kinesio taping over exercise in the treatment of patellofemoral pain syndrome. Acta Orthop Traumatol Turc. 2011;45(5):335–41.
52. Chang WD, Chen FC, Lee CL, Lin HY, Lai PT. Effects of Kinesio taping versus McConnell taping for patellofemoral pain syndrome: a systematic review and meta-analysis. Evid Based Complement Alternat Med. 2015;2015:471208.

Newest Surgical Treatments for Patellofemoral Osteochondral Lesions

6

Ignacio Dallo and Alberto Gobbi

6.1 Introduction

The articular cartilage has limited intrinsic healing potential attributed to the presence of a few specialized and undifferentiated cells with low mitotic activity and a lack of vessels that can promote tissue repair. Therefore, once an injury occurs, surgical intervention is necessary to maximize the chances of articular cartilage repair. A good cartilage repair will lead to good long-term functional outcomes and avoid subsequent cartilage degeneration that could otherwise lead to osteoarthritis (OA) [1].

Chondral or osteochondral lesions are frequently found during knee arthroscopy. In a study of 1000 patients who underwent arthroscopy, the prevalence of osteochondral defects was 61%, while 17% of them were located in the patellofemoral (PF) joint (11% patella, 6% trochlea) [2].

A review of a Polish registry found that more than half of patients undergoing knee arthroscopy had chondral defects, with 5.2% having Outerbridge Grade III or IV lesions. Of these, 37.5% were in the patella alone [3]. Curl et al., in a review of 31.516 knee arthroscopies, found over 53.000 hyaline cartilage lesions in over 19.000 patients; most of the lesions found were grade III defects in the patella [4].

Several surgical techniques for the regeneration of the articular cartilage have been proposed. Among them, two-step procedures like autologous chondrocyte implantation (ACI) and matrix-induced autologous chondrocyte implantation (MACI) have been shown to provide good results, promoting the formation of new hyaline-like cartilage tissue [5–7]; while other techniques, such as microfracture, result in fibrous cartilage and less durable repair [8]. One-step cell-based procedures are an attractive treatment option, given the potential for cost savings and the requirement for the patient to undergo only one surgical procedure instead of two.

Multipotent stem cells sourced from bone marrow aspirate concentrate (BMAC) combined with a biologic scaffold have demonstrated good to excellent clinical outcomes at long-term follow-up as with ACI [9–11].

This chapter describes the newest surgical procedures that can be used to treat chondral knee injuries in a wide range of patient age and lesion sizes in the patellofemoral compartment. Optimization of PF biomechanics through concomitant bony and/or soft tissue procedures will maximize the results when cartilage repair or regeneration is indicated.

I. Dallo (✉)
Orthopaedic Arthroscopic Surgery International (O.A.S.I.) Bioresearch Foundation, Gobbi N.P.O, Milan, Italy

A. Gobbi
O.A.S.I. Bioresearch Foundation, Milan, Italy
e-mail: gobbi@cartilagedoctor.it

© ISAKOS 2022
J. L. Koh et al. (eds.), *The Patellofemoral Joint*, https://doi.org/10.1007/978-3-030-81545-5_6

6.2 Two Step Procedures: (ACI-MACI)

Histologic studies have shown that autologous chondrocyte implantation (ACI) results in hyaline-like Type II collagen [12]. While femoral lesions had acceptable outcomes, the patients with patellar lesions showed poor results [13]. However, malalignment was not addressed in this study. When bony alignment and other comorbidities were corrected concurrently or in a staged fashion, results by the different authors were shown to be similar to femoral lesions of the knee [14–16].

However, the apparent complexity of this technique, needing the sacrifice of periosteal tissue, the uncertain distribution of chondrocyte solution, and complications such as periosteal patch hypertrophy and arthrofibrosis prompted the scientific community to develop second-generation ACI [6, 17].

Studies of matrix-induced autologous chondrocyte implantation (MACI) for patellofemoral osteochondral lesions show comparable results to the tibiofemoral location when appropriate concomitant procedures are performed [18]. Filardo et al. have reported the outcomes differences between patellar and trochlear lesions when treated by MACI [19]. Some studies have shown better results with MACI compared to ACI for the PF joint [20, 21].

However, essentially it remains a two-step procedure, including an arthroscopic biopsy and subsequent implantation of the cultured chondrocytes. Apart from donor site morbidity, the risks of two surgical procedures, and the limited quantity of cartilage that could be harvested, the total cost of surgeries, scaffold, and in vitro culture still represent the major limitation of this technique [22, 23].

6.3 One Step Procedures

6.3.1 Cell-based Scaffold Implant (HA-BMAC)

The evolution of the cartilage repair technique leads to the development of new scaffolds that allowed cell proliferation but did not avoid the chondrocyte harvest and cultivation.

Performing a one-step procedure avoids the two-step surgical procedures and reduces the operation costs by approximately five times.

Bone marrow aspirate concentrate (BMAC) contains bone marrow stem cells (BMSCs) and growth factors that are a promising option for cartilage repair and regeneration because of their differentiation potential to cartilage [24–27]. Bone marrow-derived stem cells (BMSCs) interact with a non-woven scaffold, the HYAFF 11, that supports cellular adhesion, migration, and proliferation the synthesis of extracellular matrix components under static culture conditions [28–30]. Nejadnik et al. compared the clinical outcomes of patients treated with first-generation ACI and patients treated with autologous BMSCs. Concluding that BMSCs are as effective as chondrocytes for articular cartilage repair [31]. We compared patients treated with matrix-induced autologous chondrocyte implantation (MACI) with patients treated with BMSCs combined with the same scaffold in our institution. We did not notice, at three years, follow-up, any significant statistical differences between the two groups, concluding that these techniques were viable and effective [32]. It has been shown in many clinical studies that the hyaluronic acid-based scaffold with activated bone marrow aspirate concentrate (HA-BMAC) technique is a valuable method for the treatment of full-thickness cartilage lesions of the knee [33]. Different cartilage lesions can be treated, from minor injuries to significant defects up to 22 cm^2 [34]. The HA-BMAC technique has proven to be effective in treatment for patients over 45 years of age [10].

6.3.1.1 The HA-BMAC Technique

The whole procedure is performed under general anesthesia. The patient is positioned supine for standard knee arthroscopy. The ipsilateral iliac crest is prepared and exposed to bone marrow aspiration. Examination of the knee under anesthesia is done to recognize the concomitant pathologies addressed during the surgery. All cartilage lesions are then identified during diagnostic arthroscopy. It is necessary to choose whether

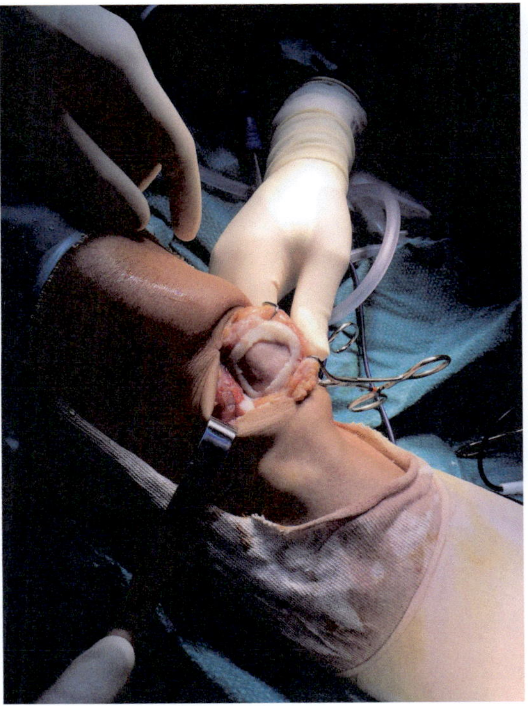

Fig. 6.1 The patellar chondral defect was prepared to obtain perpendicular edges and the circumferential borders of the lesion, which needs to be perpendicular to the subchondral bone

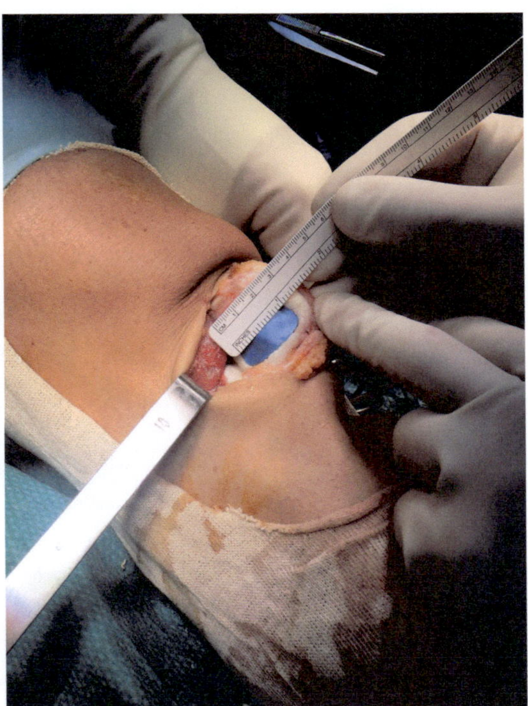

Fig. 6.2 The lesion dimensions have to be measured to prepare the matching implant from the three-dimensional hyaluronic acid-based scaffold (Hyalofast, Anika Therapeutics, Srl, Abano Terme, Italy) based on the template

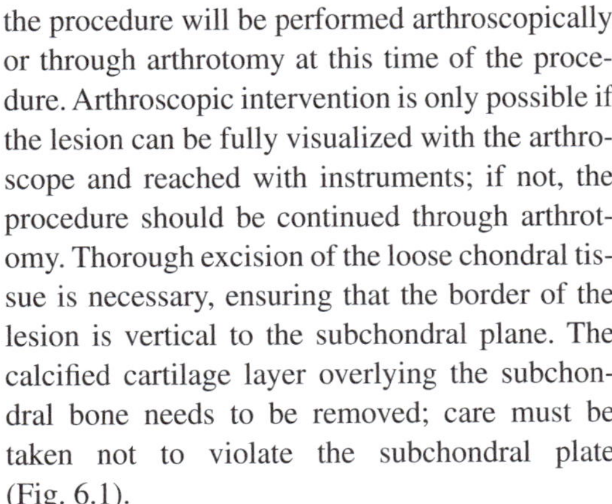

the procedure will be performed arthroscopically or through arthrotomy at this time of the procedure. Arthroscopic intervention is only possible if the lesion can be fully visualized with the arthroscope and reached with instruments; if not, the procedure should be continued through arthrotomy. Thorough excision of the loose chondral tissue is necessary, ensuring that the border of the lesion is vertical to the subchondral plane. The calcified cartilage layer overlying the subchondral bone needs to be removed; care must be taken not to violate the subchondral plate (Fig. 6.1).

BMAC preparation is started after the lesion is prepared. Approximately 60 mL of bone marrow from the ipsilateral iliac crest is harvested using a dedicated aspiration kit. The aspirate is centrifuged with a commercially available system to obtain the concentrated bone marrow (Angel, Cytomedix, Gaithersburg, MD). The dimensions of the lesion have to be measured to prepare the matching implant from the three-dimensional hyaluronic acid-based scaffold (Hyalofast, Anika Therapeutics, Srl, Abano Terme, Italy) (Fig. 6.2). It is also possible to cut an aluminum foil model before to check if it fits the lesion and then cut the scaffold according to the aluminum foil model (Fig. 6.3). When the scaffold is ready, BMAC is activated with the batroxobin enzyme (Plateltex Act, Plateltex SRO, Bratislava, Slovakia). The activation process is necessary for BMAC to form a clot, which is then applied onto the prepared scaffold comprising a sticky implant that is easy to apply into the lesion.

According to the chosen approach, previously prepared HA-BMAC is then implanted into the lesion. If an open technique is selected, the surgeon should apply HA-BMAC directly into the lesion (Fig. 6.4). If needed, fibrin glue is added to secure the graft further. The knee is then flexed and extended to check the graft stability. If the

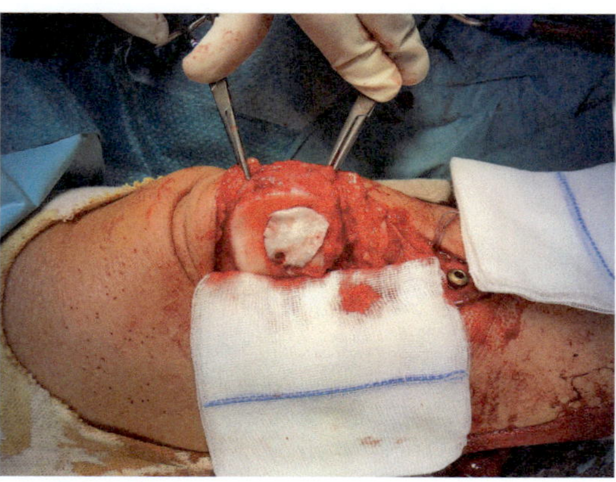

Fig. 6.3 Presentation of the three-dimensional hyaluronic acid-based scaffold (Hyalofast, Anika Therapeutics, Srl, Abano Terme, Italy) onto the lesion according to the dimensions of the template model

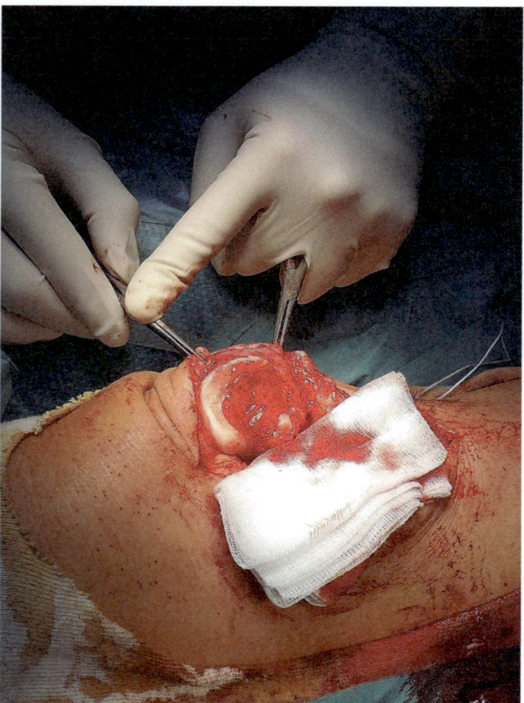

Fig. 6.4 The prepared BMAC was then placed on the hyalofast scaffold. After few minutes, the activated BMAC was absorbed by the scaffold, creating a sticky implant that is easy to apply onto the lesion. Approved, all pictures are from O.A.S.I. Bioresearch Foundation Gobbi NPO, Milan, Italy

surgeon chooses an arthroscopic approach, fluid needs to be completely drained, and the lesion should be inspected under arthroscopy after fluid drainage to ensure that the circumferential border is stable. The scaffold is introduced into the joint via the working portal through a valveless cannula and using a grasper, the implant is placed gently filling the cartilage defect. A hook can be used to press-fit the scaffold into the lesion. The crucial part of the procedure is to check the implant stability. The joint is moved a couple of times while the scaffold is observed with the arthroscope. If needed, fibrin glue is applied to improve implant stability. The working portals are sutured, a drain should not be inserted into the joint [26, 35].

6.3.2 Allograft Cell-based Scaffold Implant (DeNovo—Cartiform—BioCartilage—IMPACT)

DeNovo NT Natural Tissue Graft (Zimmer Biomet) is an off-the-shelf human tissue allograft, consisting of juvenile hyaline cartilage pieces with viable chondrocytes with promising preliminary outcomes. The immature chondrocytes have been shown to have increased metabolic and proliferative activity compared to adult chondrocytes and are intended to repair articular cartilage lesions in a one-step procedure. [36] One study showed significant improvements in MRI scores and clinical outcomes at over 2 years of follow-up [37]. A more recent study showed that lesion fill at 6, 12, and 24 months was 82%, 85%, and 75%, respectively [38]. A clinical study in patellofemoral cartilage lesions showed significant improvements in KOOS scores at 8 months [39]. A prospective trial showed improvements in radiographic appearance, histology, and clinical scores at 2 years follow-up [40]. However, randomized controlled trials and long-term data are needed.

Cartiform (Osiris Therapeutics, Inc) is a cryopreserved viable chondral allograft. It is currently available in a 2 cm^2 size to treat smaller lesions. The main advantage is that the cells remain viable, up to 70%, at 2 years [41]. This can facilitate surgical planning and elective scheduling, rather than waiting for a fresh, stored graft traditionally. There are very few clinical data of this technique; only one case report is published [42].

BioCartilage (Arthrex, Naples, FL) is a new product containing dehydrated micronized allogeneic cartilage scaffold implanted with platelet-rich plasma and fibrin glue added over a contained microfracture-treated defect, can be used in the patella and trochlea for small lesions. There are limited clinical studies of short- or long-term outcomes [43, 44].

IMPACT (D. Saris's Team, Utrecht) is another promising one-step cell-based cartilage regeneration technique [45]. With this procedure, the authors prepare the cartilage defect in the standard fashion. The cartilage is then partially digested to separate the chondrons and the extracellular matrix. These are then combined with a precise ratio of allogeneic MSCs. The findings of the first study in 35 patients demonstrate good short-term clinical, MRI, and histological results. The authors conclude that allogeneic MSCs can be a safe cell source to augment or facilitate tissue regeneration through paracrine mechanisms and cellular communication in a clinical setting. [46].

6.3.3 Cell-free Scaffolds (MaioRegen)

Cell-free scaffolds have been developed with the aim of promoting and inducing tissue regeneration. To date, there are few clinical studies on patellofemoral osteochondral lesions. MaioRegen (Fin-Ceramica Faenza SpA, Faenza, Italy) is a nanostructured 3-layer biomimetic scaffold with a porous composite structure. The device mimics the entire osteochondral structure with a cartilaginous type I collagen-based layer with a smooth surface, an intermediate tidemark-like layer consisting of a combination of type I collagen and hydroxyapatite, and a bottom layer composed of a mineralized blend of type I collagen and hydroxyapatite. Only two clinical studies using the MaioRegen scaffold for patellofemoral chondral lesions showed good clinical and MRI outcomes. Kon et al. [47] in 11 patellas and 7 trochleas at 5 years follow-up. Berruto et al. reported only one trochlear groove from a cohort of 49 knees at 2 years. [48].

6.3.4 Scaffold-free Tissue-Engineered Construct (TEC)

The tissue-engineered construct (TEC) technique is an autologous 3-dimensional (3D) biologic structure made through simple-cell culture methods of synovial MSCs. The TEC contains an extracellular matrix, synthesized by the cells, composed of fibrillar collagen (type I–III). The construct is pliable and highly adherent to normal cartilage because of the adhesion molecules, such as fibronectin and vitronectin, that are present on it. Shimomura et al. [49] published the first in humans "early proof of concept" trial in five patients, one of them was in the trochlear groove. They reported positive clinical and morphologic outcomes across all patients involved, without any significant adverse events at two years follow-up.

6.4 Concomitant Surgical Procedures

A diagnostic arthroscopy should be performed to assess chondral surfaces, patella tracking, and tibiofemoral ligamentous stability. The contralateral limb should be examined and compared to the diseased limb. The abnormal biomechanical factors need to be evaluated for correction to optimize the patellofemoral osteochondral lesions treatment. The stabilization soft tissue procedures (MPFL, MPTL reconstruction) are indicated for patellar instability. For patients with malalignment and distal/lateral patella or trochlea chondral lesion(s), antero-medialization (AMZ) of the tibial tubercle (TT) is indicated. Trochleoplasty is reserved for advanced trochlear dysplasia without significant chondrosis [50].

Fact Box 6.1: Key Surgical Points of HA-BMAC Technique

- Complete exposure of the patellofemoral cartilage lesion is critical.
- Use traction methods as needed to provide a comfortable working space.
- An aluminum foil template is used to measure the prepared cartilage defect.
- Place the HA-based scaffold against the subchondral bone on either side.
- Fibrin glue is applied to improve implant stability.
- All arthroscopic techniques can be performed only in cases where the entirety of the defect is appreciated.
- Cycle the knee under arthroscopic visualization to confirm the graft seating within de defect.
- A drain should not be inserted into the joint.

Fact Box 6.2: Contraindication of HA-BMAC Surgical Technique

- Patients older (>60 years).
- Obese (BMI > 30).
- Severe tri-compartmental OA (ICRS Grade 4).

Patients with untreated malalignment (varus/valgus >5°) or knee instability.

Multiple intra-articular steroids injections.

- Hip disorders leading to abnormal gait.
- Rheumatic diseases, Bechterew's Syndrome, and Chondrocalcinosis and Gout.

Fact Box 6.3: Key Points in Clinical Considerations of PF Cartilage Injuries

- Asymptomatic OC lesions should be observed but not aggressively treated.
- Identify and address patellofemoral instability, malalignment, meniscus, and ligaments deficiency, to provide an optimal environment for cartilage restoration.
- An individualized approach based on the patient's goals, lesion characteristics, and surgeon preferences should be made.
- The single-stage HA-BMAC technique is a durable and cost-effective cartilage repair procedure.

Resources/Websites

- ICRS International Cartilage Regeneration & Joint Preservation Society: https://cartilage.org
- O.A.S.I Bioresearch Foundation: https://kneecartilagedoctor.com/about/dr-alberto-gobbi/
- Patellofemoral Foundation: https://www.patellofemoral.org
- Patellofemoral Study Group: https://ipsg.org

Clinical Vignettes: Studies Cartilage Repair in Patellofemoral Joint					
Author, year	Treatment	*N*	Age	Follow up (years)	Results
Gomol et al. 2014	ACI	110	33	4	Good to Excellent Cl
Kon et al. 2014	MioRegen	18	34.9	5	Signif. Improv. Cl/MRI
Berruto et al. 2014	MioRegen	1	39	2	Signif. Improv. Cl/MRI
de Windt et al. 2017	IMPACT	6	30.8	1.5	Signif. Improv Cl/MRI
Shimomura et al. 2018	TEC	1	35	2	Signif. Improv Cl/MRI
Gobbi et al. 2019	HA-BMAC	23	48.5	8	Good to Excellent Cl

Take Home Messages

- An individualized approach based on the patient's goals and the surgeon's preferences is crucial.
- Pathological background factors such as malalignment, meniscus deficiency, or ligament laxity have to be addressed to provide an optimal environment for cartilage repair.
- Single-step cartilage repair eliminates the need for a two-step procedure, thereby reducing the cost and morbidity to the patient.
- HA-BMAC procedure provides good to excellent clinical outcomes at long-term follow-up in small or large lesions, single or multiple injuries, various compartments.
- Results may be comparatively more successful in younger patients.
- We need future studies on cartilage repair based on biological and imaging biomarkers testing the inflammatory and degenerative environment of a joint to estimate survival and success better.

References

1. Mankin HJ. The response of articular cartilage to mechanical injury. J Bone Joint Surg Am. 1982;64(3):460–6.
2. Hjelle K, et al. Articular cartilage defects in 1,000 knee arthroscopies. Arthroscopy. 2002;18(7):730–4.
3. Widuchowski W, Widuchowski J, Trzaska T. Articular cartilage defects: study of 25,124 knee arthroscopies. Knee. 2007;14(3):177–82.
4. Curl WW, et al. Cartilage injuries: a review of 31,516 knee arthroscopies. Arthroscopy. 1997;13(4):456–60.
5. Gobbi A, et al. Patellofemoral full-thickness chondral defects treated with Hyalograft-C: a clinical, arthroscopic, and histologic review. Am J Sports Med. 2006;34(11):1763–73.
6. Kon E, et al. Arthroscopic second-generation autologous chondrocyte implantation compared with microfracture for chondral lesions of the knee: prospective nonrandomized study at 5 years. Am J Sports Med. 2009;37(1):33–41.
7. Battaglia M, et al. Validity of T2 mapping in characterization of the regeneration tissue by bone marrow derived cell transplantation in osteochondral lesions of the ankle. Eur J Radiol. 2011;80(2):e132–9.
8. Gobbi A, Karnatzikos G, Kumar A. Long-term results after microfracture treatment for full-thickness knee chondral lesions in athletes. Knee Surg Sports Traumatol Arthrosc. 2014;22(9):1986–96.
9. Gobbi A, Whyte GP. One-stage cartilage repair using a hyaluronic acid-based scaffold with activated bone marrow-derived mesenchymal stem cells compared with microfracture: five-year follow-up. Am J Sports Med. 2016;44(11):2846–54.
10. Gobbi A, et al. One-step surgery with multipotent stem cells and Hyaluronan-based scaffold for the treatment of full-thickness chondral defects of the knee in patients older than 45 years. Knee Surg Sports Traumatol Arthrosc. 2017;25(8):2494–501.
11. Gobbi A, Whyte GP. Long-term clinical outcomes of one-stage cartilage repair in the knee with hyaluronic acid-based scaffold embedded with mesenchymal stem cells sourced from bone marrow aspirate concentrate. Am J Sports Med. 2019;47(7):1621–8.
12. Brittberg M, et al. Rabbit articular cartilage defects treated with autologous cultured chondrocytes. Clin Orthop Relat Res. 1996;326:270–83.
13. Brittberg M, et al. Treatment of deep cartilage defects in the knee with autologous chondrocyte transplantation. N Engl J Med. 1994;331(14):889–95.
14. Vasiliadis HS, et al. Malalignment and cartilage lesions in the patellofemoral joint treated with autologous chondrocyte implantation. Knee Surg Sports Traumatol Arthrosc. 2011;19(3):452–7.
15. Gillogly SD, Arnold RM. Autologous chondrocyte implantation and anteromedialization for isolated patellar articular cartilage lesions: 5- to 11-year follow-up. Am J Sports Med. 2014;42(4):912–20.
16. Gomoll AH, et al. Autologous chondrocyte implantation in the patella: a multicenter experience. Am J Sports Med. 2014;42(5):1074–81.

17. Peterson L, et al. Autologous chondrocyte transplantation. Biomechanics and long-term durability. Am J Sports Med. 2002;30(1):2–12.
18. Ebert JR, et al. A comparison of 2-year outcomes in patients undergoing tibiofemoral or patellofemoral matrix-induced autologous chondrocyte implantation. Am J Sports Med. 2017;45(14):3243–53.
19. Filardo G, et al. Treatment of "patellofemoral" cartilage lesions with matrix-assisted autologous chondrocyte transplantation: a comparison of patellar and trochlear lesions. Am J Sports Med. 2014;42(3):626–34.
20. Macmull S, et al. The role of autologous chondrocyte implantation in the treatment of symptomatic chondromalacia patellae. Int Orthop. 2012;36(7):1371–7.
21. Nawaz SZ, et al. Autologous chondrocyte implantation in the knee: mid-term to long-term results. J Bone Joint Surg Am. 2014;96(10):824–30.
22. Henderson I, et al. Autologous chondrocyte implantation for treatment of focal chondral defects of the knee--a clinical, arthroscopic, MRI and histologic evaluation at 2 years. Knee. 2005;12(3):209–16.
23. Minas T, Peterson L. Advanced techniques in autologous chondrocyte transplantation. Clin Sports Med. 1999;18(1):13–44. v–vi
24. Caplan AI. Review: mesenchymal stem cells: cell-based reconstructive therapy in orthopedics. Tissue Eng. 2005;11(7–8):1198–211.
25. Caplan AI. Mesenchymal stem cells: the past, the present, the future. Cartilage. 2010;1(1):6–9.
26. Dimarino AM, Caplan AI, Bonfield TL. Mesenchymal stem cells in tissue repair. Front Immunol. 2013;4:201.
27. Huselstein C, Li Y, He X. Mesenchymal stem cells for cartilage engineering. Biomed Mater Eng. 2012;22(1–3):69–80.
28. Pasquinelli G, et al. Mesenchymal stem cell interaction with a non-woven hyaluronan-based scaffold suitable for tissue repair. J Anat. 2008;213(5):520–30.
29. Lisignoli G, et al. Chondrogenic differentiation of murine and human mesenchymal stromal cells in a hyaluronic acid scaffold: differences in gene expression and cell morphology. J Biomed Mater Res A. 2006;77(3):497–506.
30. Facchini A, et al. Human chondrocytes and mesenchymal stem cells grown onto engineered scaffold. Biorheology. 2006;43:471–80.
31. Nejadnik H, et al. Autologous bone marrow-derived mesenchymal stem cells versus autologous chondrocyte implantation: an observational cohort study. Am J Sports Med. 2010;38(6):1110–6.
32. Gobbi A, et al. Matrix-induced autologous chondrocyte implantation versus multipotent stem cells for the treatment of large patellofemoral chondral lesions: a nonrandomized prospective trial. Cartilage. 2015;6(2):82–97.
33. Gobbi A, Karnatzikos G, Sankineani SR. One-step surgery with multipotent stem cells for the treatment of large full-thickness chondral defects of the knee. Am J Sports Med. 2014;42(3):648–57.
34. Gobbi A, et al. One-step cartilage repair with bone marrow aspirate concentrated cells and collagen matrix in full-thickness knee cartilage lesions: results at 2-year follow-up. Cartilage. 2011;2(3):286–99.
35. Whyte GP, Gobbi A, Sadlik B. Dry arthroscopic single-stage cartilage repair of the knee using a hyaluronic acid-based scaffold with activated bone marrow-derived mesenchymal stem cells. Arthrosc Tech. 2016;5(4):e913–8.
36. Bonasia DE, et al. Cocultures of adult and juvenile chondrocytes compared with adult and juvenile chondral fragments: in vitro matrix production. Am J Sports Med. 2011;39(11):2355–61.
37. Tompkins M, et al. Preliminary results of a novel single-stage cartilage restoration technique: particulated juvenile articular cartilage allograft for chondral defects of the patella. Arthroscopy. 2013;29(10):1661–70.
38. Grawe B, et al. Cartilage regeneration in full-thickness patellar chondral defects treated with Particulated juvenile articular allograft cartilage: an MRI analysis. Cartilage. 2017;8(4):374–83.
39. Buckwalter JA, et al. Clinical outcomes of patellar chondral lesions treated with juvenile particulated cartilage allografts. Iowa Orthop J. 2014;34:44–9.
40. Farr J, et al. Clinical, radiographic, and histological outcomes after cartilage repair with Particulated juvenile articular cartilage: a 2-year prospective study. Am J Sports Med. 2014;42(6):1417–25.
41. Geraghty S, et al. A novel, cryopreserved, viable osteochondral allograft designed to augment marrow stimulation for articular cartilage repair. J Orthop Surg Res. 2015;10:66.
42. Hoffman JK, Geraghty S, Protzman NM. Articular cartilage repair using marrow stimulation augmented with a viable chondral allograft: 9-month postoperative histological evaluation. Case Rep Orthop. 2015;2015:617365.
43. Fortier LA, et al. BioCartilage improves cartilage repair compared with microfracture alone in an equine model of full-thickness cartilage loss. Am J Sports Med. 2016;44(9):2366–74.
44. Wang KC, et al. Arthroscopic Management of Isolated Tibial Plateau Defect with microfracture and micronized allogeneic cartilage-platelet-rich plasma adjunct. Arthrosc Tech. 2017;6(5):e1613–8.
45. de Windt TS, et al. Early health economic modelling of single-stage cartilage repair. Guiding implementation of technologies in regenerative medicine. J Tissue Eng Regen Med. 2017;11(10):2950–9.
46. de Windt TS, et al. Allogeneic MSCs and recycled autologous Chondrons mixed in a one-stage cartilage cell transplantation: a first-in-man trial in 35 patients. Stem Cells. 2017;35(8):1984–93.
47. Kon E, et al. Clinical results and MRI evolution of a nano-composite multilayered biomaterial for osteochondral regeneration at 5 years. Am J Sports Med. 2014;42(1):158–65.

48. Berruto M, et al. Treatment of large knee osteochondral lesions with a biomimetic scaffold: results of a Multicenter study of 49 patients at 2-year follow-up. Am J Sports Med. 2014;42(7):1607–17.
49. Shimomura K, et al. First-in-human pilot study of implantation of a scaffold-free tissue-engineered construct generated from autologous synovial mesenchymal stem cells for repair of knee chondral lesions. Am J Sports Med. 2018;46(10):2384–93.
50. Sherman SL, Thomas DM, Farr J II. Chondral and osteochondral lesions in the patellofemoral joint: when and how to manage. Ann Joint. 2018; https://doi.org/10.21037/aoj.2018.04.12.

Part II

Patellofemoral Instability: Case-Based Evaluation and Treatment

7 Patellofemoral Instability in the Pediatric Patient with Open Physes: A 11-Year-Old Girl with Trochlear Dysplasia

Alexandra H. Aitchison, Daniel W. Green, Jack Andrish, Marie Askenberger, Ryosuke Kuroda, and Geraldo Schuck de Freitas

7.1 Case Presentation

Patient is a premenarchal 11.0-year-old female who presents with recurrent patella dislocations of the left knee. She has had three prior atraumatic dislocations in the past year. The latest one was sustained while playing a tennis video game. She previously completed two rounds of physical therapy and has also worn a knee brace. She is in the fifth grade and plays club soccer. Her past medical and surgical history are negative. Her family history is positive for an older sister who has had one patellar dislocation.

The patient is 155 cm tall and weighs 43 kg with a BMI of 17.7. Clinical exam reveals 5 degrees of hyperextension and 130 degrees of flexion in both knees with no effusion. She has positive J-signs bilaterally. Her left knee is positive for patella apprehension and a laterally subluxable patella with weak MPFL endpoint, with approximately 3 out of 4 quadrants lateral patella subluxation with stress testing in extension. Her contralateral knee exhibits 2 out of 4 quadrants of lateral patellar translation with a firm end point and no apprehension. She is ligamentous lax with a Beighton score of seven out of nine (no genu recurvatum).

Radiographical assessment of X-rays performed after the third dislocation (Figs. 7.1, 7.2, and 7.3) reveals patella alta and a positive crossing sign on the lateral X-ray. There is a small bony fragment in the soft tissues adjacent to the medial aspect of the left patella consistent with an avulsion fracture from a lateral patella dislocation. Patellofemoral alignment is normal bilaterally. Standing alignment X-rays revealed open growth plates, and no evidence of pathologic genu valgum or limb length discrepancy (Fig. 7.4). Radiograph of the left hand and wrist reveals a bone age of 11 according to the atlas of Greulich and Pyle (Fig. 7.5).

Radiographical assessment of an MRI performed on 9/17/2016 demonstrates trochlear dysplasia with a sulcus angle of 142.5° (Dejour type A) and an elevated TT-TG of 21 mm, indicating a patellar tracking abnormality (Fig. 7.6). There is bone marrow edema in the medial aspect of the patella and lateral femoral condyle scarring and

A. H. Aitchison · D. W. Green (✉)
Department of Orthopaedic Surgery, Hospital for Special Surgery, New York, NY, USA
e-mail: GreenDW@hss.edu

J. Andrish
Department of Orthopaedic Surgery, Cleveland Clinic, Cleveland, OH, USA

M. Askenberger
Department of Women's and Children's Health, Karolinska Institute, Stockholm, Sweden

R. Kuroda
Orthopaedic Surgery, Kobe University Graduate School of Medicine, Kobe, Japan
e-mail: kurodar@med.kobe-u.ac.jp

G. S. de Freitas
.scola de Medicina, Universidade Federal do Rio Grande do Sul, Porto Alegre, Brazil

© ISAKOS 2022

J. L. Koh et al. (eds.), *The Patellofemoral Joint*, https://doi.org/10.1007/978-3-030-81545-5_7

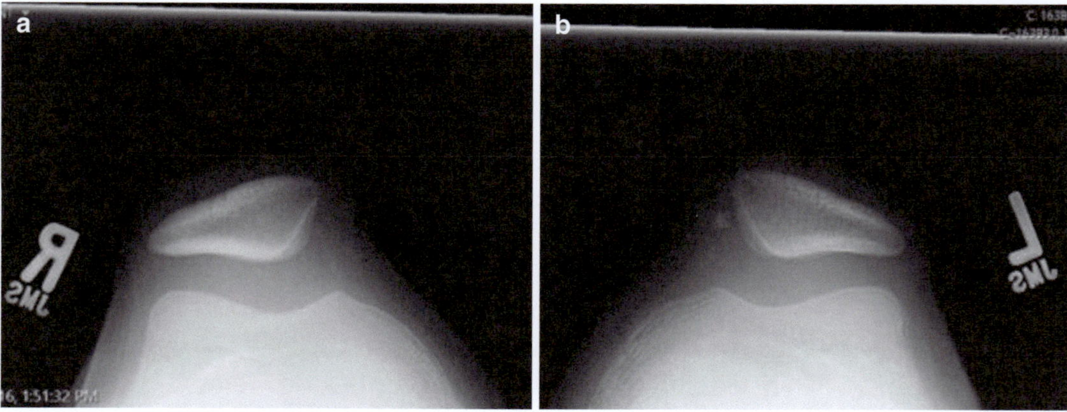

Fig. 7.1 Merchant view radiographs of both knees prior to surgery

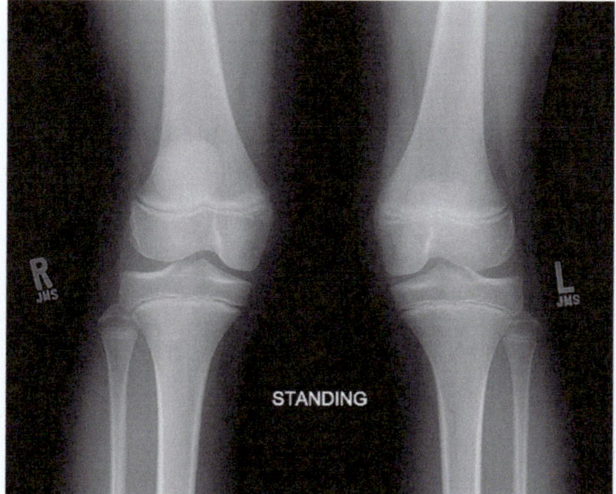

Fig. 7.2 AP view radiographs of both knees prior to surgery

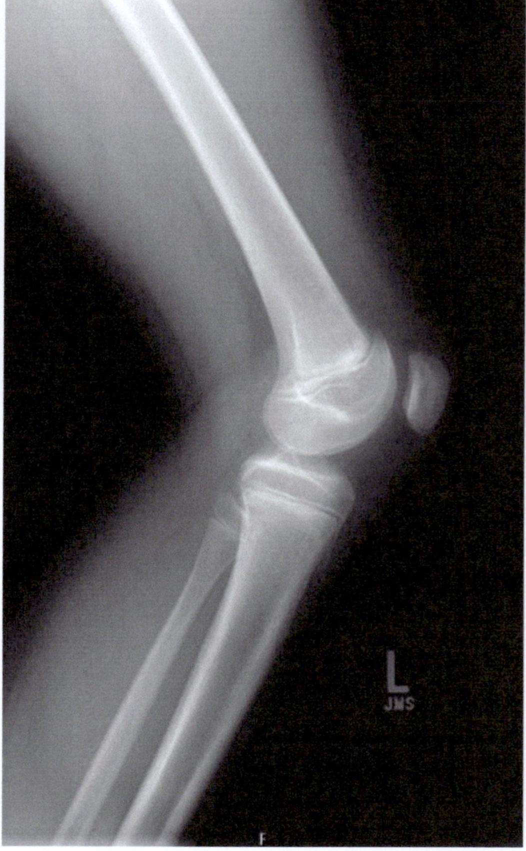

Fig. 7.3 Lateral radiograph of the left knee prior to surgery

tearing of the MPFL at the patella insertion consistent with a recent patellar dislocation. There is mild fraying of the free edge of the body of the lateral meniscus without discrete tear. There is no evidence of significant articular cartilage wear.

7.2 Evaluation and Treatment by Presenting Physician (Daniel Green)

Due to recurrent patellar dislocations, ligament laxity, and skeletal immaturity, the patient is indicated for an epiphyseal MPFL reconstruction with a two-limb hamstring tendon autograft harvested via a small incision posterior harvest. Our technique for MPFL reconstruction is to harvest the gracilis or semitendinosus tendon and use a doubled two-limbed hamstring autograft (Fig. 7.7). We use suture anchors and short osse-

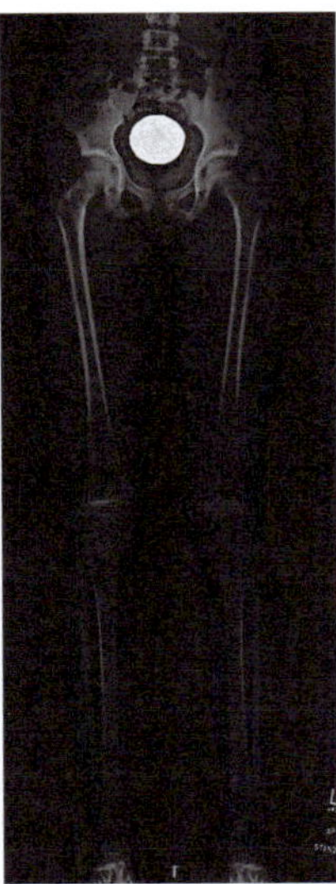

Fig. 7.4 Preoperative standing hip to ankle EOS film

ous sockets to avoid the increased risk of patellar fractures associated with osseous tunnels (Video: https://www.ncbi.nlm.nih.gov/pmc/articles/PMC3716230/bin/mmc1.mov) [1]. This technique requires the creation of a femoral tunnel and therefore caution must be taken in patients with an open distal femoral physis. Use of intraoperative fluoroscopic guidance and thorough understanding of patient anatomy is important to avoid violation of the femoral physis [2]. In a cadaveric study, Nguyen et al. determined that when reaming the femoral tunnel, the drill should be angled distally and anteriorly approximately 15–20 degrees in the sagittal and coronal planes to avoid injury to the physis [3].

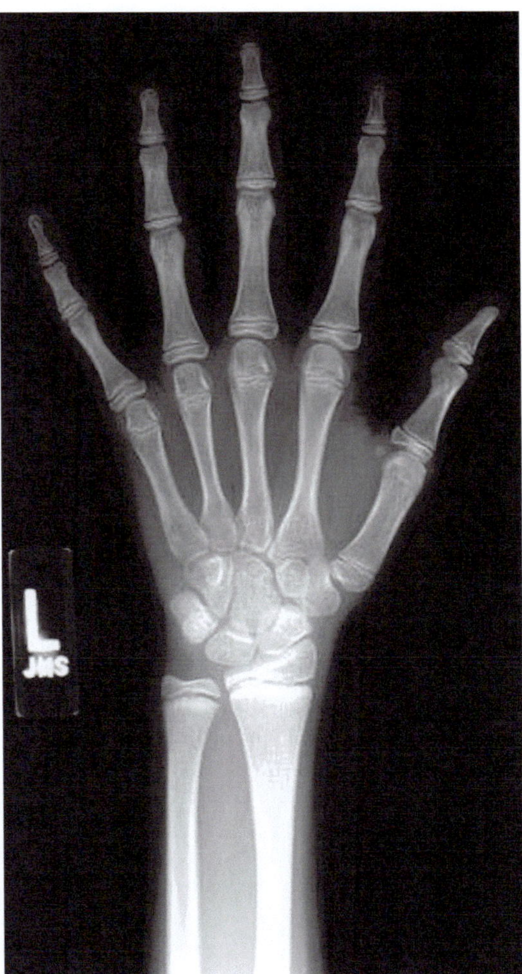

Fig. 7.5 Left hand and wrist radiograph for bone age assessment prior to surgery

Fact Box: MPFL Reonstruction [1]

Advantages

- Anatomic placement is performed below the distal femoral growth plate of the graft.
- No long tunnels are necessary.
- Aperture fixation is used, which may be stronger than fixation with soft tissue alone.
- The minimized effect of longitudinal growth on the isometrics of the reconstructed graft.
- The hamstring tendon is stronger than the original MPFL and allows for robust reconstruction.
- Can be performed with relatively small incisions.
- Early range of motion and weight bearing are allowed.

Disadvantages

- Before final graft fixation, precise positioning and assessment of graft length are required.
- Intraoperative fluoroscopy is necessary.

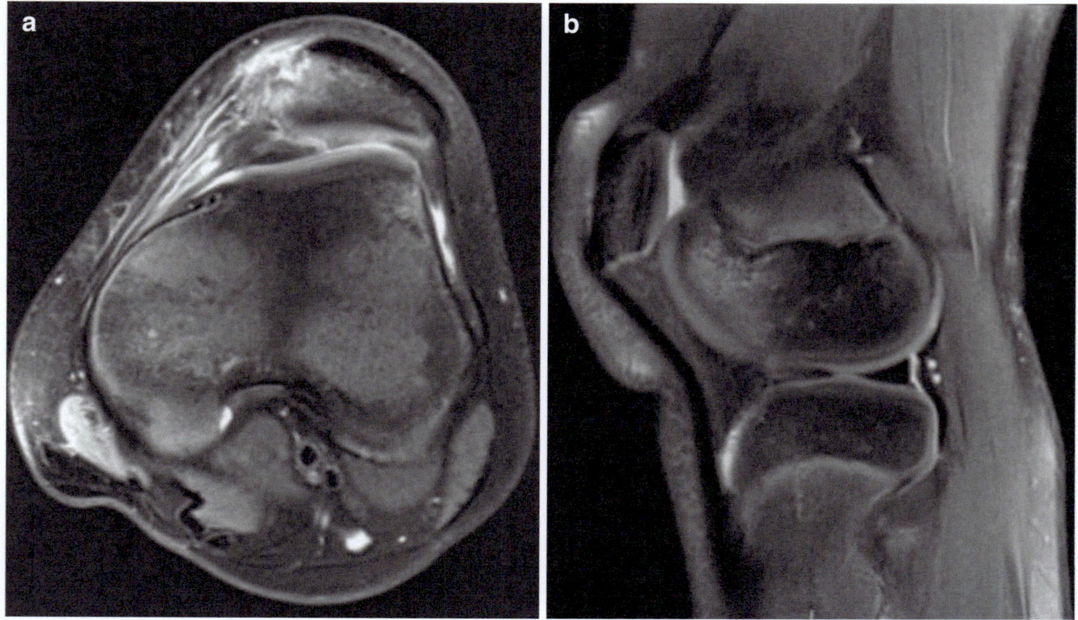

Fig. 7.6 Axial (left) and sagittal (right) slices of left knee MRI prior to surgery

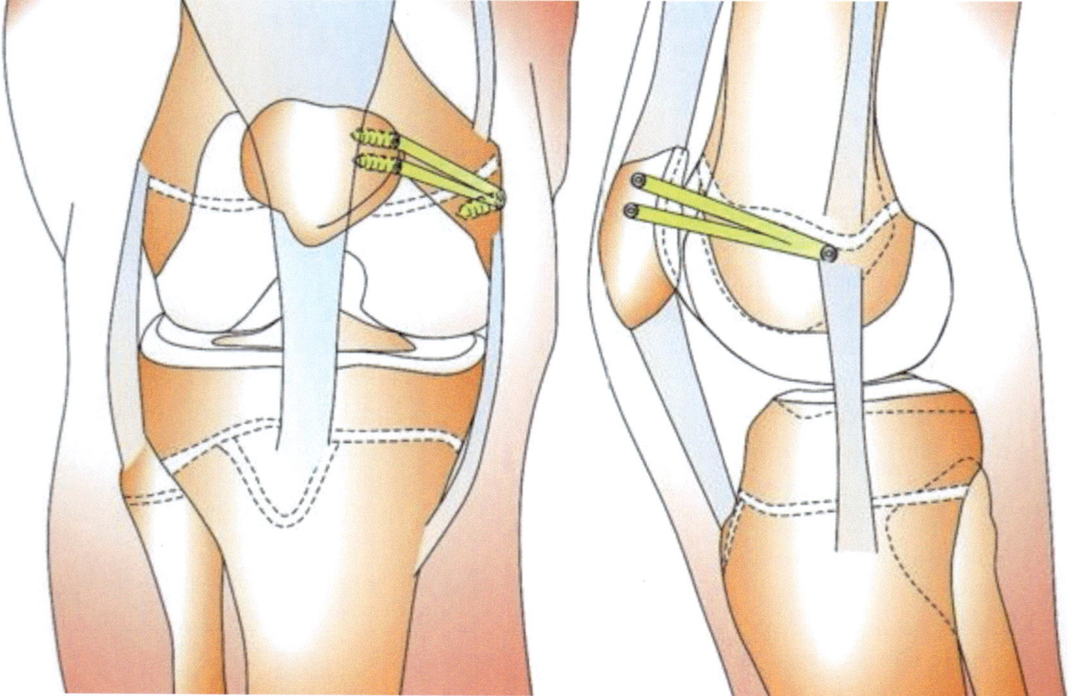

Fig. 7.7 Schematic representation of a double two-limbed free hamstring graft MPFL reconstruction

In a study of 21 skeletally immature patients who underwent MPFL reconstruction with a doubled gracilis autograft, Nelitz et al. found all patients had favorable outcomes [4]. There were no subsequent dislocations, and there was marked improvement in Kujala scores at a mean follow-up of 2.8 years. Ladenhauf et al. also demonstrated favorable outcomes after using a free doubled hamstring autograft in 41 patients, 23 of whom were skeletally immature [5]. Using the described technique, the two ends of the hamstring autograft were docked in the superior medial patella using short sockets. The femoral socket was carefully placed distal to the physis under fluoroscopic guidance with anteroposterior and lateral imaging. At a mean follow-up of 16 months, no patients experienced subsequent dislocations or growth disturbances.

Fact Box: Treating Patellofemoral Instability in the Skeletally Immature

- The distal femoral physis does not close until around 14 years of age in females and 16 years of age in males.
- In order to preserve the patellofemoral joint, immature patients with patellofemoral instability should NOT have their MPFL reconstruction delayed until reaching skeletal maturity.
- MPFL reconstruction can also be performed safely with osteochondral fracture fixation.
- MPFL reconstruction and implant mediated guided growth can be performed concomitantly to address any angular deformity that may be contributing to instability.

7.3 Commentary and Treatment Recommendation from US Surgeon (Jack Andrish)

7.3.1 General Comments

Patella dislocations in youth can be classified in various ways. They can occur with minimal trauma such as a non-contact pivot or jump (atraumatic). They can occur during sport from a contact injury (traumatic). They can be present at birth (congenital) or acquired after birth. Or they can be permanently dislocated (fixed) or present with every flexion/extension movement of the knee (habitual). This case represents the most common clinical presentation of patella instability in children and youth, atraumatic. And when we encounter an atraumatic patella dislocation in children, most often there are variations in anatomy that predispose to patella dislocations [6]. These pathoanatomies include trochlea dysplasia, patella alta, or patellofemoral malalignment secondary to increased Q-angle, increased tibial tuberosity–trochlea groove distance, torsional abnormalities such as medial femoral torsion and/or external tibial torsion, or genu valgum. Another risk factor is hyper-elasticity. The two most common pathoanatomies in children and youth associated with recurrent patella dislocations are first, trochlea dysplasia and second, patella alta. The importance of recognizing the presence or absence of these anatomic risk factors is to enable the physician to better counsel the family and, if needed, to design the surgical treatment to directly address the predisposing pathoanatomies [7].

As we examine this case, we see that the lateral radiograph demonstrates a positive "crossing sign" depicting some degree of trochlea dysplasia. The axial image seen on the MRI reveals a normal inclination of the lateral trochlea facet, but a very shallow groove. Furthermore, the MRI confirms the disruption of the medial patellofemoral ligament (MPFL) which is the primary soft-tissue restraint to lateral translation of the patella. It also confirms what is suggested by the presence of the small avulsion fracture from the medial boarder of the patella, seen on the Merchant view. The lateral radiograph as well as the lateral view on the MRI demonstrates normal patella height (no patella alta) and the full-length radiograph of the lower extremities demonstrates normal femoral-tibial alignment. The physical examination revealed hyper-elasticity.

With the presence of several important risk factors for recurrence of patella dislocation, I agree with the recommendation for surgical intervention. The presence of open growth plates does limit some of our possible surgical options. We cannot eliminate the inherent hyper-elasticity, but we can address the incompetence of the torn MPFL by augmenting a repair/imbrication of her medial retinaculum (part of my choice) with an augmentation of the native MPFL with a tendinous reconstruction. The use of either a semitendinosus or gracilis tendon autograft will provide a soft-tissue restraint that is 10 times stronger than the native MPFL. The fact that the graft is also significantly stiffer than the native MPFL, however, makes the technical proficiency of the surgical reconstruction critical in order to avoid over-constraint and destructive patellofemoral contact forces. Other graft choices such as quadriceps tendon have also demonstrated success. Trochleoplasty is not an option because of the open growth plates, but MPFL reconstruction has been a significant addition that has demonstrated the ability to compensate for the presence of mild to moderate trochlea dysplasia.

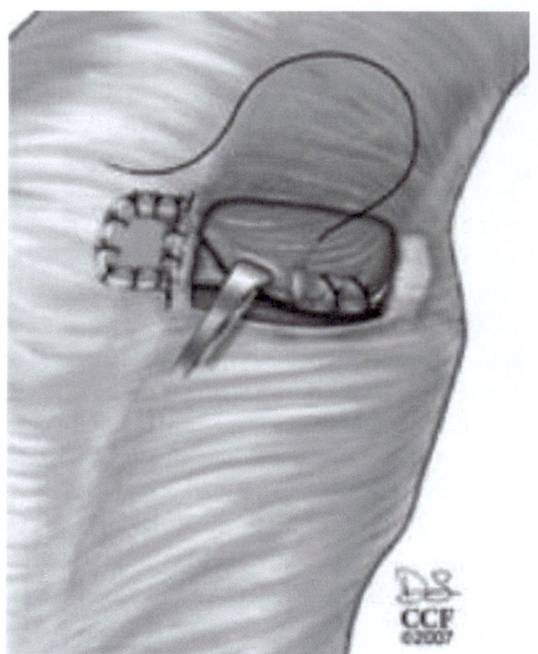

Fig. 7.8 The medial retinaculum is exposed and a 2-centimeter strip of the superficial and intermediate layers is detached from the medial boarder of the superior half of the patella and reflected posteriorly off of the underlying joint capsule. In so doing this is sharply detached from the distal boarder of the tendon of the vastus medialis oblique (VMO). This isolated medial retinaculum which contains the native MPFL is then advanced through a small button-hole incision in the remaining medial parapatellar fascia, re-tensioned, and secured to the prepatellar fascia. The tendon of the VMO is then advanced over the re-tensioned medial retinaculum

7.3.2 Treatment Recommendation

For the management of recurrent patella dislocations not associated with trochlea dysplasia or hyper-elasticity or failures of prior surgical attempts, I prefer to perform an imbrication of the medial retinaculum which includes the native MPFL (Fig. 7.8), but in the case presented here with trochlea dysplasia and hyper-elasticity, I would include a tendinous MPFL reconstruction and my choice for autograft would be semitendinosus [8]. There are various methods of fixation both to the patella as well as the femur. I think we all agree that making large drill hole for the patella fixation is problematic as that can lead to patella stress fracture. Fixation is secured to the proximal half of the medial boarder of the patella with suture anchor and/or soft tissue alone to the remaining pre- and peri-patella fascia. The choice for the femur is more technical. As noted by Dr. Green, if a docking technique is chosen, the direction of the drill must be aimed slightly distal and slightly anterior in order to avoid possible injury to the distal femoral physis. Other methods used for the skeletally immature patient that avoid possible injury to the femoral physis involve either using the tendon insertion of the adductor magnus about the adductor tubercle or looping under and around the superficial medial collateral ligament adjacent to the medial epicondyle (my choice as depicted in Fig. 7.9). In any case it is important that we understand the strain patterns placed upon the tendons used for reconstruction in order to avoid over-constraint which could result in stiffness and/or excess patellofemoral joint stress. For example, the technique of using the superficial MCL as the femoral anchor results in a graft that requires increased length during extension and less tension during flexion. In that sense, this is more

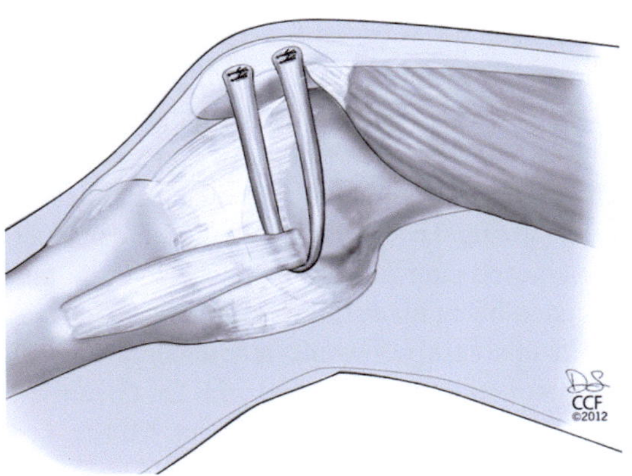

Fig. 7.9 The origin of the superficial medial collateral ligament (MCL) is exposed adjacent to the medial femoral epicondyle. A pathway is made deep to this ligament and a free semitendinosus graft is looped around the MCL which is now used as the femoral anchor for the MPFL reconstruction. It is important not to over-tension the reconstruction by making the final fixation to the patella with the knee in full extension. During fixation, proximal traction is applied to the central quadriceps tendon in order to simulate active quadriceps contraction

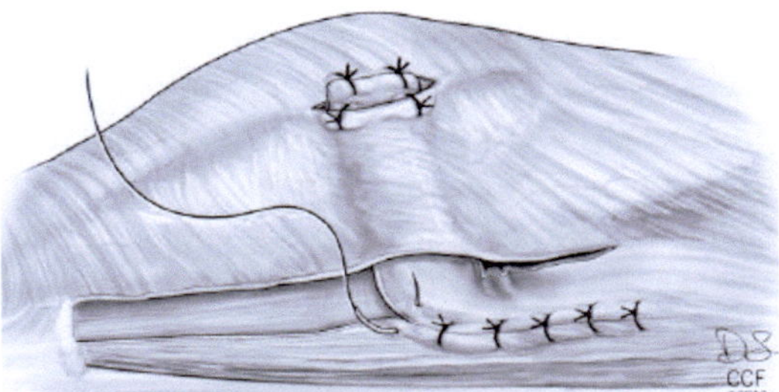

Fig. 7.10 This lateral reconstruction uses the anterior half of the central iliotibial band. This reconstruction is intended to simulate the deep transverse lateral retinaculum, normally the thickest and most stout portion of the lateral retinaculum

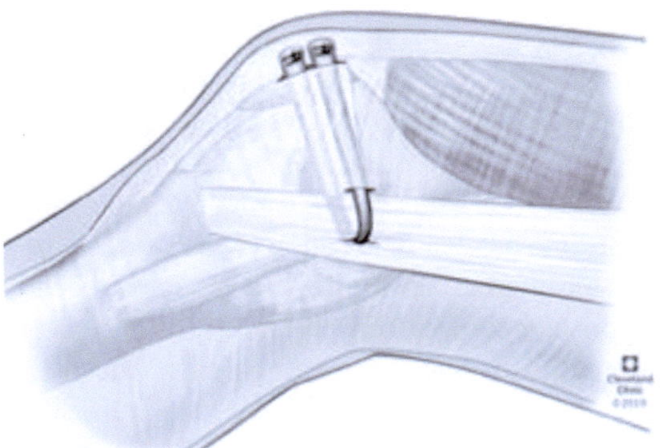

Fig. 7.11 For the hyper-elastic patient with medial as well as lateral instability, a free semitendinosus graft (either autograft or allograft) can provide enhanced stability. The graft is placed overlying the joint capsule and, if present, deep to the superficial oblique lateral retinaculum

like the strain pattern of the native MPFL, but it also can lead to iatrogenic medial patella subluxation during active knee extension. In order to avoid this complication, during final fixation of the graft (to the patella) the knee should be in full extension and proximal traction should be placed upon the quadriceps tendon to simulate active extension.

My final thoughts about this case are for the consideration of the hyper-elasticity. We should always remember that the stability of the extensor mechanism includes the lateral retinaculum as well as the medial retinaculum [9]. For a knee that is unstable medially as well as laterally, I include imbricating and/or even reconstructing the lateral retinaculum as depicted in Fig. 7.10. For a documented pathologic hyperelasticity syndrome, my graft choice for the medial reconstruction as well as the lateral reconstruction (Fig. 7.11) includes semitendinosus allograft.

Regardless of techniques chosen, my postoperative management avoids bracing; allows early weight bearing as tolerated and advancement along the principles of pelvi-femoral rehabilitation. [6–9]

7.4 Recommendation from EU (Marie Askenberger)

For me it is essential to gather all imaging measurements from the MRI before making a surgical decision, viewing the whole MRI series including measurements of trochlear anatomy, patellar height, the TT-TG distance, lateral patellar tilt, MPFL injury, as well as ruling out osteochondral/chondral injuries.

When treating children, it is important to have well informed parents and children. Patient's history including heredity and physical activity, the results from clinical examination, radiographs, and MRI are all taken in consideration. For first-time dislocations, this helps the treating physician to evaluate the risk of redislocation and to better counsel the family of treatment choices, promote physiotherapy to possibly avoid surgery, or chose the necessary surgery based on individual patient factors. Guided growth should be considered in a case with open growth plates and valgus knee with limb malalignment. Almost always in this age (if no chondral/osteochondral injuries are present that needs surgery), we recommend physiotherapy combined with a soft knee brace as first choice treatment. We follow the patient after a first-time dislocation, with a report from the physiotherapist with final results including strength and functional stability tests before returning to sports, or if instability persists despite adequate rehabilitation, then surgery is considered.

The incidence of anatomic patellar instability risk factors is generally high among children with first-time patellar dislocation [10]. A clinical prediction model based on MRI variables recently published showed that the risk of redislocation increased to 78.5% after a first-time dislocation if the patient is skeletal immature have a trochlear dysplasia with Sulcus Angle ≥154° and Patellar alta with Insall-Salvati ratio ≥1.3 [11]. Recently studies have also showed that failures after MPFL reconstruction are mainly due to:

- failure to consider additional bony risk factors (high-grade trochlear dysplasia, severe patellar alta, elevated TT-TG distance, rotational malalignment).
- Intraoperative technical errors/decisions (non-anatomical femoral fixation, patellar fractures after drilling in small patellae).
- Inappropriate patient selection [12, 13].

In this case we have a child (skeletal immature) that wants to be physical active and has tried physiotherapy without satisfaction which leads us to recommend surgery. Information should be given to the family, explaining how the hypermobility effects the joint combined with an insufficient MPFL, a low-grade trochlear dysplasia, and elevated TT-TG distance. For patient and family education, we use a knee-model as well as images from the MRI to achieve as clear of understanding as possible. When informing the family, we would say that an MPFL reconstruction is necessary and probably enough, but with a high TT-TG distance (21 mm), there is a small chance that we might have to add a medialization (medializing the patellar tendon in some manner) to achieve better patellar tracking, with the final decision being made intraoperatively.

In this case we would recommend a MPFL reconstruction, in agreement with Dr. Green.

The patient has an insufficient MPFL, systemic hypermobility (which in itself cannot be treated surgically), a low-grade trochlear dysplasia (not high grade and is skeletally immature, therefore no indication for trochleoplasty), and an elevated TT-TG distance. In this scenario I would use inter-operative criteria to evaluate the need to treat the elevated TT-TG by temporarily fixating the MPFL, and evaluate tracking. If intraoperative evaluation reveals that patellar tracking is lateralized, a patellar tendon medialization utilizing a modified Roux-Goldthwait or other method of patellar tendon medialization, should be added.

7.5 Preferred Graft and Method for MPFL Reconstruction

As said before there are many options to reconstruct the MPFL. For MPFL reconstruction we prefer using the Quadriceps Tendon (QT) with the graft still attached to the patella as previously described by Fink, which avoids drilling in small patellae (risk for fracture) [14, 15]. Biomechanically, Herbort et al. found that the characteristics of QT graft as described below are very similar to the native MPFL (maximum load, yield load, and stiffness) [16].

The minimal QT method is performed with a 3 cm long transverse skin incision over the superomedial pole of the patella. QT is exposed and a special set of tendon harvesting instruments are used. The QT is harvested with a minimum of 8 cm length, 10 mm wide, and 3 mm thick, leaving intact the distal attachment to the patella. When the tendon is taken centrally and flipped 90° medially there is sufficient medial prepatellar

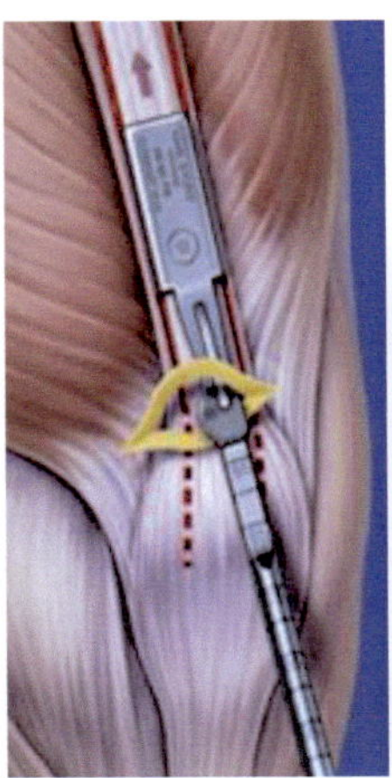
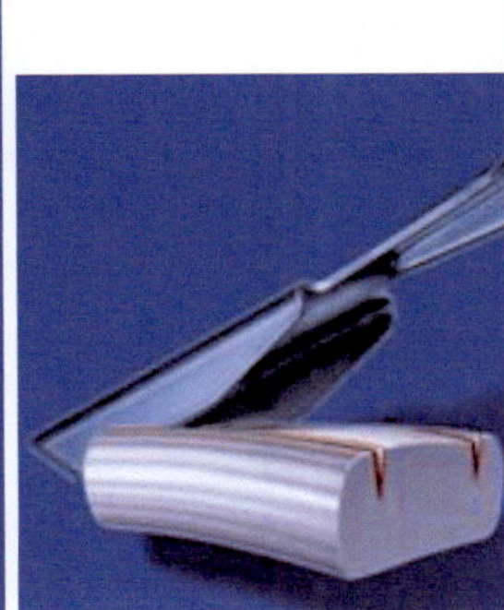
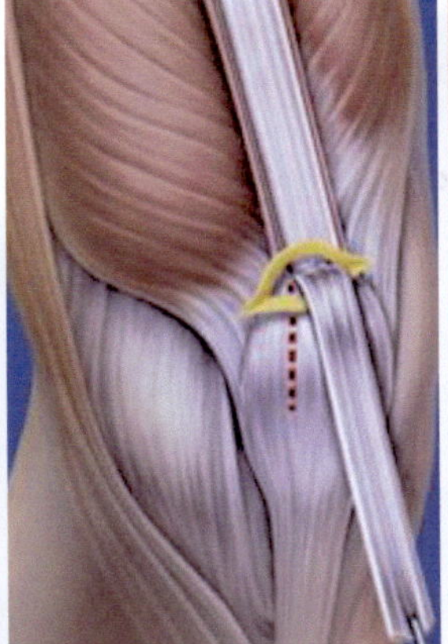
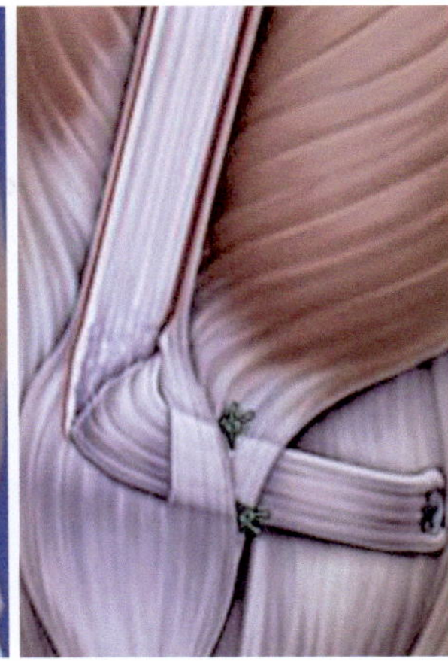

Fig. 7.12 The minimal QT method is performed with a 3 cm long transverse skin incision over the superomedial pole of the patella. QT is exposed and the tendon harvesting instruments are used. The QT is harvested with a minimum of 8 cm length, 10 mm wide, and 3 mm thick, leaving intact the distal attachment to the patella. When the tendon is taken centrally and flipped 90° medially there is sufficient medial prepatellar tissue to pass the graft under this tissue, adding to stability and positioning the neoattachment of the MPFL in a deeper plane at the medial patellar border. The QT is secured proximal and medial at the patella with sutures. The graft is then tunneled and passed between the joint capsule and the vastus medialis (VMO) towards the femoral incision. This technique requires the creation of a femoral tunnel to secure the graft, the use of intraoperative fluoroscopic guidance for femoral fixation location, and thorough understanding of patient anatomy to avoid violation of the femoral physis. (Karl STORTZ, Tuttlingen Germany)

tissue to pass the graft under this tissue, adding to stability and positioning the neoattachment of the MPFL in a deeper plane at the medial patellar border (Fig. 7.12).

The QT is secured proximal and medial at the patella with sutures. The graft is then tunneled and passed between the joint capsule and the Vastus Medialis Oblique (VMO) towards the femoral incision. This technique requires the creation of a femoral tunnel to secure the graft, the use of intraoperative fluoroscopic guidance for femoral fixation location, and thorough understanding of patient anatomy to avoid violation of the femoral physis. An additional insertion over the adductor tubercle is performed. The desired femoral insertion is determined with a guide pin (Schottles point), remaining below the physis using fluoroscopic guidance. The graft is temporarily fixated to test for isometry prior to drilling the tunnel, with a maximum length of 20 mm, distal to the physis and angled about 15–20° anterior and distal to the entry point to avoid damage to the physis, intra-condylar notch, and distal femoral cartilage [3]. The graft is secured with a cannulated Bio Composite SwiveLock 4.75 mm with the knee in about 30° of flexion. Patellar tracking and graft tension are controlled before final fixation, to avoid over tightening the graft. Postoperatively, we allow full weight bearing with a locked knee brace for 4 weeks for the younger patients <15 years old, and physiotherapy starts with ROM 0–90° during this period. Thereafter, free range of motion and physiotherapy to gain full strength and knee control. The results of the procedure should be judged with strength test and patient reported outcome measures at follow-up.

Nelitz has described another method using the quadriceps tendon as a pedicled QT graft with favorable results in children, excluding those children with high-grade trochlear dysplasia (type B-D according to Dejour). The study included 25 children with recurrent patellar instability, average age at the time of operation was 12.8 years and average follow-up was 2.6 years. There were no redislocations, significantly improved postoperative Kujala scores and improved Tegner activity scores with this patient selection and technique [17].

The surgical method for patellar stabilization should be based on a thorough preoperative evaluation including history, physical exam, and imaging risk factors to individualize the treatment. It is also important to align outcome expectations with the chosen method of surgical stabilization and timeline for return to activities. In few cases it can be a planned stepwise stabilization depending on the skeletal maturation, an MPFL-R can be a good option to gain stability; but when present with severe additional risk factors, a later surgery when mature should be discussed and performed if necessary.

7.6 Recommendation from APKAS Country (Ryosuke Kuroda)

Previous studies have shown that skeletally immature patients are at high risk for having recurrent dislocation compared with adult patients [18, 19]. Particularly, younger patients with other risk factors such as trochlear dysplasia, patellar alta, and increased TT-TG have significantly high odds ratios of having redislocation after the first-time dislocation as seen in this patient [11, 20–22]. Therefore, we recommend surgical treatment for this patient.

In the preoperative assessments for the present patient, the risk/predisposing factors that need to be considered are trochlear dysplasia and the TT-TG distance of 21 mm. Other osseous abnormalities including mal-rotational and valgus limb alignment were not exhibited.

Considering the surgical invasiveness, MPFL reconstruction would be the most reasonable choice of procedure for this patient. Although the shape of the trochlea is flat in the proximal part, the distal part of the sulcus in flexion is maintained. Therefore, it seems the degree of the dysplasia will not be severe after a MPFL reconstruction is performed. The increased TT-TG distance of 21 mm can be a concern, but additional tibial tuberosity transfer is impossible to perform due to the presence of an open tibial physis. In addition, we have obtained favorable outcomes even in patients with increased TT-TG of >20 mm after MFPL reconstruction [23]. Therefore, TT-TG of 21 mm alone may not be absolute indication for tibial tuberosity transfer when performing MPFL reconstruction, at least for the present patient. Other soft-tissue distal realignment procedure such as Roux-Goldthwait will also likely be unnecessary in this case based on our experience.

J-sign is another concern to address since a recent study showed that severe J-sign was a significant risk factor for graft laxity after MPFL reconstruction [24]. However, a surgical method to prevent J-sign has not been established yet, and the presence of J-sign in bilateral knees suggests that it is not critical to be treated. Based on these assessments, we would plan for an isolated MPFL reconstruction.

7.6.1 Our Preferred Surgical Technique

7.6.1.1 Graft Choice

Since allograft is not regularly used in Japan, as in most of the Asian countries, our primary graft choice is semitendinosus tendon. Gracilis can be an option. However, the length of the gracilis is less than 20 cm and the length may not be enough in some patients.

7.6.1.2 Surgical Procedure

The surgical procedure is similar to that presented by Dr. Green. For graft fixation, we use two metal suture anchors for the patellar side and one or two anchors for the femoral side. The identification of the femoral graft attachment site is one of the most critical points during surgery. Care must be taken not to violate the physis while obtaining proper length change pattern of the graft. In adult patients, the Schöttle's point is often used to identify the femoral attachment site [25]. In skeletally immature patients, the Schöttle's point is closed to the physis and it overlaps with the physis in the lateral view fluoroscopy, while the position is still distal to the physis in the anteroposterior view. Although placement of the graft in the anatomical position with a proper graft length change pattern without breaking the physis is ideal, sometimes it is not easy to accomplish both of them. We believe that safety is the most important and the position can be compromised or modified while prioritizing the safety (preserve the physis) and the function of the graft.

We first carefully choose the position under fluoroscopy and insert a guide wire slightly distal to the physis. Then the length change during range of motion (ROM) was checked. If the graft change pattern shows a physiological or favorable pattern that nearly isometric or slightly long in extension and short in flexion, then the suture anchor is inserted into the position [26]. The center of the folded graft is fixed at the position after decortication. The fixation was further augmented by covering it with the surrounding periosteum. For the patellar side, two ends of the graft are fixed with the suture anchors and are further covered with the periosteum.

7.6.1.3 Postoperative Rehabilitation

Progressive ROM exercise is performed 1 week after surgery. Weight bearing is allowed as tolerated. Knee brace was applied during walking until the patients have extension lag <5 degrees in the straight leg raise test. Jogging is permitted 3 months after surgery and sports activity is 6 months after surgery.

7.6.1.4 MPFL Reconstruction with Suture Anchors

Advantages:

- Easy to avoid violation of the physis.
- Low risk of patellar fracture.
- Functional graft placement.

Disadvantages:

- Fixation strength is relatively weak compared with other trans-osseous/bone-socket techniques especially at femoral side.
- Metal implants are used.

7.7 Recommendation from SLARD Country (Geraldo Schuck De Freitas)

7.7.1 General Comments

The patellar instability and patella dislocation are very complex disorders of the knee extensor mechanism that has been a challenger to knee surgeons, as we can see by the enormous diversity of suggested treatments. More than a hundred different procedures have been described to stabilize the patellofemoral joint. Regarding pediatric patients, the stakes are particularly high. In these patients, patellar instability can occur in the setting of severe multiple or syndromic birth defects. Early treatment is required to whom that exhibit the most severe forms.

As with anterior cruciate ligament surgery in children, specific pediatric techniques have been developed for the reconstruction of MPFL, however, the ideal sequence has yet to be determined. The role of MPFL procedures in personalized treatment strategies for pediatric patients remains uncertain.

Currently, there is a general consensus that the MPFL femoral insertion in children is epiphyseal, a few millimeters away from the growth plate [27, 28]. The MPFL is not isometric, and it will be stretched during the knee is full extended and it will relax in flexion when the patella is engaged in the trochlear groove [29]. Many adult orthopedic surgeons believe that the role of MPFL is crucial [30]; however, in the most severe forms of patellar instability seen in pediatrics, the contribution of MPFL has yet to be completely clear.

Other important subject to be considered is related to the trochlear and patellar shape. Nietosvaara et al. documented changes in the thickness, in an ultrasound study, of the trochlear cartilage between 12 and 18 years of age [31]. The cartilaginous trochlea of an abnormal knee is flatter than the bone trochlea, which makes the cartilaginous trochlear angle a better parameter for separating normal and abnormal knees. The positive correlation between the angle of the trochlea and the height of the patella when the knee is extended supports a role for developmental disorders, with decreased forces being applied to the trochlea in patients with high patella. We have to be aware about other bone factors that may increase the risk of patellar instability, such as excessive anterior femoral torsion and genu valgus, which develop secondarily. These factors change during growth.

Patients with occasional patellar dislocation have the quadriceps only slightly shortened, and the only manifestation may be the high patella, with limited involvement in the trochlear groove, excessive mobility and, in some cases, inclination of the patella in the extension. In patients with severe patellar instability the muscle factor is responsible for permanent or frequent dislocation, besides it has received less attention since the development of MPFL reconstruction techniques became more frequent.

Occasional dislocation of the patella can result in acute traumatic distention of MPFL, while chronic distention of MPFL can occur in severe dislocation. "One difference between adults is that, in children, the MPFL avulsion site in the acute phase is usually located in the patella (61% of cases in a study by Kepler et al. [31]) although almost half of the lesions are complex and involve multiple sites [32].

Ascension of the vast medial oblique (VMO), edema, and some muscle injuries were demonstrated in 56% of cases in MRI study [32]. The rise of the VMO is a marker of severe MPFL injury and can promote recurrent patellar dislocation.

MPFL helps to control patellar tilt and translation. MPFL distension can be assessed clinically based on the physical tests such as patellar seizure. Tilt and patellar translation can be measured on magnetic resonance or computed tomography images (taken with the knee extended and the quadriceps contracted or relaxed).

Knee hypermobility manifests as a genuine recurvatum with high patella and excessive external rotation of the patella in the knee and should be seen as a secondary factor of instability.

MPFL reconstruction was introduced only very recently in pediatric practice. The creation of a femoral tunnel, as recommended in adults, is not recommended in patients with open growth plates: growth would result in migration of the metaphyseal tunnel towards the diaphysis. It would be difficult to create an epiphyseal tunnel due to the proximity of the growth plate and the absence of reliable anatomical landmarks. We prefer techniques that involve femoral fixation in soft tissues; these techniques also avoid tunnel-related complications.

7.7.2 Case Discussion

In the case at hand, we usually consider that children with constitutional hypermobility syndrome with a Beighton hypermobility score greater than 6/9 should be investigated for cardiac, skeletal, and ocular abnormalities [33, 34].

At the physical examination we find the patella can slide laterally is 3 out for 4 quadrants on the left knee and there is soft MPFL endpoint, these findings indicate that MPFL is stretched. I use to measure the angle of the patella with the horizontal plane by the patellar tilt test. These maneuvers help to assess MPFL distension [35]. I believe there is little to be gain in children with use of many other sings, such as pain due to compression of the patella in the trochlear groove, crepitation during sliding of the patella, etc.

Analyzing all images and their measurements, we see a patient with skeletal immature, and presence of many anatomical risk factors. The lateral radiograph demonstrates trochlear dysplasia with clear "crossing sign," and patella alta. We can see disruption of the MPFL, shallow trochlea, and small avulsion of the patellar medial border at the MRI.

Besides the patient being very active, and we could see all these anatomical abnormalities, I would consider in a first take the conservative treatment as long as it possible, based on the followings considerations:

- Usually girls reach their skeletal maturity around 13 years old. So, it is very tempting to try to maintain the treatment with physiotherapy, using brace and keep them away of physical activities with risk of dislocation of the patella, for only 2 years more.

- There are anatomical abnormalities, such as trochlea dysplasia, increased TT-TG and patella Alta, which potentially tend to get worse until the end of growth and which cannot be adequately addressed with surgical gestures only in soft tissues. We should consider great possibility to correct these anatomical disorders in the near future.
- The reconstruction of MPFL in this situation would lead to the need to use non-isometric techniques, in order to eliminate the risk of growth plate injury, in addition to creating a risk of jeopardizing future MPFL reconstruction if it would be necessary.
- Regarding to the reconstruction of the MPFL at this time, it would imply an increased risk of causing joint over-constraint, since the graft to be used is often stronger than the natural MPFL. In addition, it has less expandability than growth will require.
- Finally, if we do isolate MPFL reconstruction, we are going to perform surgery to prevent patellar dislocation, but we are not treating the instability. Once the other important anatomical disorders, such as trochlea dysplasia, increased TT-TG and patella alta, would not be addressed.

If the conservative treatment failed, as it did in the present case, then surgical treatment is the treatment of choice. In this case I agree with the indication of Dr. Green and also recommend isolated MPFL reconstruction.

In this case, I would prefer a technique of MPFL reconstruction that use the center part of quadriceps tendon, leaving the distal arm inserted to the patella. I think using this graft brings some important advantages, such as keeping the hamstrings for future surgery, if necessary, and in the same way, if we do not drill the patella at the MPFL insertion point, we will avoid potential complications. Besides, keeping the patellar bone undisrupted could be useful for future surgeries if it would be necessary. In relation to femur attachment, an osteoperiosteal tunnel is created under the adductor magnus at its distal femoral insertion (Fig. 7.13). The free end of the graft is then passed through this tunnel, folded over itself, and fixed with stitches (Figs. 7.14 and 7.15). This fixation is performed with the knee flexed 60°, but the wounds are not closed before fully testing flexion and extension (Figs. 7.16, 7.17, 7.18, and 7.19) [36].

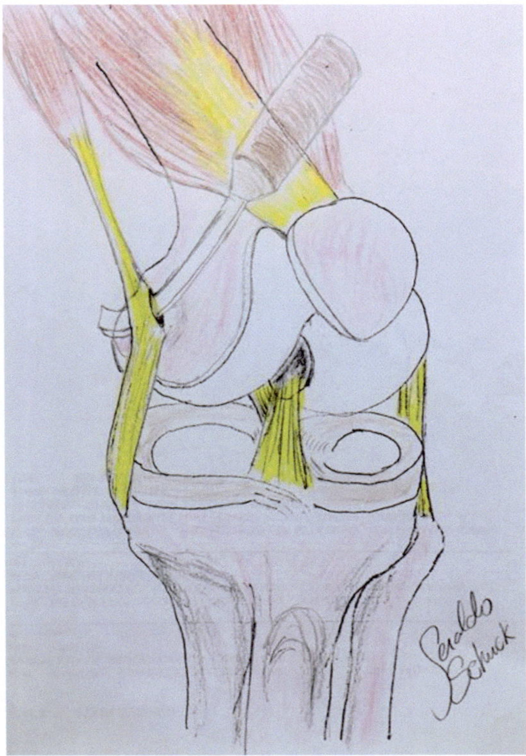

Fig. 7.13 Illustration depicting the formation of the subperiosteal tunnel under the insertion of the adductor Magnus

Thus, in avoiding creating a femoral tunnel, we decrease the risk of growth plate injury. We finalize the stabilization procedures by carefully closing the medial retinaculum and repositioning and stitching the VMO in the best anatomical position. Using this technique, we also produce the fixation without implants. This soft-tissue fixation would allow more distensibility of the graft fixation, avoiding over-constraint of patellofemoral joint until the end of growth.

Other option would be the Chassaing technique [37]. The gracilis tendon is passed through the entire thickness of the posterior portion of the medial retinaculum, close to its location of the

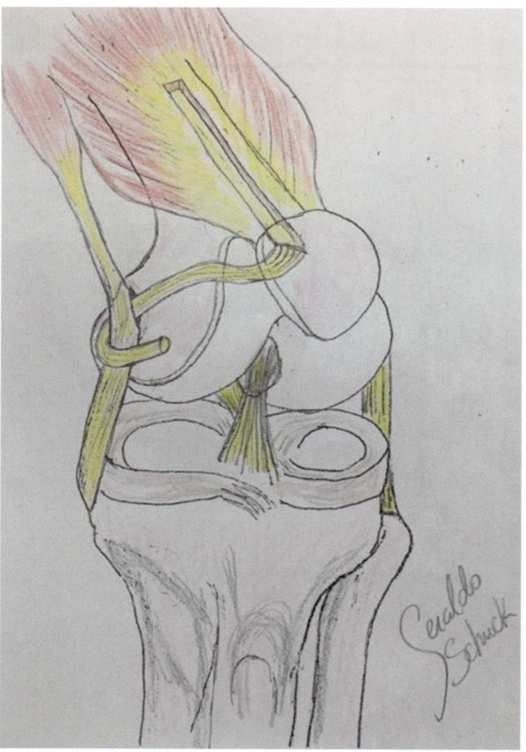

Fig. 7.14 Illustration depicting the use of the central third of quadriceps tendon as a graft for MPLF reconstruction by passing the free arm of the graft into the periosteal tunnel and folding it on to itself

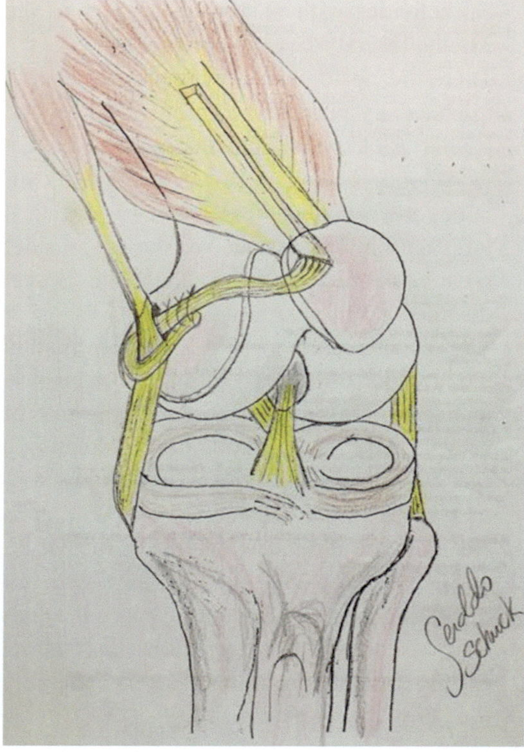

Fig. 7.15 After passing the graft and folding it back on to itself, femoral fixation is achieved by using stitches to avoid the use of hardware for attaching

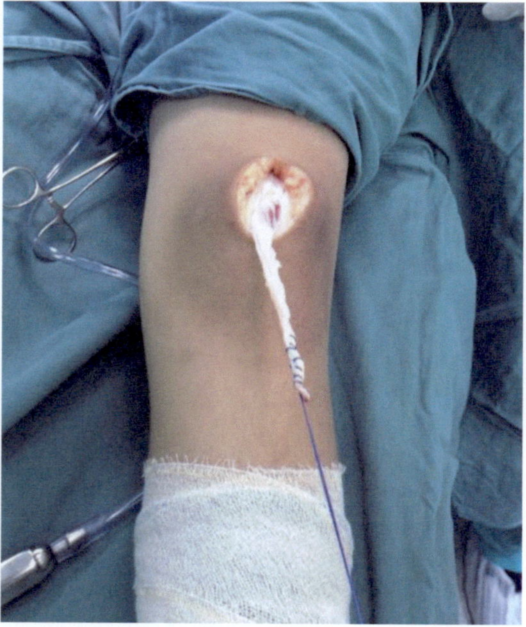

Fig. 7.16 The surgical approach showing the incision at the superior patella exposing the quadriceps tendon. Next, the third central part of the quadriceps tendon is harvested leaving the distal part attached to the patella

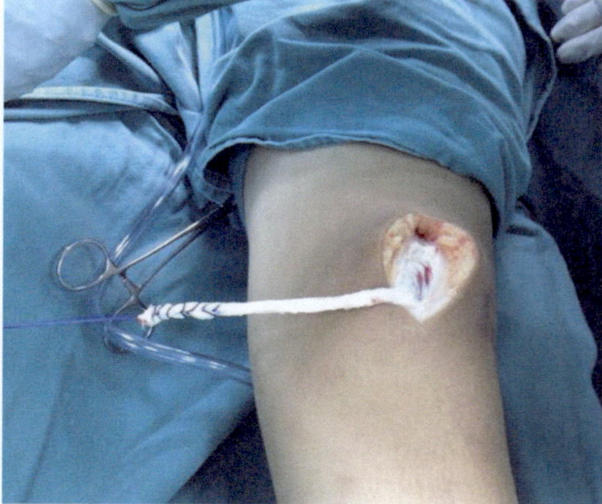

Fig. 7.17 The quadriceps tendon graft is taken down to the center of the patella, utilizing the thick periosteal in front of the patella. This way, adequate length is achieved and it is positioned at the isometric patella point

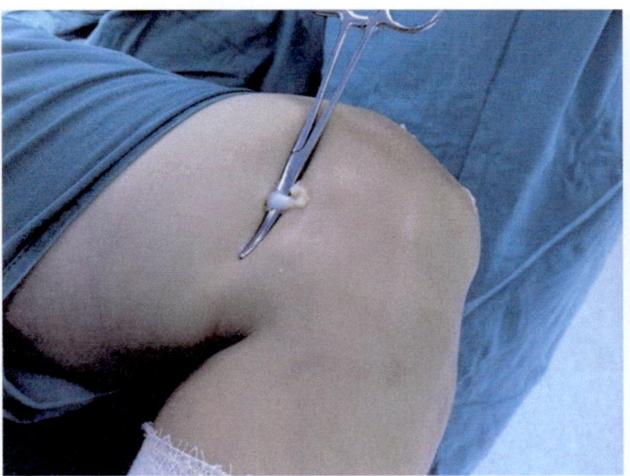

Fig. 7.18 Performing the approach on the medial femoral epicondyle (MFE), we can easily find the insertion of the adductor Magnus tendon at the adductor tuberosity, just proximal and posterior to MFE. Then we perform the subperiosteal tunnel under the insertion of the long adductor, and we pass a clamp through the tunnel to later capture the graft

femoral fixation, where the tendon folds again into a U-shape. The gracilis tendon is anchored in the patella subperiosteally.

Postoperatively, regardless of MPFL reconstruction technique chosen, we allow weight bearing with aid of crutches as tolerated. Physiotherapy starts on the day after, encouraging free range of motion. The muscles strengthening begins as soon as possible, and we avoid using any bracing.

7.7.3 General Rules

An individually adapted surgical procedure must be performed in a single stage that combines one or more of the techniques. The staged execution of some techniques before the end of growth and waiting to perform others after skeletal maturity is not advisable. When the child is close to the end of growth, the best strategy may be to wait a few months to allow the use of simpler and more comprehensive techniques, such as trochleoplasty with deepening of the grooves or medialization/distalization of the tibial tuberosity.

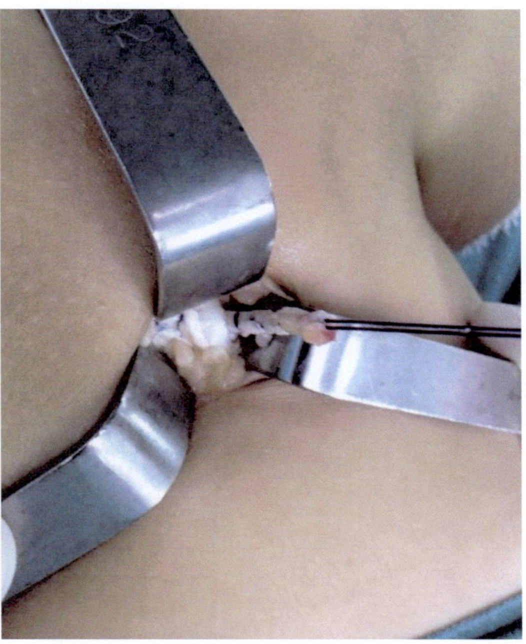

Fig. 7.19 After passing the free arm of the quadriceps graft through the subperiosteal tunnel, we fold it back over itself, attaching it to itself with stitches

7.8 Postoperative Course and Outcomes (Daniel Green)

Immediately following surgery, the patient was placed in a postoperative knee brace locked in extension and will be partial weight bearing on crutches. She was prescribed a CPM machine and cold therapy to use at home. Two weeks after surgery she began a physical therapy program to work on ROM and strengthening. The patient began to wean crutches at 6 weeks and was transitioned to abort brace at 8 weeks. After completion of a strength and conditioning program, she was cleared and returned to sports around 8 months postoperatively. She had mild pain with extension at her one-year visit that resolved without intervention.

At her final follow-up, 24 months postoperatively, the patient was still doing well with no complaints of pain or feelings of instability. There were no dislocations or subluxation epi-

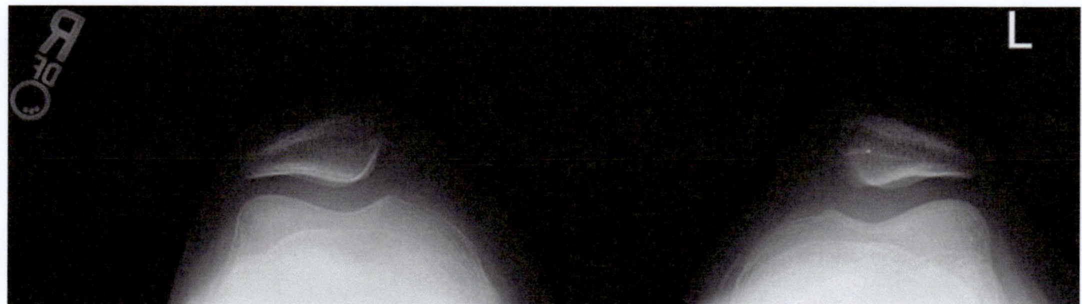

Fig. 7.20 Postoperative merchant X-ray of both knees

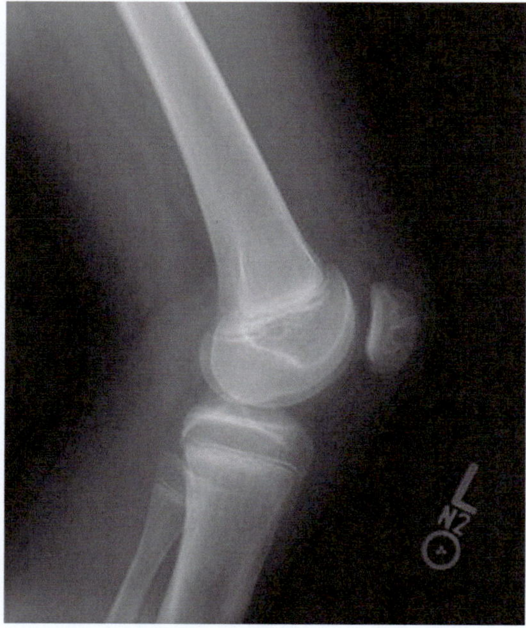

Fig. 7.21 Postoperative lateral X-ray of the left knee

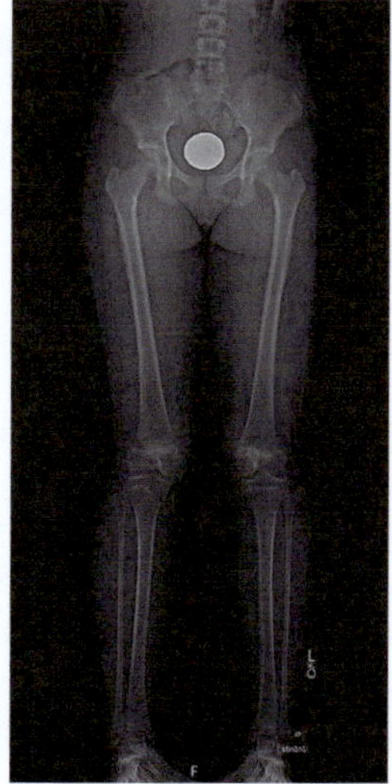

Fig. 7.22 Postoperative standing hip to ankle EOS film

sodes since her surgery. She successfully returned to sports and has been playing basketball and soccer with no limitations in her knee. On clinical exam she had a mild bilateral J-sign with well centered patellae and good MPFL endpoint. She had full range of motion (0–130 degrees) bilaterally with no evidence of swelling. Her knee radiographs revealed no osseous abnormalities and satisfactory patellofemoral alignment (Figs. 7.20 and 7.21). Standing hip to ankle radiographs showed good alignment and no leg length discrepancy (Fig. 7.22). MRI of the left knee demonstrated continued patellofemoral dysplasia with an intact medial patellofemoral ligament reconstruction (Figs. 7.23 and 7.24). There is chondral heterogeneity without fibrillation or defect over the patella, which is mildly laterally subluxed.

Take Home Message

Treating patellofemoral instability in a skeletally immature patient with trochlear dysplasia should not be delayed until the physes close. If a patient has risk factors for recurrent instability and/or

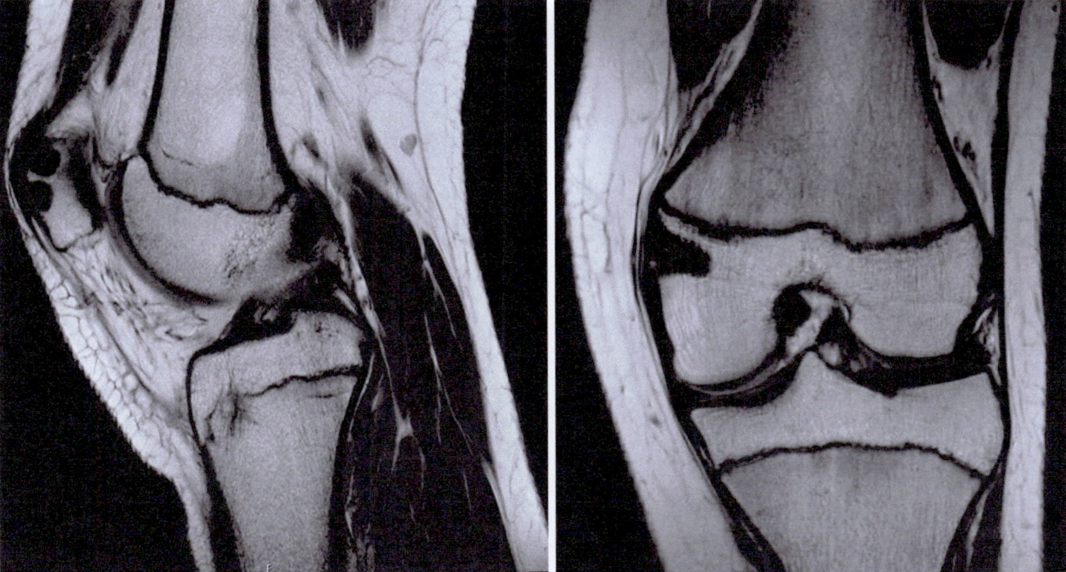

Fig. 7.23 Postoperative sagittal and coronal MRI

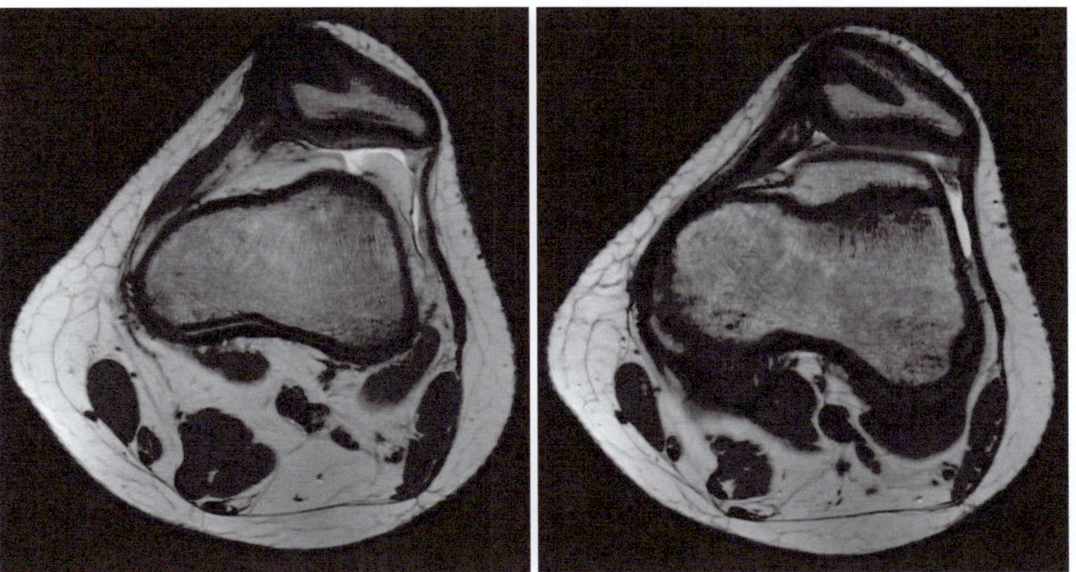

Fig. 7.24 Postoperative axial MRI

conservative management has failed, there are options for effective surgical treatment that will not harm the open physis. The general recommendation in this case is to perform a MPFL reconstruction with either a semitendinosus or quadriceps tendon autograft and use either soft tissue fixation or small sockets to avoid patellar fracture. If indicated, these types of patients may also benefit from imbrication of the medial and or lateral retinaculum, and a soft-tissue distal realignment procedure such as a Roux-Goldthwait procedure.

Resources

- http://www.patellofemoral.org/pfoe/
- https://www.patellofemoral.org/education
- https://www.arthroscopytechniques.org/knee-patellofemoral
- https://www.hss.edu/sports-patellofemoral-center.htm

References

1. Schlichte LM, Sidharthan S, Green DW, Parikh SN. Pediatric management of recurrent patellar instability. Sports Med Arthrosc. 2019; https://doi.org/10.1097/JSA.0000000000000256.
2. Balcarek P, Walde TA. Accuracy of femoral tunnel placement in medial patellofemoral ligament reconstruction: the effect of a nearly true-lateral fluoroscopic view. Am J Sports Med. 2015;43(9):2228–32. https://doi.org/10.1177/0363546515591265.
3. Nguyen CV, Farrow LD, Liu RW, Gilmore A. Safe drilling paths in the distal femoral epiphysis for pediatric medial patellofemoral ligament reconstruction. Am J Sports Med. 2017;45(5):1085–9. https://doi.org/10.1177/0363546516677795.
4. Nelitz M, Dreyhaupt J, Reichel H, Woelfle J, Lippacher S. Anatomic reconstruction of the medial patellofemoral ligament in children and adolescents with open growth plates: surgical technique and clinical outcome. Am J Sports Med. 2013;41(1):58–63. https://doi.org/10.1177/0363546512463683.
5. Ladenhauf HN, Berkes MB, Green DW. Medial patellofemoral ligament reconstruction using hamstring autograft in children and adolescents. Arthrosc Tech. 2013;2:e151–4. https://doi.org/10.1016/j.eats.2013.01.006.
6. Andrish J. Surgical options for patellar stabilization in the skeletally immature patient. Sports Med Arthrosc. 2017;25(2):100–4. https://doi.org/10.1097/JSA.0000000000000145.
7. Andrish J. Recurrent patellar dislocation. In: Fulkerson JP, editor. AAOS Monograph: Common Patellofemoral Problems. American Academy of Orthopaedic Surgeons; 2005. p. 43–55.
8. Andrish J. Surgical options for patellar stabilization in the skeletally immature patient. Sports Med Arthrosc. 2007;15(2):82–8. https://doi.org/10.1097/JSA.0b013e31805752d0.
9. Sanchis-Alfonso V, Montesinos-Berry E, Monllau JC, Andrish J. Deep transverse lateral retinaculum reconstruction for medial patellar instability. Arthrosc Tech. 2015;4(3):e245–9. https://doi.org/10.1016/j.eats.2015.02.003.
10. Askenberger M, Janarv PM, Finnbogason T, Arendt EA. Morphology and anatomic patellar instability risk factors in first-time traumatic lateral patellar dislocations. Am J Sports Med. 2017; https://doi.org/10.1177/0363546516663498.
11. Arendt EA, Askenberger M, Agel J, Tompkins MA. Risk of Redislocation after primary patellar dislocation: a clinical prediction model based on magnetic resonance imaging variables. Am J Sports Med. 2018; https://doi.org/10.1177/0363546518803936.
12. Parikh SN, Nathan ST, Wall EJ, Eismann EA. Complications of medial patellofemoral ligament reconstruction in young patients. Am J Sports Med. 2013; https://doi.org/10.1177/0363546513482085.
13. Nelitz M, Williams RS, Lippacher S, Reichel H, Dornacher D. Analysis of failure and clinical outcome after unsuccessful medial patellofemoral ligament reconstruction in young patients. Int Orthop. 2014; https://doi.org/10.1007/s00264-014-2437-4.
14. Fink C, Veselko M, Herbort M, Hoser C. MPFL reconstruction using a quadriceps tendon graft. Knee. 2014;21(6):1175–9. https://doi.org/10.1016/j.knee.2014.05.006.
15. Peter G, Hoser C, Runer A, Abermann E, Wierer G, Fink C. Medial patellofemoral ligament (MPFL) reconstruction using quadriceps tendon autograft provides good clinical, functional and patient-reported outcome measurements (PROM): a 2-year prospective study. Knee Surg Sport Traumatol Arthrosc. 2019; https://doi.org/10.1007/s00167-018-5226-6.
16. Herbort M, Hoser C, Domnick C, et al. MPFL reconstruction using a quadriceps tendon graft. Part 1: Biomechanical properties of quadriceps tendon MPFL reconstruction in comparison to the Intact MPFL. A human cadaveric study. Knee. 2014; https://doi.org/10.1016/j.knee.2014.07.026.
17. Nelitz M, Dreyhaupt J, Williams SRM. Anatomic reconstruction of the medial patellofemoral ligament in children and adolescents using a pedicled quadriceps tendon graft shows favourable results at a minimum of 2-year follow-up. Knee Surg Sport Traumatol Arthrosc. 2018;26(4):1210–5. https://doi.org/10.1007/s00167-017-4597-4.
18. Lewallen L, Mcintosh A, Dahm D. First-time patellofemoral dislocation: risk factors for recurrent instability special focus section 303. J Knee Surg. 2015;28:303–10. https://doi.org/10.1055/s-0034-1398373.
19. Christensen TC, Sanders TL, Pareek A, Mohan R, Dahm DL, Krych AJ. Risk factors and time to recurrent ipsilateral and contralateral patellar dislocations. Am J Sports Med. 2017;45(9):2105–10. https://doi.org/10.1177/0363546517704178.
20. Lewallen LW, McIntosh AL, Dahm DL. Predictors of recurrent instability after acute patellofemoral dislocation in pediatric and adolescent patients. Am J Sports Med. 2013; https://doi.org/10.1177/0363546512472873.
21. Balcarek P, Oberthür S, Hopfensitz S, et al. Which patellae are likely to redislocate? Knee Surg Sport Traumatol Arthrosc. 2014; https://doi.org/10.1007/s00167-013-2650-5.

22. Sanders TL, Pareek A, Hewett TE, Stuart MJ, Dahm DL, Krych AJ. High rate of recurrent patellar dislocation in skeletally immature patients: a long-term population-based study. Knee Surg Sport Traumatol Arthrosc. 2018; https://doi.org/10.1007/s00167-017-4505-y.
23. Matsushita T, Kuroda R, Oka S, Matsumoto T, Takayama K, Kurosaka M. Clinical outcomes of medial patellofemoral ligament reconstruction in patients with an increased tibial tuberosity–trochlear groove distance. Knee Surg Sport Traumatol Arthrosc. 2014; https://doi.org/10.1007/s00167-014-2919-3.
24. Zhang ZJ, Zhang H, Song GY, Zheng T, Feng H. A pre-operative grade 3 J-sign adversely affects short-term clinical outcome and is more likely to yield MPFL residual graft laxity in recurrent patellar dislocation. Knee Surg Sport Traumatol Arthrosc. 2019; https://doi.org/10.1007/s00167-019-05736-4.
25. Schöttle PB, Schmeling A, Rosenstiel N, Weiler A. Radiographic landmarks for femoral tunnel placement in medial patellofemoral ligament reconstruction. Am J Sports Med. 2007; https://doi.org/10.1177/0363546506296415.
26. Matsushita T, Araki D, Hoshino Y, et al. Analysis of graft length change patterns in medial patellofemoral ligament reconstruction via a fluoroscopic guidance method. Am J Sports Med. 2018; https://doi.org/10.1177/0363546517752667.
27. Nelitz M, Dornacher D, Dreyhaupt J, Reichel H, Lippacher S. The relation of the distal femoral physis and the medial patellofemoral ligament. Knee Surg Sport Traumatol Arthrosc. 2011; https://doi.org/10.1007/s00167-011-1548-3.
28. Kepler CK, Bogner EA, Hammoud S, Malcolmson G, Potter HG, Green DW. Zone of injury of the medial patellofemoral ligament after acute patellar dislocation in children and adolescents. Am J Sports Med. 2011; https://doi.org/10.1177/0363546510397174.
29. Thaunat M, Erasmus PJ. The favourable anisometry: an original concept for medial patellofemoral ligament reconstruction. Knee. 2007; https://doi.org/10.1016/j.knee.2007.08.008.
30. Lind M, Jakobsen BW, Lund B, Christiansen SE. Reconstruction of the medial patellofemoral ligament for treatment of patellar instability. Acta Orthop. 2008; https://doi.org/10.1080/17453670710015256.
31. Nietosvaara Y. The femoral sulcus in children: an ultrasonographic study. J Bone Jt Surg Ser B. 1994; https://doi.org/10.1302/0301-620x.76b5.8083274.
32. Seeley M, Bowman KF, Walsh C, Sabb BJ, Vanderhave KL. Magnetic resonance imaging of acute patellar dislocation in children: patterns of injury and risk factors for recurrence. J Pediatr Orthop. 2012; https://doi.org/10.1097/BPO.0b013e3182471ac2.
33. Remvig L, Jensen DV, Ward RC. Are diagnostic criteria for general joint hypermobility and benign joint hypermobility syndrome based on reproducible and valid tests? A review of the literature. J Rheumatol. 2007;
34. Chotel F, Peltier A, Kohler RBJ. What should we think about or look at, in case of patella dislocation in children. La Patella. 2012;15:33–40.
35. Tanner SM, Garth WP, Soileau R, Lemons JE. A modified test for patellar instability: the biomechanical basis. Clin J Sport Med. 2003; https://doi.org/10.1097/00042752-200311000-00001.
36. Ellera Gomes JL, Stigler Marczyk LR, De César PC, Jungblut CF. Medial patellofemoral ligament reconstruction with semitendinosus autograft for chronic patellar instability: a follow-up study. Arthrosc J Arthrosc Relat Surg. 2004; https://doi.org/10.1016/j.arthro.2003.11.006.
37. Chassaing V, Trémoulet J. Medial patellofemoral ligament reconstruction with gracilis autograft for patellar instability. Rev Chir Orthop Reparatrice Appar Mot. 2005; https://doi.org/10.1016/S0035-1040(05)84331-4.

8 A 12-Year-Old Girl with Recurrent Patellar Dislocation and Multiple Risk Factors Including Genu Valgum

Shital N. Parikh, Jacob R. Carl, Andrew Pennock, Javier Masquijo, and Franck Chotel

8.1 Case Presentation

A premenarchal 12-year-old female presented with recurrent patella dislocations of the left knee. She had two prior atraumatic patellar dislocations in the past year. The first patellar dislocation was sustained while she was dancing at home, lost her balance, and had a fall. The patella had to be reduced in the ED under sedation. The latest dislocation was sustained about seven months after the first episode. She was walking and accidentally bumped into someone, causing her to fall. The patella dislocated and spontaneously reduced. Prior to her latest episode, she had completed 28 sessions of formal physical therapy and had also worn a knee brace. She was in seventh grade and participated in competitive dancing. Her past medical and surgical history and family history were negative.

The patient was 163 cm tall and weighed 51.5 kg with a BMI of 19.3. Prior to her latest dislocation, the clinical exam showed 5° of hyperextension and 130° of flexion in both knees. She had mildly positive J-signs bilaterally. There was mild effusion in the left knee. The left knee was positive for patella apprehension and a laterally subluxable patella with weak Medial Patellofemoral Ligament (MPFL) endpoint, with approximately 3 out of 4 quadrants lateral patellar subluxation with stress testing in extension. Her contralateral knee exhibited 2 out of 4 quadrants of lateral patellar translation with a firm end point and positive apprehension. She had no physiologic hyperlaxity; Beighton score was five out of nine (no genu recurvatum). She had a normal gait with foot progression angle of 15° external. She had mild valgus standing alignment. She had a normal rotational profile in the prone position with hip internal rotation of 55° and external rotation of 40°.

Radiographical assessment of X-rays performed after the second dislocation (Figs. 8.1 and 8.2) revealed open growth plates, patella alta, and a positive crossing sign on the lateral X-ray. Caton-Deschamp index was 1.4. There was a bony fragment on the medial aspect of the left

S. N. Parikh (✉)
Orthopaedic Sports Medicine, Cincinnati Children's Hospital, Cincinnati, OH, USA

Orthopaedic Surgery, University of Cincinnati School of Medicine, Cincinnati, OH, USA
e-mail: Shital.Parikh@cchmc.org

J. R. Carl
Staff Pediatric Orthopaedic Surgeon, Shriners Hospital for Children-Spokane, Spokane, WA, USA

A. Pennock
360 Sports Medicine, Portland, OR, USA

Rady Children Hospital, San Diego, CA, USA

Orthopaedic Surgery, University of California San Diego, La Jolla, CA, USA

J. Masquijo
Department of Pediatric Orthopaedics, Sanatorio Allende, Córdoba, Nueva Córdoba, Argentina

F. Chotel
Orthopaedic Surgery, Claude Bernard University, Lyon, France

© ISAKOS 2022
J. L. Koh et al. (eds.), *The Patellofemoral Joint*, https://doi.org/10.1007/978-3-030-81545-5_8

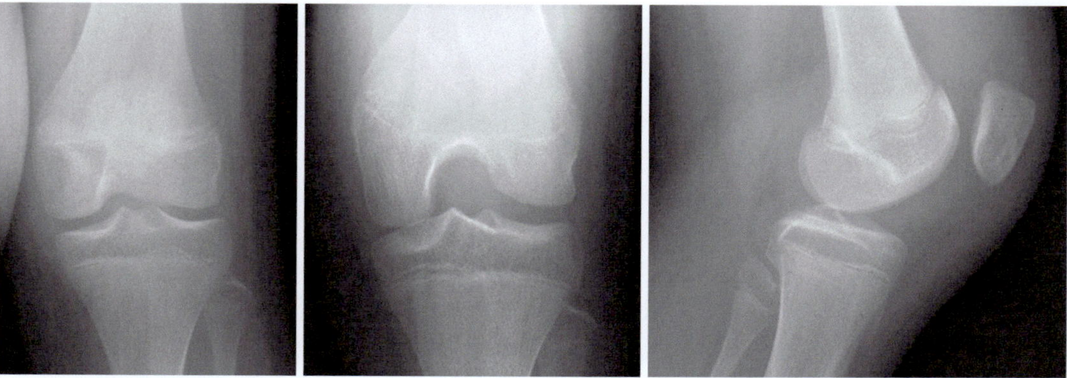

Fig. 8.1 Anteroposterior, notch, and lateral radiographic views of the left knee. They show open physis, crossing sign on lateral view suggesting trochlear dysplasia and mild patella alta

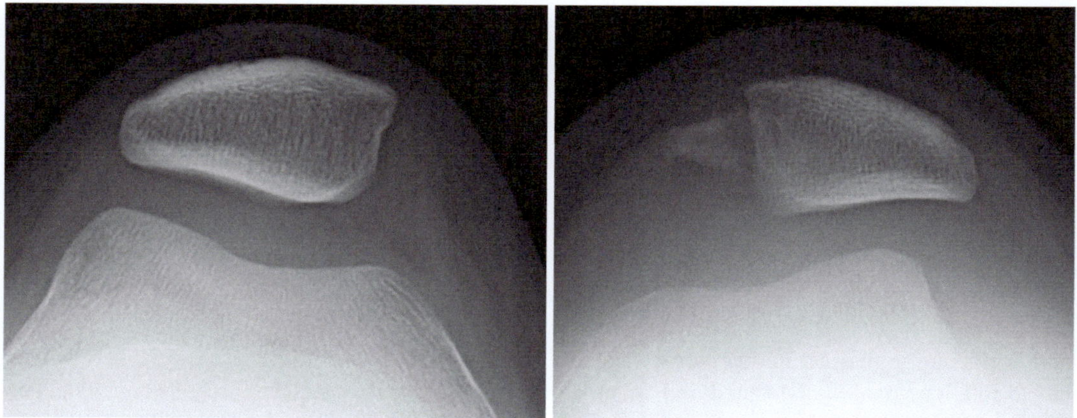

Fig. 8.2 Axial view of both patella shows the medial avulsion fracture from left patella, along with patellar tilt and subluxation

patella consistent with an avulsion fracture from a lateral patella dislocation. Standing alignment X-rays, which were bilateral full-length weight bearing radiographs from hip to ankle, revealed mild genu valgum, and no limb length discrepancy (Fig. 8.3). The valgus angle, which is the angle formed by the anatomic (diaphyseal) axis of femur and tibia, measured 10°. The mechanical lateral distal femoral angle was 80° and the mechanical medial proximal tibial angle was 89°. The mechanical axis from the center of the head of the femur to the center of the ankle passed through the lateral tibiofemoral knee compartment (+2 quadrant). Radiograph of the left hand and wrist showed a bone age of 12 years according to the atlas of Greulich and Pyle (Fig. 8.4).

MRI was obtained (Figs. 8.5, 8.6, and 8.7). It demonstrated trochlear dysplasia with a flat sulcus on an axial sect. 3 cm above the tibiofemoral joint line, and a trochlear bump on sagittal section (Dejour type B). TT-TG distance was 12.6 mm. Patellar tilt was 25°. There was no bone marrow edema on the medial aspect of the patella or the lateral femoral condyle but there was an MPFL tear on the patellar side consistent with a past patellar dislocation. There was mild articular cartilage wear on the medial aspect of the patella but there was no full thickness cartilage defect.

8.2 Treatment (SNP, JRC)

Due to recurrent patellar dislocations, skeletal immaturity, multiple anatomic risk factors (trochlear dysplasia, patella alta, patellar tilt, and genu

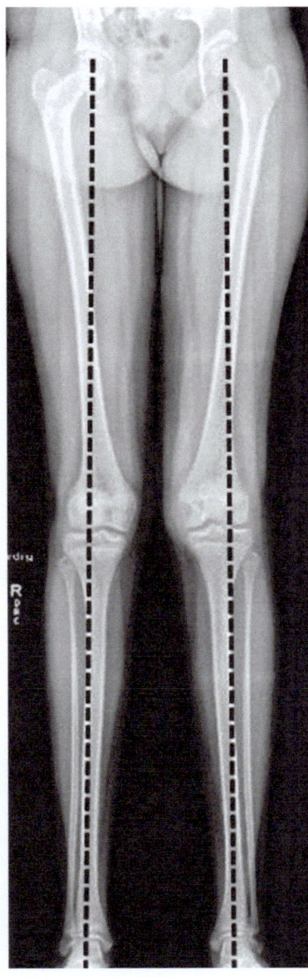

Fig. 8.3 Bilateral, full-length, standing radiograph shows the mechanical axis of the lower limb (dotted line) passing through the lateral compartment of the knee

valgum), and failure of trial of conservative treatment, the patient was indicated for surgical treatment. An epiphyseal MPFL reconstruction is the cornerstone and our preferred surgical treatment for recurrent patellar instability in a skeletally immature patient. Correction of mildly increased genu valgum by simultaneous distal femoral medial hemi-epiphysiodesis (also known as guided growth or growth modulation) was considered as this was the least morbid procedure to correct an anatomic risk factor as compared to surgical techniques for correction of other anatomic risk factors like patella alta and trochlear dysplasia [1, 2].

At surgery, examination under anesthesia showed a dislocatable patella with the knee at 0° and 30° flexion. The patellar tilt was correctable

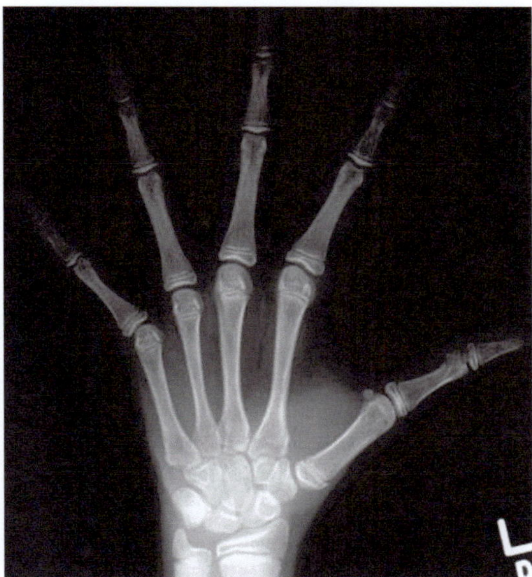

Fig. 8.4 Left hand radiograph for assessment of skeletal age

and the patella could be everted to neutral and hence a formal lateral retinacular release or lengthening was not performed. Knee arthroscopy was performed using standard portals and loose chondral flaps from medial aspect of patella were debrided (Fig. 8.8). The medial distal femur hemi-epiphysiodesis was then performed using Metaizeau technique. A transphyseal 6.5 mm cannulated cancellous screw was inserted percutaneously in a proximal-lateral to distal-medial direction to temporarily stop the growth of medial femur (Fig. 8.9). The screw should cross the physis at medial one-third, lateral two-third junction on an AP view and is in the midline on lateral view. The screw was placed prior to MPFL reconstruction to avoid inadvertent damage to the MPFL graft from later screw placement. The epiphyseal MPFL reconstruction involved gracilis tendon harvest through a posteromedial incision, patellar fixation in a 3.5 mm bone tunnel without implants and femoral fixation below the level of the distal femoral physis using interference screw fixation. The femoral tunnel was established under fluoroscopic guidance by placement of a beath pin directed parallel and slightly inferior to the distal femoral physis. On a lateral view, the beath pin was directed slightly anteri-

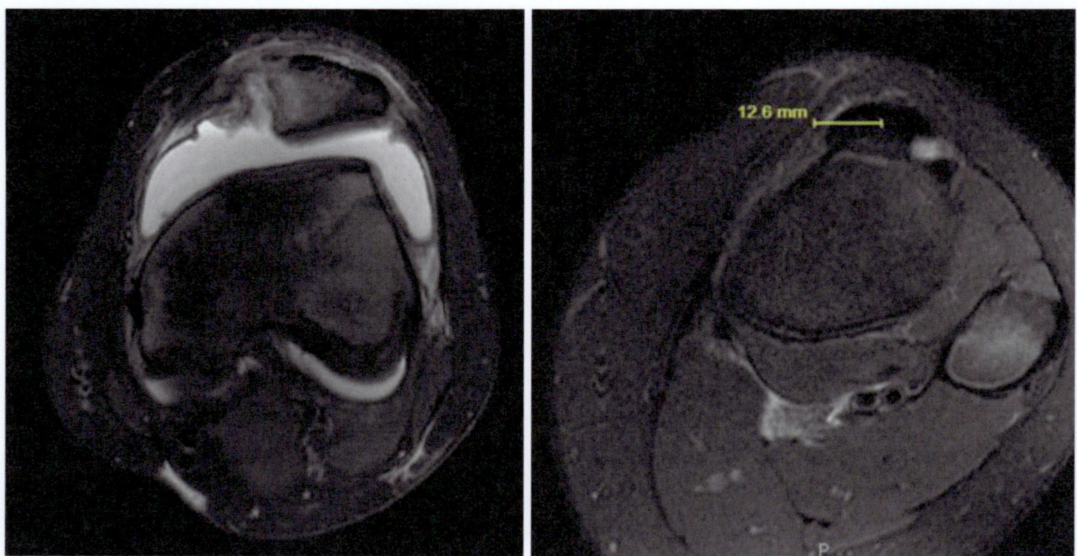

Fig. 8.5 T2 weighted, axial views of MRI show effusion, avulsion of medial patella, flat trochlea, and measurement of TT-TG distance

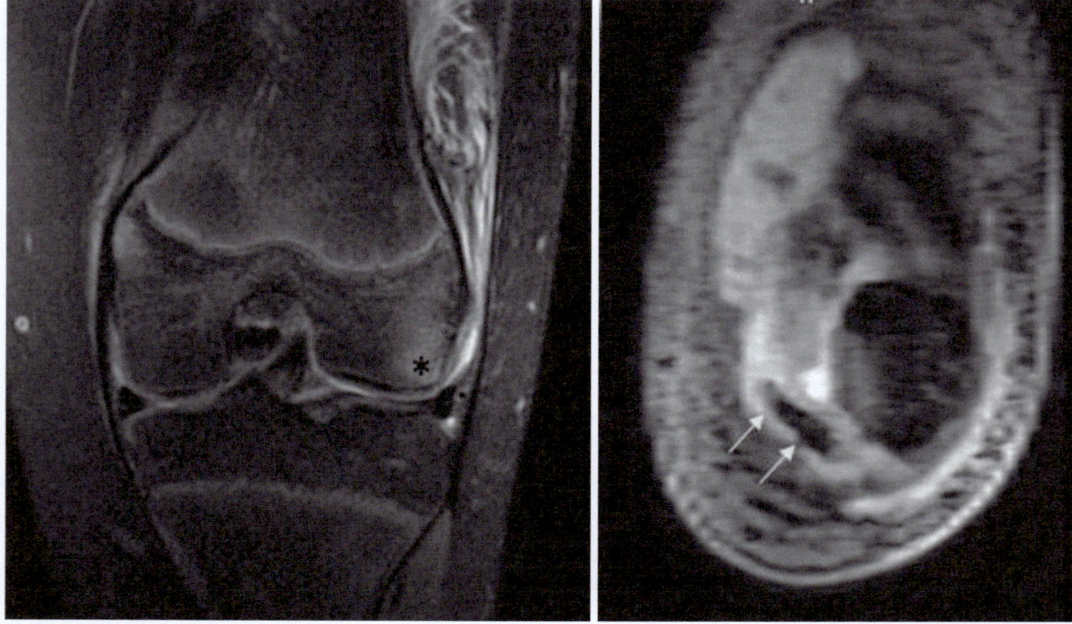

Fig. 8.6 T2 weighted, coronal views of MRI show the bruise (*) over the lateral femoral condyle, open physis, and inferomedial avulsion fracture (arrows) of patella

orly, between the transphyseal screw and the intercondylar notch. The beath pin was over-drilled and then the graft was pulled into the femoral tunnel using a pull-through technique. The surgical technique and early results of this combined procedure have been published.

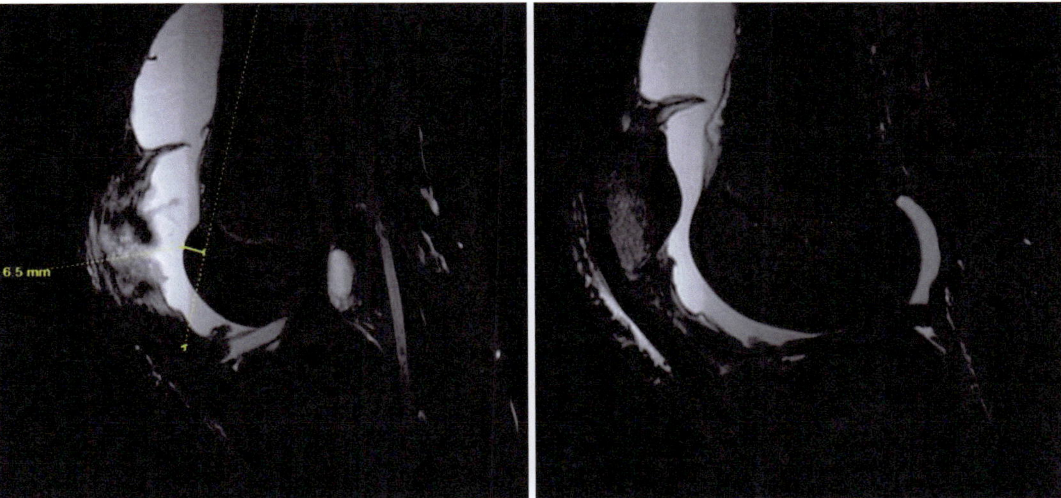

Fig. 8.7 T2 weighted, sagittal views of MRI show the measurement of trochlear bump, effusion, and patellofemoral engagement

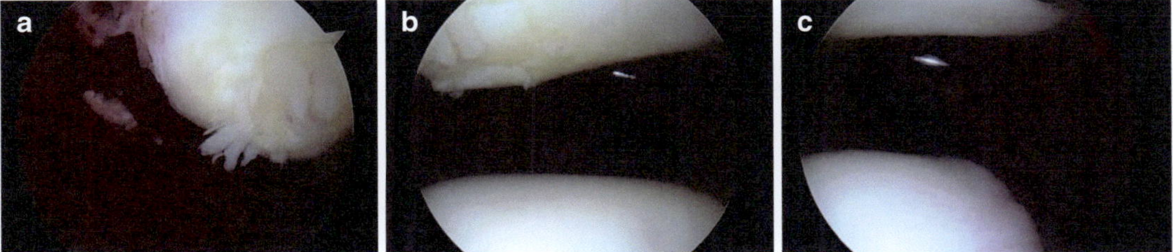

Fig. 8.8 Arthroscopic evaluation of patellofemoral joint shows cartilage fraying over the medial (**a**) aspect of patella. (**b**) The trochlea is flat. (**c**) The patella is subluxed laterally but there is no cartilage injury over the lateral aspect of patella

Fact Box: Epiphyseal MPFL Reconstruction

Advantages

Gracilis graft harvest from posteromedial incision is a cosmetic approach and avoids saphenous neuropraxia.

The gracilis tendon helps to keep the patellar tunnel size to a minimum.

No implants are used on the patellar side and tunnel size is restricted to 3.5 mm to avoid patellar fracture.

Femoral tunnel placement performed below the distal femoral growth plate of the graft allows for anatomic reconstruction.

Pull-through technique is used to allow for final fixation of the graft on the femoral side.

Early range of motion and weight bearing are allowed.

Transphyseal screw is an acceptable method for growth modulation and does not interfere with epiphyseal MPFL reconstruction.

Disadvantages

Intraoperative fluoroscopy is necessary.

Position of beath pain below the distal femoral physis and posterior to the transphyseal screw requires precise placement.

Though fluoroscopy is used during femoral tunnel placement, there is a small risk of damage to the distal femoral physis.

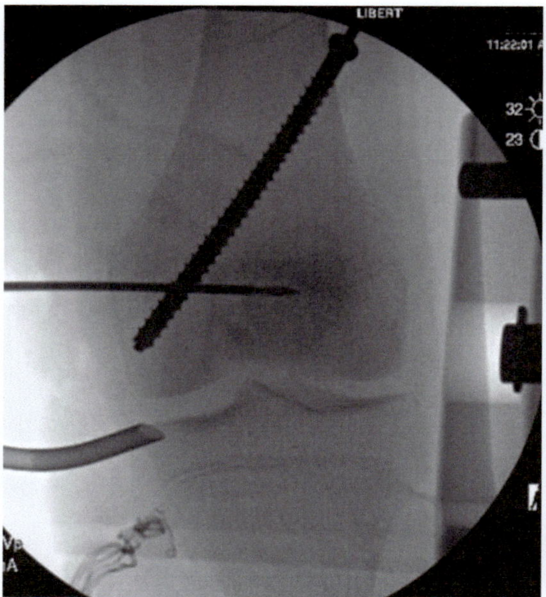

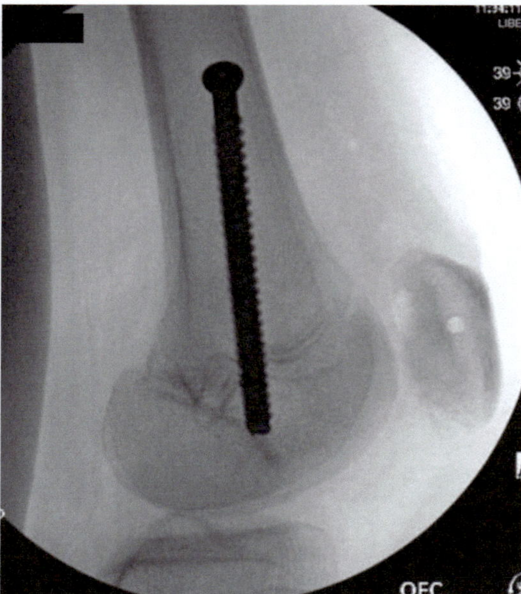

Fig. 8.9 A 6.5 mm transphyseal screw is first inserted under fluoroscopic image guidance. It should intersect the physis at its medial one-third, lateral two-third junction on anteroposterior view and at its center on lateral view. The guide pin for MPFL femoral tunnel is inserted below the level of the distal femoral physis. On lateral view, the femoral tunnel is placed just anterior to the posterior femoral cortical line and posterior to the transphyseal screw. The tunnel in the patella is seen on the lateral view

Serial follow-up with full-length radiographs is required to avoid overcorrection of valgus deformity during growth modulation.

Second surgery for removal of transphyseal screw may be required unless the valgus correction coincides with completion of growth.

8.3 Post-operative Course and Outcome

Immediately following surgery, the patient was placed in a knee immobilizer and was weight bearing as tolerated on crutches. She was prescribed cold therapy to use at home. Three days after surgery she began a physical therapy program to work on ROM and quadriceps activation. The patient began to wean from the knee immobilizer and crutches at 3–4 weeks. After completion of a strength and conditioning program and isokinetic assessment (Biodex), she was cleared and returned to full activities without restrictions around 6 months post-operatively.

She achieved genu valgum correction and minor overcorrection at about 6 months after the index surgery (Fig. 8.10). The transphyseal screw was removed in the operating room. She subsequently had recurrent patellar instability on the contralateral side and underwent an isolated MPFL reconstruction. At her final follow-up, 3 years post-operatively, the patient was doing well with no complaints of pain or feelings of instability. There were no dislocations or subluxation episodes since her surgery. She successfully returned to dance and has been playing basketball and soccer with no limitations in her knee. On clinical exam she had a mild bilateral J sign with well centered patellae and good MPFL endpoint. She had full range of motion (−5 to 130°) bilaterally with no evidence of swelling. She had positive apprehension sign on both sides. Her knee radiographs revealed no osseous abnormalities and satisfactory patellofemoral alignment (Fig. 8.11). Standing hip to ankle radiographs showed neutral alignment on the left side (compared to slight overcorrection prior to screw removal) and no leg length discrepancy (Fig. 8.12).

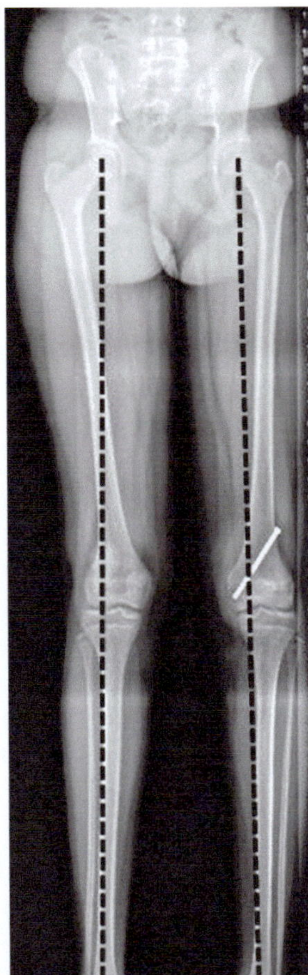

Fig. 8.10 Full-length, standing radiograph, 6 months after transphyseal screw placement, shows that the mechanical axis of the lower limb (dotted line) on the left side is now passing through the medial compartment of the knee

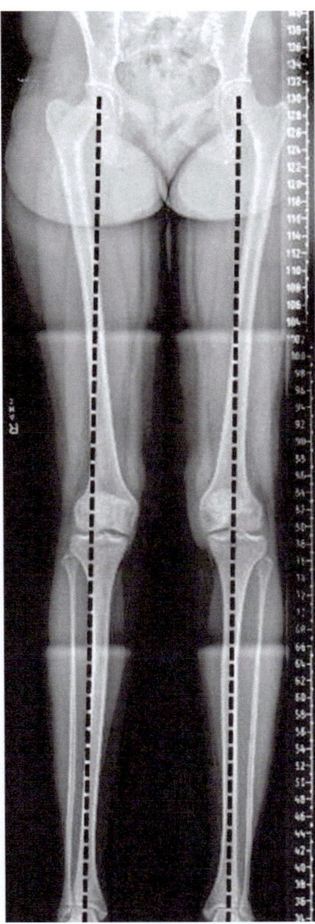

Fig. 8.12 Full-length, standing radiograph, 3 years after index surgery, shows the maintained correction of genu valgum on the left side. The right side show mild valgus that is unchanged; MPFL reconstruction was performed without growth modulation as patient did not have sufficient growth remaining at the time of surgery

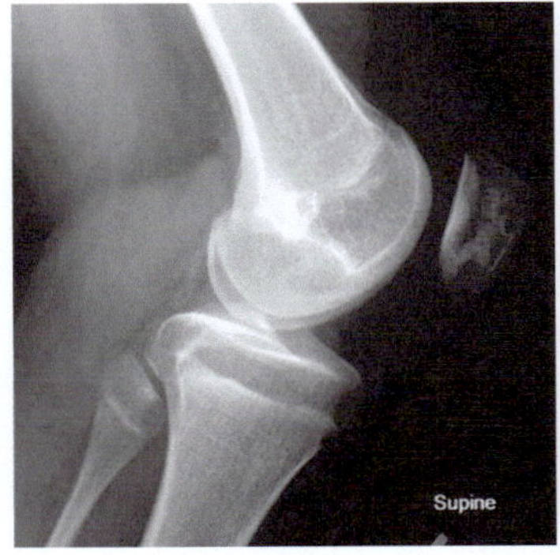

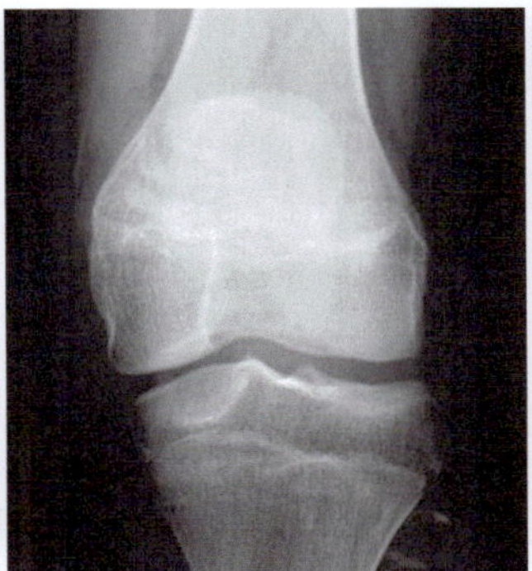

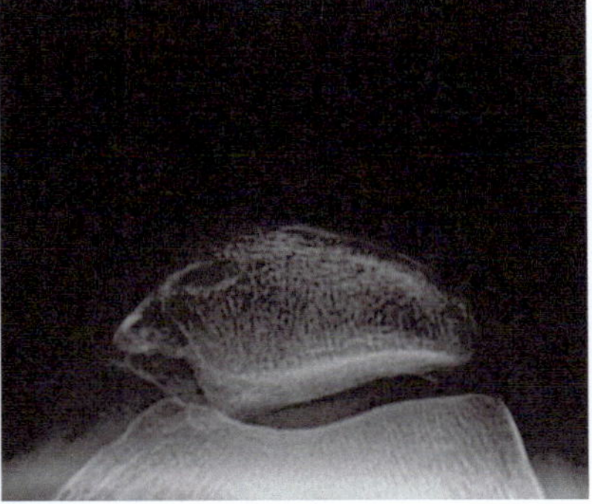

Fig. 8.11 Follow-up radiographs show adequate positioning of the MPFL tunnels, removal of transphyseal screw, no degenerative changes, adequate position of the patella and sequelae of medial avulsion fracture of patella

8.4 Perspective and Treatment Recommendation from USA (Pennock)

This case highlights many of the challenges of treating skeletally immature patients with patella instability. Typically, these younger patients have multiple risk factors for patella inability and in the current case this included trochlear dysplasia, patella alta, genu valgum, mildly increased lateral offset of the TT-TG (normal is approximately 10–12 mm), VMO atrophy, and increased femoral version (normal prone internal rotation is 30–40°). One of the primary treatment challenges for this young population is that many of the surgical options that would be utilized in patients with closed growth plates are not available in patients with open physes. For example, in these younger patients the tibial tubercle cannot be osteotomized to address a pathologic TT-TG or to correct significant patella alta. Over the last several decades though, advancements in surgical technique have enabled surgeons to work around these issues.

Physeal sparing MPFL reconstructions have been developed and have been shown to have promising outcomes with respect to stabilizing the patella and avoiding growth plate abnormalities. To date, the ideal graft choice has not been identified. Many surgeons prefer to use hamstring tendon autograft (as used in this case), but other surgeons have moved to soft tissue allograft since it reduces surgical time, it is more cosmetic, and it avoids harvest site morbidity. Unlike ACL reconstructions in younger patients, allograft tissue for MPFL reconstructions has not been associated with higher rates of graft failure which may in part be due to the extra-articular nature of the reconstruction. Fixation of the graft on the femur is another important consideration in patients with open growth plates. As acknowledged in the case above, care must be taken placing the femoral screw distal and parallel to the physis. In this slightly "non-anatomic" position, the graft is less isometric than when the graft is placed in the anatomic position at Schottle's point. The surgeon must be aware that it is easy to over-tension the graft when it is placed in this distal position which can lead to loss of flexion. A simple pearl to avoid this complication is to "tension" or pull the slack out of the graft with the knee in hyperflexion prior to final fixation of the graft. After the slack is pulled out of the graft, the knee is straightened into 30–60 degrees of flexion and the graft is secured, but care is taken to not further tighten the graft.

It is also important for the surgeon to recognize that an MPFL reconstruction alone cannot adequately address all skeletally immature patients with patella instability. Fortunately, other concomitant procedures can and should be selectively utilized in patients with open physes. This case example highlights the utilization of growth modulation in a patient with an underlying valgus deformity. Limb alignment correction can be achieved using the Metaizeau technique as demonstrated in this case or an alternative option would be to carefully place a guided growth plate that can be positioned just proximal or superficial to the graft. In slightly older patients with open growth plates, but minimal growth remaining, surgery can be timed such that a definitive hemi-epiphysiodesis can be performed concurrently with the MPFL reconstruction or the femoral screw can even be placed across the medial physis. Another important consideration in these patients is the position of the patella relative to the trochlear groove. In patients with mild patella alta (as seen in this case), an MPFL reconstruction is a great surgical option because the distal pull of the graft leads to mild reduction in the patient's patella alta. For patients with severe patella alta with open growth plates, a patella tendon imbrication can be performed by folding the tendon on top of itself and sewing it together. Surgical solutions have also been proposed to address significant elevations of the TT-TG. Historical procedures like the Roux-Goldthwait involve transferring a portion of the patella tendon to a more medial position on the proximal tibia. Recently, a complete transfer of the patellar tendon has been described for skeletally immature patients (under the age of 12 years) with severe patellar maltracking that can be performed in conjunction with other soft tissue procedures. Another important surgical consideration is the role that femoral version plays on patellar mechanics. With mild rotational abnormalities, as seen in this case, physical therapy targeting the core and proximal musculature can be utilized, but in cases of severe rotational abnormalities, derotational femoral osteotomies can be considered. In these cases, the surgeon may choose to stage the procedure addressing the bony abnormality first and coming back during a second procedure to remove any symptomatic implants and to address the soft tissue abnormalities.

Establishing realistic post-operative expectations is important for these patients and families. As in this case where the screw had to be removed on the one side and an MPFL reconstruction was necessary on the other side, it is not infrequent that a second procedure will be necessary on the ipsilateral knee or the contralateral knee. In particular when the patient reaches skeletal maturity, bony surgery in the form of a tubercle osteotomy or even a trochleoplasty for severe cases of trochlear dysplasia may be necessary. It is also important for these families to recognize the high risk of future arthritis. This may be the result of chondral injuries from the original instability events or possible over-constraint as a result of surgery. Radiographic arthritis has been shown to be a common complication 20–30 years after even a single episode of patella instability. Therefore, in these younger patients, they may only be in their 30s or 40s when they begin experiencing symptoms of patellofemoral arthritis.

With respect to this specific case, my approach would have varied slightly from that presented. First, I would not have performed a femoral nerve block. In my experience, these patients have significant quadriceps weakness often associated with profound VMO atrophy. I worry that a femoral nerve block puts the patient at risk of an iatrogenic femoral nerve injury which may further weaken a leg that has minimal muscular reserve. After finishing an examination under anesthesia, I would have performed a similar knee arthroscopy to address any chondral pathology or loose bodies. I then would have proceeded with an adult-style MPFL reconstruction with soft tissue allograft that is pulled through a 5 mm converging osseous tunnel on the medial patella and secured with a biocomposite interference screw to the femoral. My rationale for this approach is that a girl with a bone age of 12 years and an MRI with physes that are already thinning has minimal growth remaining. I have found that placing the screw in the more isometric and more anatomic position compared to the "physeal sparing" position leads to a patella that feels and tracks better after the graft is secured. Additionally, the risk of an iatrogenic growth disturbance is small and in the rare event where it occurs, the ensuing progressive varus growth will actually help correct the baseline valgus deformity. If the current patient had a bone age that was 1 or 2 years younger, I would have then proceeded with a physeal sparing MPFL reconstruction with the screw placed distal and parallel to the physis. My approach also would have varied in that I would not have utilized a growth modulation technique. There are several reasons for this. First, the patient only had a mild baseline deformity. Second, the patient had minimal growth remaining so there was limited correction potential. Third, the implants used for growth modulation often need to be removed with a second surgery which adds significant costs to the care of the patient. Finally, these implants need to be placed in the vicinity of the MPFL attachment which can potentially compromise the placement of the femoral tunnel. If the current patient had been a boy with a bone age of 12 years or a girl with a bone age of 10 years, I would have added a growth modulation procedure. With respect to the patient's trochlear dysplasia, elevated TT-TG, patella alta, and the increased femoral version, I would not address these surgically since all of these abnormalities were relatively mild (other than the trochlear dysplasia which was more severe). I would clearly counsel the family, that a future osteotomy may be necessary when she is skeletally immature to address these residual risk factors.

8.5 Perspective and Treatment Recommendation from Argentina (Masquijo)

Recurrent patellar instability, particularly in the skeletally immature patient, presents a challenging problem. The normal course of the patella within the trochlear groove is determined by a complex interplay of several anatomic components. This case presents with multiple anatomic risk factors, including genu valgum, patella alta, and trochlear dysplasia. Guided growth has a clear role in the skeletally immature patient with significant genu valgum and patellofemoral instability. My preference is to perform guided growth with a single tension-band plate placed on the medial distal femur as the initial procedure, and then the medial patellofemoral complex (MPFC) reconstruction at the time of hardware removal. I prefer the plate over transphyseal screw technique because it does not disrupt the physis, thereby minimizing the possibility of premature growth arrest, and removal is easier. After guided growth, patients are usually followed every 3 months, and the implants are removed when the mechanical axis is restored to normal. A recent study has shown that frontal leg axis correction has an effect on the Q-angle similar to medialization of the tibial tubercle, which can be beneficial in this patient with borderline TT-TG distance [3].

A reconstruction of the MPFC is my preferred approach for pediatric–adolescent patients with recurrent patellofemoral instability. I favor this technique because it attempts to reconstruct the normal anatomy of the medial quadriceps tendon femoral ligament (MQTFL) and MPFL, and avoid transosseous patellar drilling for docking which can cause fracture [4, 5]. Knee arthroscopy is performed using standard portals to address associated chondral injuries. For MPFC reconstruction, I use either a hamstring autograft or allograft. The graft is prepared for double-bundle reconstruction with the use of the whipstitch technique of the distal ends, allowing for approximately 50 to 60 mm for the reconstructed MPFL/MQTFL and 15 to 20 mm for graft placement within the femoral socket. The proximal limb should be slightly longer than the distal limb such that it can reach the quad tendon. Under fluoroscopy, a guidewire is placed distal to the medial distal femoral growth plate on the anteroposterior image and in line with the posterior femoral cortex on the lateral radiograph. A 15 mm long femoral socket distal to the distal femoral physis is created. The graft is fixed on the femur with use of an interference tenodesis screw, and then passed through the soft tissue space between the fascia and synovium to the superior half of the patella. The patella is then centralized in the trochlea with the knee placed in 30° to 40° degrees of flexion. The graft length is adjusted so that it holds the patella in the center of the trochlear groove. The patellar limb is fixed with a 3.5 mm suture anchor on the medial patellar border at the junction of the proximal 1/3 and distal 2/3. The quadriceps limb is passed through a hole in the rectus femoris and secured to itself with nonabsorbable sutures under the correct tension. Finally, with the knee in extension, patellar glide is evaluated. The graft should be positioned to allow for patellar excursion of one-fourth the width of the patella to successfully resolve maltracking without overconstraining the patella [6]. I rarely combine lateral retinacular lengthening in a patient with recurrent patellar dislocation. If under anesthesia the patella could be everted to neutral, I would avoid this additional procedure. In the author's practice, lateral retinacular lengthening is more commonly performed in the setting of habitual or irreducible patellar dislocation.

Patella alta has long been recognized as a risk factor for recurrent patellar instability. Patella alta means that the patella will engage in the trochlea later in flexion, making the patella more susceptible to lateral dislocation in early flexion. Some studies [7, 8] have shown that MPFL reconstruction can reduce the patellar height and correct patella alta. The degree of correction reported by these authors is limited (a decrease of CD index between 11% and 16%), however it can be enough to correct borderline patella alta like the current case.

Trochlear dysplasia involves an abnormality of the shape and depth of the trochlear groove and is frequently found in individuals with recurrent patellofemoral instability. Several different types of groove-deepening trochleoplasty procedures have been described in adults with documented efficacy [9]. However, for skeletally immature patients, trochleoplasty has the potential risk of injury to the distal femoral growth plate and subsequent growth disturbance. Also, there is some recent evidence that patellar stabilization in very young patients may allow for trochlear remodeling [10]. Therefore, I avoid trochleoplasty in this patient population.

This particular patient had no hyperlaxity (Beighton score 5/9). In my practice, generalized laxity is also an important factor in the surgical decision. In a hypothetical patient with similar features and an elevated Beighton score (>6/9), I would be more aggressive addressing the risk factors. In such a patient, I would perform distal realignment (tibial tubercle periosteum transfer) in addition to the MPFC reconstruction to generate the most favorable conditions for the graft.

8.6 Perspective and Treatment Recommendation from France (Chotel)

This girl has episodic patellar instability and she is close to skeletal maturity. Skeletal maturity in girls is typically associated with menarche and 13.5 years of skeletal age. Thus, at 12 years of skeletal age, she should be skeletally mature in the next 16 to 20 months. In such a situation, the Beighton score should be interpreted with caution as the cut off value for diagnosis of generalized hyperlaxity is 6/9 in children and 4/9 in adults. A 5/9 score in a 12 years old girl could be considered as generalized hyperlaxity. This is a risk factor for recurrence of instability after surgery and thus, the "à la carte" menu for surgical procedures, including osteotomies, should be considered.

The analysis of major factors for patella instability in this patient reveals: trochlear dysplasia Type B of Dejour with a significant bump on lateral X-ray, patella alta, and MPFL insufficiency (J sign and patellar tilt in extension). The authors reported that the distance TT-TG was normal and clinically, there was no tightness of the lateral retinaculum. Our treatment approach for this patient would include Dejour deepening trochleoplasty and tibial tubercle osteotomy for distalization, associated with MPFL reconstruction. This treatment would be delayed till skeletal maturity (few months) in order to adequately correct major risk factors for instability.

The valgus alignment of lower extremity without increased TT-TG distance or without rotational abnormality is only a minor risk factor for patella instability. The X-ray analysis of valgus can be misleading, considering variation in angle measurement based on rotation of the lower extremity at the time of X-ray; in the present case, it is surprising that the first post-operative X-ray reveals an overcorrection in varus (Fig. 8.10) and later, the second X-ray shows perfect axis (Fig. 8.12). With a normal BMI (no overweight), hemi-epiphysiodesis for knee valgus would only be indicated if the intermalleolar distance was more than 8–10 cm. If valgus correction was indicated, medial curettage of the distal femoral physis (Bowen's procedure) could be considered at around 12.5 or 13 years of skeletal age but not earlier. In the present case, the first author chose the option of unilateral correction combined with MPFL reconstruction. At the last follow-up, a perfect mechanical axis with unilateral correction of valgus deformity cannot be obtained without slight overcorrection in varus on the epiphysiodesis (femoral) side.

It would also be an option to perform bilateral genu valgum correction (trochlear dysplasia is nearly always bilateral in case of instability). The advantage of definitive epiphysiodesis by curettage is to avoid a second procedure for device removal. Hemi-epiphysiodesis would not be considered in the present case as the valgus is mild, and there is a risk of hypercorrection; the normal mechanical lateral distal femoral angle range is between 85 and 90°.

The first author strategy with a simple soft tissue procedure and unilateral hemi-epiphysiodesis before maturity differs from our strategy. Though it gave good functional results at 3 years follow-up, we are concerned about post-operative J sign and positive apprehension test. These points are related to MPFL insufficiency. The bad trochlear engagement in extension is due to patella alta. We believe that a perfect patella stabilization would eliminate the apprehension test.

It has been demonstrated that isolated MPFL reconstruction could decrease patellar height; however, a Caton-Deschamps index around 1.4 is too high and its correction would require distalization procedure. MPFL reconstruction with patella alta would increase the risk of an anisometric ligament when stretching in extension. Of course the patellar tendon can be shortened before maturity but bony distalization is more accurate when it is acceptable to wait for a few months. The second argument to wait till skeletal maturity is the need for deepening trochleoplasty. Deepening trochleoplasty should not be performed in skeletally immature patients due to the risk of anterior distal femoral physeal injury and resultant recurvatum deformity. This procedure is proposed for trochlear dysplasia Dejour type B and D. In the present case, and because of monosurface and single patellar slope, it would have been done mainly to remove the spur rather to recreate two trochlear facets.

Considering the chronology of procedures, the surgery would start with a medial arthrotomy, checking medial patella bony avulsion fragment and removal if necessary. The second procedure would be bony tibial tuberosity osteotomy in order to facilitate the deepening trochleoplasty. After that, the distalization of the tibial tuberosity would be achieved with 2 anterior screws, to reach a post-operative CD index around 1.1. Next, the MPFL reconstruction would be performed. Our technique for MPFL reconstruction is a mix between Fithian procedure for adult (double strand, femoral and patellar tunnels) and the Die procedure for children (the third post of medial collateral ligament acts as a reflection pulley for the reconstruction) using hamstring tendon [11]. The graft tension is a key point; it would be performed in the present case so that the translation test in extension reaches 0% (rather than 10% lateral residual translation for a patient without general hyperlaxity). The advantage of this technique is a strong graft with pyramidal shape due to double strand of the reconstruction, and the absence of femoral tunnel that can be a source of secondary tunnel enlargement. Finally, a plasty of the superficial medial retinaculum would achieve the procedure, as lateral retinaculum release is not required here. A perfect patella tracking (anti-J sign with slight medialization in full extension) is checked at the end of the procedure.

8.7 Discussion

Normal valgus angle between anatomic axis of femur and tibia is reported to be less than 6° [12]. This angle is varied based on age, sex, stature, and ethnicity of the patient; thus there is no consensus on the absolute value that differentiates physiologic valgum from pathologic genu valgum. Genu valgum has been associated with altered biomechanical forces on the lateral tibiofemoral and patellofemoral joint. It increases the Q-angle and the resultant lateral force vector across the knee joint, which causes the patella to translate laterally, leading to increased lateral patellofemoral joint contact pressure and lateral patellar instability. Thus, correction of genu valgum is desirable in the setting of management of patellar instability. The indication for growth modulation of distal femur in our patient with patellar instability and open physis was a valgus angle of 10°, mechanical axis position in the lateral compartment (normally should be just medial to the center of the knee) and a lateral distal femoral angle of 84° or less [1, 2].

Metaizeau et al. first described the technique of transphyseal screw for temporary hemiepiphysiodesis [13]. The procedure was reported to be simple, extra-capsular, and minimally invasive with rapid return to function. Compared to permanent ablation of physis, this technique offers the advantage that accurate prediction of remaining growth is not required as resumption of normal growth is expected when deformity is corrected and the screw is removed. Another technique of growth modulation has been the tension-band plate as popularized by Stevens [14]. The advantage of tension-band plate is that, unlike transphyseal screw, the implants do not cross the physis. However, position of the plate on the medial aspect of the distal femoral physis can interfere with MPFL femoral attachment and hence is not the ideal choice of implant [15].

Can correction of genu valgum be enough to treat patellar instability? Kearney and Mosca reported that normalizing mechanical axis in 15 patients (26 knees) with patellar instability and genu valgum by isolated hemi-epiphysiodesis resolved the symptoms of patellar instability in most [16]. However, only 7 of 15 patients had documented patellar dislocation preoperatively, and four of these seven continued to have instability symptoms after valgus correction. Thus isolated growth modulation may have a role but patient selection could be difficult. Moreover, it would not be desirable to have continued recurrent patellar instability episodes during gradual deformity correction, as more episodes can lead to more permanent damage to the patellofemoral joint and more functional loss.

The safety and outcomes of MPFL reconstruction in skeletally immature patients have been evaluated [17]. In a recent systematic review and meta-analysis of 7 studies that entailed 132 MPFL reconstructions (126 patients) in skeletally immature patients with mean age of 13.2 years (range, 6–17 years) and mean postoperative follow-up of 4.8 years (range, 1.4–10 years), there was significant improvement in patient-reported outcomes and re-dislocation rates. All of the grafts used were autograft, with gracilis tendon (60.6%) being the most common. Methods of femoral fixation included interference screw (39.4%), suture anchor (38.6%), and soft tissue pulley around the medial collateral ligament or adductor tendon (21.9%). Pooled Kujala scores improved from 59.1 to 84.6 after MPFL reconstruction. The total reported complication rate was 25% and included 5 redislocations (3.8%) and 15 subluxation events (11.4%). No cases of premature physeal closure were noted, and there were 3 reports of donor site pain (2.3%). Neither autograft choice nor method of femoral fixation influenced recurrent instability or overall complication rates.

In a 1994 landmark article, four major risk factors for patellar instability were identified [18]. These included trochlear dysplasia, patella alta (CDI > 1.2), patellar tilt >20°, and TT-TG distance >20 mm. Three of the four major risk factors were present in our patient but were not surgically addressed. Of all the anatomic risk factors for patellar instability, trochlear dysplasia is considered the most important risk factor. All signs of trochlear dysplasia on lateral radiograph in adults (crossing sign, supratrochlear bump, and double contour sign) have been described in skeletally immature patients [19]. Trochlear dysplasia is present at birth and its shape is considered to be predominantly genetically determined. However, the proximity of primary and secondary physis of the distal femur to the trochlea raises some questions as to the role and contribution of the physis on the growth and development of trochlear dysplasia [20]. This is supported by reports of remodeling of trochlea after patellar stabilization in patients less than 10–11 years age [10]. From a treatment perspective, the close relationship between the anterior aspect of distal femoral physis and the trochlea would preclude a groove-deepening trochleoplasty in skeletally immature patients. A recent study has described safety of trochleoplasty in adolescent patients nearing skeletal growth completion [21]. Alternative to groove-deepening trochleoplasty in skeletally immature patient is the lateral facet elevating trochleoplasty, though there are no outcome studies supporting such an approach in this patient population. Thus trochlear dysplasia is evaluated for its prognostic value but it is not surgically addressed in skeletally immature patients.

Patella alta is a known risk factor for patellar instability as a high-riding patella would not engage in the trochlea during early flexion, leaving the patella unconstrained and unstable. Thus, correction of patella alta is desirable. Patellar height is typically evaluated on a lateral radiograph. There are several methods used to measure patellar height; the two frequently used methods include Insall-Salvati ratio and Caton-Deschamps index. On MRI, the patellotrochlear overlap index has been described. The important thing to consider in skeletally immature patients is that the patella ossifies from proximal to distal. Hence on a radiograph, the unossified distal-most aspect of patella would not be visible. This would decrease the measurement of patellar length and increase the measurement between the lower end of patella and tibia. This would cause a spurious elevation of patellar height measurement in children. The normative age-appropriate values of Caton-Deschamps index in children should be considered. If patella alta is present, its surgical correction in adults would be in the form of tibial tubercle osteotomy and distalization. Such surgery would be contraindicated in skeletally immature patients with open tibial tubercle apophysis, as damage to the apophysis could lead to genu recurvatum. Patellar tendon shortening or transposition has been described, as an alternative to tibial tubercle osteotomy, to obviate the risk of growth disturbances in skeletally immature patients. In the last decade, however, several studies have shown that isolated MPFL reconstruction can lower the patellar height and correct patella alta to some extent [7]. This is largely due to the vector of MPFL from patellar insertion to femoral insertion, which is medial and inferior. Thus, higher the patella, larger is the MPFL vector which helps to lower the patellar height. This, along with surgical complications of lowering patellar height, has led to decreased enthusiasm to address patella alta in skeletally immature patients [22].

Patellar tilt more than 20°, which cannot be corrected passively, has been described as pathologic. It indicates tight lateral structures, including tight lateral retinaculum. A lateral retinacular release or retinacular lengthening is indicated to address pathologic patellar tilt. In our patient, though the patellar tilt on MRI was >20°, it was passively correctable to almost neutral in the operating room. Hence, a lateral retinacular release or lengthening was not indicated. An unindicated or ill-performed lateral retinacular release could lead to iatrogenic medial instability and should be avoided.

8.8 Summary

The current chapter highlights various considerations while addressing recurrent patellar instability in a skeletally immature patient. All surgeons from different parts of the world do agree that surgical treatment is required and MPFL reconstruction should be part of the surgical algorithm. There are variations in the surgical technique for MPFL reconstruction including graft choice, patellar fixation, femoral fixation, and graft tensioning. The main difference, however, is the surgical management of anatomic risk factors. The factors that should be corrected at the time of index procedure ranges from none to genu valgum to all major risk factors. Knowledge is related to assessment of anatomic risk factors, their relative contribution to patellar instability, and the surgical technique for their correction continues to evolve.

Acknowledgment None of the authors has received any outside funding or grants in support of their research for preparation of this work. Neither authors nor members of their immediate family received payments or other benefits or a commitment or agreement to provide such benefits from a commercial entity.

References

1. Parikh SN, Redman C, Gopinathan NR. Simultaneous treatment for patellar instability and genu valgum in skeletally immature patients: a preliminary study. J Pediatr Orthop B. 2019 Mar;28(2):132–8.
2. Shah A, Parikh SN. Medial patellofemoral ligament reconstruction with growth modulation in children with patellar instability and genu Valgum. Arthrosc Tech. 2020 Mar 31;9(4):e565–74.
3. Flury A, Jud L, Hoch A, Camenzind RS, Fucentese SF. Linear influence of distal femur osteotomy on the Q-angle: one degree of varization alters the Q-angle by one degree. Knee Surg Sports Traumatol Arthrosc. 2020;9
4. Parikh SN, Wall EJ. Patellar fracture after medial patellofemoral ligament surgery: a report of five cases. J Bone Joint Surg Am. 2011;93(17):e97(1–8).
5. Spang RC, Tepolt FA, Paschos NK, Redler LH, Davis EA, Kocher MS. Combined reconstruction of the medial patellofemoral ligament (MPFL) and medial quadriceps tendon-femoral ligament (MQTFL) for patellar instability in children and adolescents: surgical technique and outcomes. J Pediatr Orthop. 2019 Jan;39(1):e54–61.
6. Elias JJ, Jones KC, Lalonde MK, Gabra JN, Rezvanifar SC, Cosgarea AJ. Allowing one quadrant of patellar lateral translation during medial patellofemoral ligament reconstruction successfully limits maltracking without overconstraining the

patella. Knee Surg Sports Traumatol Arthrosc. 2018 Oct;26(10):2883–90.

7. Lykissas MG, Li T, Eismann EA, Parikh SN. Does medial patellofemoral ligament reconstruction decrease patellar height? A preliminary report. J Pediatr Orthop. 2014 Jan;34(1):78–85.
8. Fabricant PD, Ladenhauf HN, Salvati EA, Green DW. Medial patellofemoral ligament reconstruction improves radiographic measures of patella alta in children. Knee. 2014 Dec;21(6):1180–4.
9. Longo UG, Vincenzo C, Mannering N, et al. Trochleoplasty techniques provide good clinical results in patients with trochlear dysplasia. Knee Surg Sports Traumatol Arthrosc. 2018;26(9):2640–58.
10. Rajdev NR, Parikh SN. Femoral trochlea does not remodel after patellar stabilization in children older than 10 years of age. J Pediatr Orthop B. 2019;28(2):139–43.
11. Chotel F, Bérard J, Raux S. Patellar instability in children and adolescents. Orthop Traumatol Surg Res. 2014;100(1 Suppl):S125–37.
12. Heath CH, Staheli LT. Normal limits of knee angle in white children--genu varum and genu valgum. J Pediatr Orthop. 1993;13(2):259–62.
13. Métaizeau JP, Wong-Chung J, Bertrand H, Pasquier P. Percutaneous epiphysiodesis using transphyseal screws (PETS). J Pediatr Orthop. 1998;18(3):363–9.
14. Stevens PM. Guided growth for angular correction: a preliminary series using a tension band plate. J Pediatr Orthop. 2007;27(3):253–9.
15. Bachmann M, Rutz E, Brunner R, Gaston MS, Hirschmann MT, Camathias C. Temporary hemiepiphysiodesis of the distal medial femur: MPFL in danger. Arch Orthop Trauma Surg. 2014;134(8):1059–64.
16. Kearney SP, Mosca VS. Selective hemiepiphyseodesis for patellar instability with associated genu valgum. J Orthop. 2015;12(1):17–22.
17. Shamrock AG, Day MA, Duchman KR, Glass N, Westermann RW. Medial patellofemoral ligament reconstruction in skeletally immature patients: a systematic review and meta-analysis. Orthop J Sports Med. 2019;7(7):2325967119855023.
18. Dejour H, Walch G, Nove-Josserand L, Guier C. Factors of patellar instability: an anatomic radiographic study. Knee Surg Sports Traumatol Arthrosc. 1994;2(1):19–26.
19. Lippacher S, Reichel H, Nelitz M. Radiological criteria for trochlear dysplasia in children and adolescents. J Pediatr Orthop B. 2011 Sep;20(5):341–4.
20. Parikh SN, Rajdev N. Trochlear dysplasia and its relationship to the anterior distal femoral Physis. J Pediatr Orthop. 2019 Mar;39(3):e177–84.
21. Nelitz M, Dreyhaupt J, Williams SRM. No growth disturbance after Trochleoplasty for recurrent patellar dislocation in adolescents with open growth plates. Am J Sports Med. 2018 Nov;46(13):3209–16.
22. Bartsch A, Lubberts B, Mumme M, Egloff C, Pagenstert G. Does patella alta lead to worse clinical outcome in patients who undergo isolated medial patellofemoral ligament reconstruction? A systematic review. Arch Orthop Trauma Surg. 2018 Nov;138(11):1563–73.

9 Patellofemoral Instability in the Young Male Soccer Player with First-Time Dislocation, Closed Physes, Normal Trochlea: 17-Year-Old

Petri Sillanpää, Julian Feller, and Laurie A. Hiemstra

9.1 Case

9.1.1 History

A 17-year-old male, who was physically active and soccer player at high level with no history of knee complaints: He had a contact injury resulting in a fall and obvious knee-twisting injury. He felt sudden pain in his right knee, and he had to leave the court immediately. He could not describe how he felt, despite acute pain. Walking was possible only with crutches and physical examination revealed knee hemarthrosis.

P. Sillanpää (✉)
Pihlajalinna Koskisairaala Hospital, Tampere, Finland
e-mail: petri.sillanpaa@tuni.fi

J. Feller
School of Allied Health, Human Services and Sport, La Trobe University, Melbourne, Vic, Australia

OrthoSport Victoria, Epworth HealthCare, Melbourne, Vic, Australia
e-mail: jfeller@osv.com.au

L. A. Hiemstra
Banff Sport Medicine, Banff, AB, Canada
e-mail: hiemstra@banffsportmed.ca

9.1.2 Physical Examination

There were significant pain and tenderness around the medial side of the knee. Knee was heavily swollen. Aspiration was done with 50 ml blood. Difficulties were there in full extension, and flexion was at 80 degrees. Passive ROM was 0–110. He was able to stand on the injured leg. Stable knee joint was found in varus and valgus stress, and anteroposterior laxity seemed normal, but pain caused muscle tension and difficulty evaluating in physical examination. Patellar movement made significant pain on the medial patellar side and acute lateral patellar dislocation injury was suspected; concomitant injuries were hard to define, though.

9.1.3 Radiographic Examination

X-rays of the right knee at the date of the injury were reviewed. They showed no significant fractures, despite suspicion of small osteochondral fracture on lateral view. Patella was located centrally, and patellar height was normal. MRI was required to confirm the location of the suspected osteochondral fracture, as well as to confirm the diagnosis (Figs. 9.1, 9.2 and 9.3).

© ISAKOS 2022
J. L. Koh et al. (eds.), *The Patellofemoral Joint*, https://doi.org/10.1007/978-3-030-81545-5_9

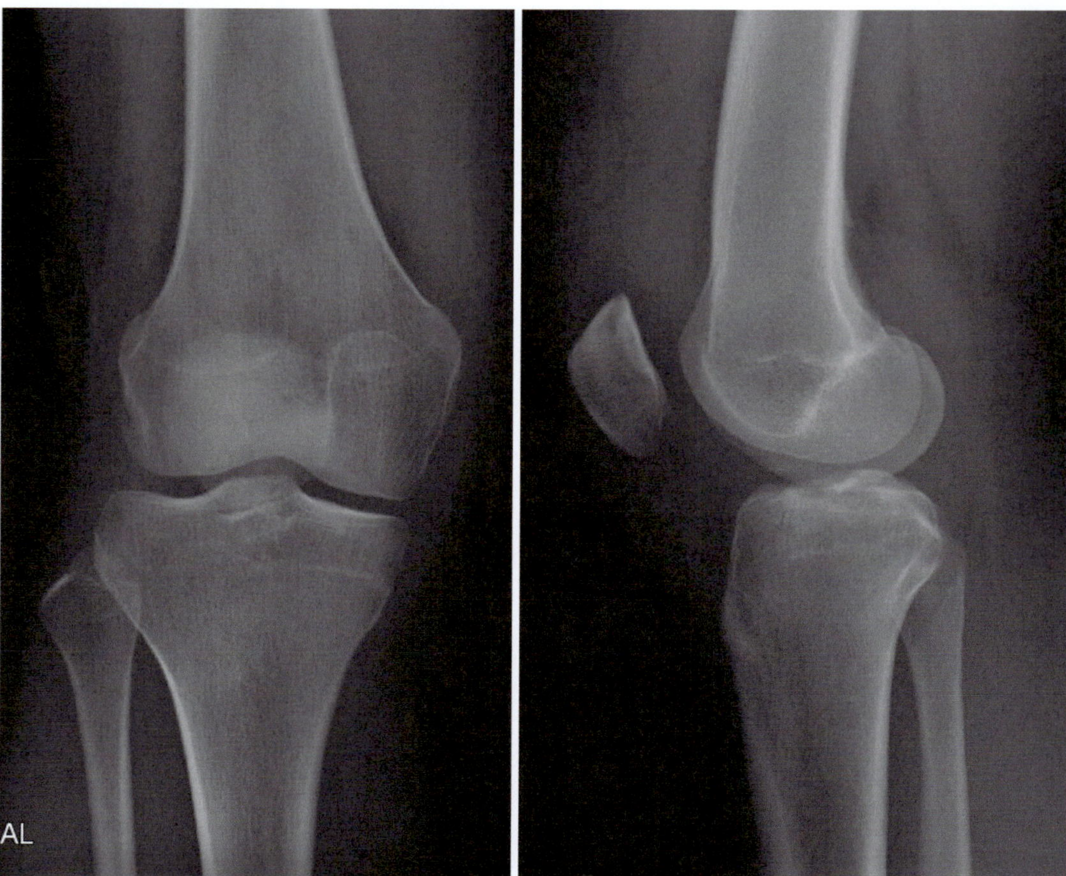

Fig. 9.1 Radiographs at the time of injury

9.1.4 Proposed Treatment

Surgery was necessary to fixate the osteochondral fracture, which was considered as significantly affecting the lateral femoral condyle articulating surface. In a case of primary patellar dislocation without an osteochondral fracture amenable for fixation, the proposed treatment would have been nonoperative. An arthroscopic evaluation was performed, and the osteochondral fracture was fixated arthroscopically with two bioabsorbable nails. As the patella was significantly unstable, a medial patellofemoral ligament (MPFL) reconstruction was performed simultaneously. The gracilis tendon was harvested and the free ends of the graft were placed into 3.5 mm blind-end tunnels at the medial patella. The graft was located between the layers two and three at the medial side of the knee and the loop of the graft was fixated at the femoral 5 mm tunnel. Tunnel position was confirmed to be anatomical with true lateral fluoroscopic view, and before fixation of the femoral side, MPFL graft behavior was checked to be anatomical not resulting in any graft tightening at any point of knee flexion cycle. Femoral fixation was performed with a 5 mm bioabsorbable interference screw.

9.2 Rehabilitation and Outcome

After surgery, free knee range of motion was allowed as tolerated by pain, and no brace was used. Full weight bearing was permitted in straight leg and partial weight bearing for 2 weeks for flexed knee. Crutches were necessary for 4 weeks, until pain and muscle control allowed controlled gait pattern. Full extension

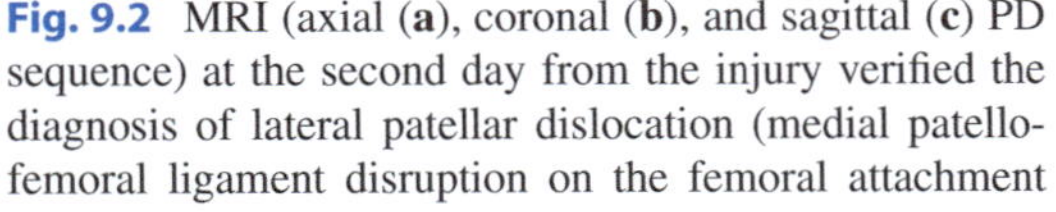

Fig. 9.2 MRI (axial (**a**), coronal (**b**), and sagittal (**c**) PD sequence) at the second day from the injury verified the diagnosis of lateral patellar dislocation (medial patellofemoral ligament disruption on the femoral attachment and lateral femoral condyle bone bruises) and revealed osteochondral fracture at the lateral femoral condyle. No trochlea dysplasia or patella alta was seen

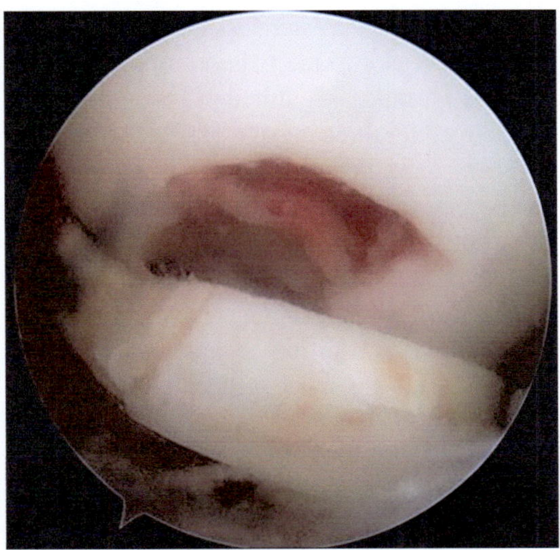

Fig. 9.3 Arthroscopic view of the displaced lateral femoral condyle osteochondral fracture. Size of the fragment was 9 mm of width and 18 mm anteroposterior length

should be gained as soon as possible and 90 degrees flexion at 3 weeks. The patient could walk normally at 4–5 weeks and return to near-normal daily activities including stairs by 6–8 weeks. He could run at 3 months and return to play soccer at 5 months. In follow-up, the osteochondral fracture healed and the patient had no complaints, pain, or instability at 6 months postoperatively.

Take-Home Message

In a case of first-time patellar dislocation, the medial stabilizing structures, MPFL being the most important, have evident, though limited, spontaneous healing capacity. Nonoperative treatment is therefore a viable option, if no significant concomitant injuries or risk factors for recurrent instability exist. When load-bearing articular cartilage has fractured, the fracture should be fixated if possible to preserve articulating joint surface and simultaneous patellar stabilizing surgery (MPFL reconstruction) performed.

Key Points

1. First-time patellar dislocation results in MPFL injury and osteochondral fractures are frequently seen.
2. MRI is necessary to confirm the diagnosis and to assess any concomitant osteochondral fractures.
3. Indication for surgery is an osteochondral fracture amenable for fixation.

9.3 Perspective

Julian Feller

jfeller@osv.com.au

9.3.1 First-Time Dislocation with OC Fragment

This is an interesting case, not only because it highlights important factors in decision-making, but also because some aspects are a little unusual. The important factors in the decision-making process are the patient's age, the contact mechanism of injury, the lack of risk factors for recurrent patellar dislocation, and the presence of a displaced osteochondral fragment.

Both the contact mechanism and the absence of risk factors put this patient at low risk of recurrent patellar instability, probably in the order of less than 15%, based on the findings of a recent systematic review [1].

The displaced osteochondral fragment raises the question of surgical intervention. Given the location of the fragment on MRI scans (anterolateral) it needs to be addressed as it is likely to become symptomatic. However, it is not uncommon for an osteochondral fragment to become adherent to the lateral aspect of the lateral femoral condyle and not cause any symptoms. This can be seen at arthroscopy as part of a patellar stabilization procedure in the setting of recurrent patellar instability.

Once a decision has been made to perform an arthroscopy, consideration needs to be given to whether the fragment is suitable for reduction and fixation. In this case, the single arthroscopic view suggests an unusual site for the osteochondral lesion in that anteriorly it seems quite central on the lateral femoral condyle in a mediolateral plane. Typically, these lesions are lateral and involve the lateral margin of the condyle.

When deciding whether to retain or remove such a fragment, an assessment of the potential for healing needs to be made. On the positive side, this patient is relatively young, but the fragment appears to be essentially chondral with no or minimal bone on the deep surface, at least anteriorly. This reduces the chance of it healing. The size of the lesion probably tips the balance in favor of an attempt at repair. It would not need to have been much smaller for me to have removed it.

Accessing these lesions arthroscopically can be challenging. The more anterior they are, the harder it is. The site reflects the degree of knee flexion at the time of patellar dislocation/reduction with greater knee flexion leading to a more posterior lesion. I am impressed that in this particular case the fragment could be reduced and fixed arthroscopically. It is not uncommon to need a limited anterolateral arthrotomy to achieve this. With regard to the fixation used, this is the same as I would use, although I would perhaps have aimed for three nails.

Then comes the question of whether a medial patellofemoral ligament (MPFL) repair or reconstruction is warranted. I think there is general agreement that if the MPFL is to be addressed, the results of reconstruction are superior to repair or plication, even in the acute situation. In this case, the surgeon made a decision to proceed with an MPFL reconstruction. I think a reasonable case can be made either way.

On the one hand, the patient is young and the osteochondral fragment needs to be addressed surgically, so that the addition of an MPFL reconstruction might be seen to be relatively small in terms of affecting the recovery from surgery. A risk of recurrence following an MPFL reconstruction in this setting is very low and less than the assumed risk of 15% or less.

On the other hand, a <15% risk of recurrence can be seen as relatively low, particularly as the injury mechanism was contact, in keeping with the absence of risk factors for recurrence. In other words, without contact, this patella was very unlikely to dislocate in the first place. Without the osteochondral injury, this patient would most likely have been treated nonoperatively. As such, I would have strongly considered limiting my surgery to addressing the osteochondral fragment.

9.4 Perspective

Laurie A. Hiemstra

Dr. Sillanpää has presented a relatively straightforward case of a first-time patellofemoral dislocation in a young athlete with no significant pathoanatomic risk factors. This case raises several interesting issues, more related to pathways of care than to the final treatment of this athlete. Early assessment of knee injuries, with early imaging when necessary, is the crucial step that allows for optimal patient care, especially those patients that may have time-sensitive repairable osteochondral lesions.

The principal issue in this case is the early diagnosis and imaging of this young patient to allow for optimum surgical care with fixation of the cartilage injury and restoration of knee stability. This young athlete was assessed the day after his injury, an example of an excellent care pathway for an acute knee injury. In many countries, this access to an appropriate practitioner such as a sport medicine physician or orthopedic surgeon that can provide a definitive diagnosis may take weeks to months. This delay decreases the chances that a potentially fixable osteochondral fragment will remain fixable. It also introduces significant deconditioning to the athlete, increased time of missed play, and delayed return to sport.

Access to early MRI was the other key to this successful outcome. The early MRI allowed the surgeon to rule out an ACL injury which would be a common differential diagnosis with this history. The MRI confirmed the existence, location, and size of the osteochondral fracture giving the evidence to proceed with emergent surgical management. MRI is highly sensitive to detecting osteochondral injury after patellofemoral dislocation and in children and adolescents these lesions are reported in 34–62% of patients [2, 3]. Early MRI allows for the appropriate management of any associated osteochondral injury avoiding the challenges related to addressing a neglected (undiagnosed) fracture. MRI has the added benefit in patients with patellofemoral dislocation of allowing for the assessment of pathoanatomic risk factors, which may alter surgical management [4]. In many countries, MRI may not be available in a timely fashion or may not be affordable for all patients. Further research into the cost and clinical outcome benefits of emergent MRI in a first-time patellofemoral dislocation to identify the repairable chondral lesion may lend evidence to push for early imaging in this patient population.

Dr. Sillanpää's treatment was ideal for this patient. The primary goals in treating this young athlete are to return him to full function, to prevent any recurrent dislocation, and to have minimal issues in this knee in the future. It is known that patellofemoral dislocation is associated with osteoarthritis later in life [5]. Primary repair and healing of an osteochondral fracture will provide the best outcomes, conceivably saving multiple surgeries should the chondral lesion remain symptomatic or progress to further degeneration. Early return to sport with minimal deconditioning will optimize the patient's sport performance and recreational activities. Most experts would agree with treating the first-time patellofemoral dislocation nonoperatively in the setting of relatively normal pathoanatomy and no repairable osteochondral fracture. Individual circumstances may influence toward earlier stabilization. If the patient is already having an anesthetic, surgery, and rehabilitation to address the osteochondral fracture, the benefit of adding an MPFL reconstruction is potentially significant with minimal potential harm.

References

1. Huntington LS, Webster KE, Devitt BM, Scanlon JP, Feller JA. Factors associated with an increased risk of recurrence after a first-time patellar dislocation: a systematic review and meta-analysis. Am J Sports Med. 2020;14(10):2552–62.
2. Seeley MA, Knesek M, Vanderhave KL. Osteochondral injury after acute patellar dislocation in children and adolescents. J Pediatr Orthop. 2013;33(5):511–8.
3. Zaidi A, Babyn P, Astori I, et al. MRI of traumatic patellar dislocation in children. Pediatr Radiol. 2006;36(11):1163–70.
4. Balcarek P, Ammon J, Frosch S, et al. Magnetic resonance imaging characteristics of the medial patellofemoral ligament lesion in acute lateral patellar dislocations considering trochlear dysplasia, patella alta, and tibial tuberosity-trochlear groove distance. Arthroscopy. 2010;26(7):926–35.
5. Sanders TL, Pareek A, Johnson NR, et al. Patellofemoral arthritis after lateral patellar dislocation: a matched population-based analysis. Am J Sports Med. 2017;45(5):1012–7.

10 The Patellofemoral Joint: A Case-Based Approach

Ashraf Abdelkafy, John P. Fulkerson, and Ryosuke Kuroda

10.1 Introduction

The incidence of acute patellar dislocation is 5.8 per 100,000. It increases to 29.0 per 100,000 in the age group 10–17 years [1, 2]. Recurrence rate ranges from 15 to 44% after nonsurgical management [2] and 58% of patients continue to experience pain and mechanical symptoms after the initial dislocation incident and 55% fail to return to pre-dislocation sports activity [3].

10.2 History and Examination

Several patellar stability tests are available. The most important are: (1) the patellar glide test where translation to less than two quadrants is considered normal while two or more quadrants indicate laxity of the medial patellar restraints. When this is not bilateral, related to hyperlaxity and is associated with acute injury, it represents an injury to the MPFL [4]. *(2) The patellar apprehension test which is performed with the knee in full extension and a lateral force is applied to the patella. The test is positive when the patient verbally reports apprehension and pain or quadriceps muscle contraction resisting lateral patella translation. In the previous two tests, more important than the absolute translation is the comparison to the opposite uninjured side, whether there is a firm or soft endpoint, and the presence of an apprehension sign* [5].

A 21-year-old male ice hockey player sustained traumatic dislocation of his right patella when checked into the boards 5 months previously. At that time, the patella spontaneously relocated 15 minutes after the initial event but knee swelled and return to play was impossible despite intensive physical therapy after an initial 5 days of rest with compression. Upon attempts at returning to vigorous activity, the patella felt unstable with notable clicking and shifting, even with a brace on. The patient had successful return to sport after prior anteromedial tibial tuberosity osteotomy (TTO) on the contralateral knee.

10.3 Imaging

Radiographs [6] *in the form of Standard AP, lateral in 30° flexion and axial views are mandatory. AP view must be taken with patient in standing position thus the relationship of the patella to the femur will be clear and the medial and lateral joint space narrowing may be detected* [7]. *The lateral view*

A. Abdelkafy
Faculty of Medicine, Suez Canal University, Ismailia, Egypt
e-mail: ashraf.abdelkafy@med.suez.edu.eg

J. P. Fulkerson (✉)
Yale University, New Haven, CT, USA
e-mail: patelladoc@aol.com

R. Kuroda
Orthopaedic Surgery, Kobe University Graduate School of Medicine, Kobe, Japan
e-mail: kurodar@med.kobe-u.ac.jp

© ISAKOS 2022
J. L. Koh et al. (eds.), *The Patellofemoral Joint*, https://doi.org/10.1007/978-3-030-81545-5_10

must be taken in the lateral decubitus or standing with the knee flexed to 30° to place the patellar tendon under tension, and also with the posterior femoral condyles overlapped. From the lateral view, the patellar morphology, thickness, height [Caton-Deschamps (CD) and Insall-Salvati indices (IS)], and trochlear dysplasia can be assessed. Axial view is performed tangentially with the knee flexed 30°. In the axial view, the congruence angle may be measured and lateral patellar subluxation and tilt may be identified [8]. *CT scan is very important as it allows a true axial view of the patellofemoral joint in addition to the ease of patient positioning. With CT we can assess the TT-TG (tibial tuberosity trochlear groove) distance, lateral patellar tilt angle, and the grade of trochlear dysplasia. MRI is mandatory to assess the medial soft tissue restrains of the patellofemoral joint, the medial patellofemoral complex (MPFC) as well as the articular cartilage of the patellofemoral joint.*

30 degree knee flexion axial view of our patient demonstrated lateral translation of the right patella but no bony lesion (Fig. 10.1). MRI imaging showed evidence of distal medial patellar articular cartilage damage grade 3 with lateral subluxation. The lateral trochlea was also damaged with marginal articular disruption grade 3. TT-TG measurement was 18 mm, Caton-Deschamps index measured 1.2. Dejour B proximal trochlea (the available image in Fig. 10.1 shows more distal trochlea at 30 degrees knee flexion) No orthogonal views were available.

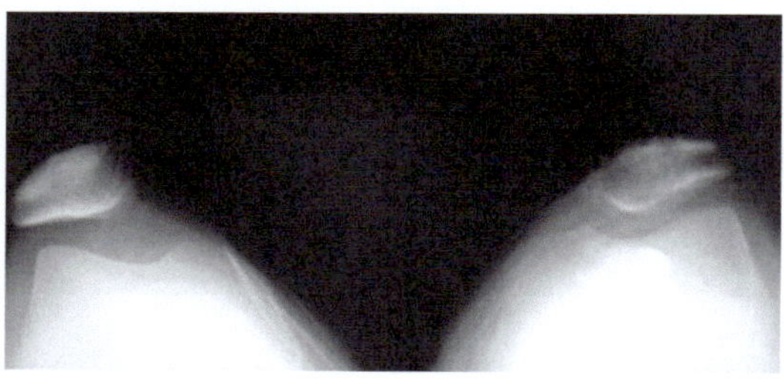

Fig. 10.1 Shows lateral translation of right patella

10.4 Diagnostic Arthroscopy

Arthroscopy after failed rehabilitation (recurrent feelings of instability but no more complete dislocation) revealed distal medial patellar and lateral trochlea articular lesions, evidence of lateral tracking, and medial patellofemoral complex (MPFC) interstitial disruption (Fig. 10.2a, b).

10.5 Options for Management

1. Resume rehabilitation and non-operative care after debridement of the lesion.
2. Isolated MPFC reconstruction [MPFL (medial patellofemoral ligament. reconstruction) or MQTFL (medial quadriceps tendon femoral ligament)].
3. MPFC reconstruction with lateral release or lengthening.
4. Medial TTO and MPFC reconstruction.

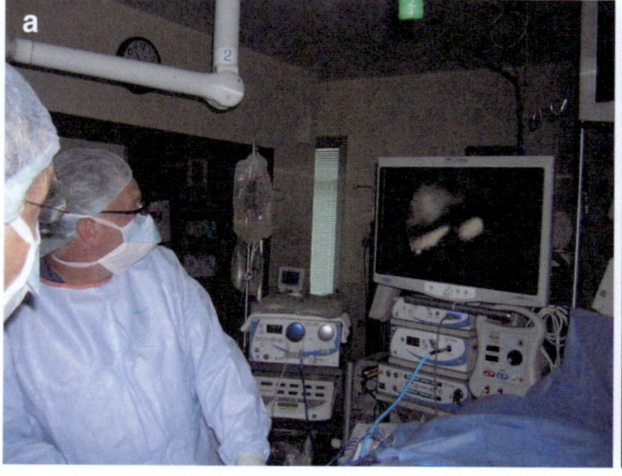

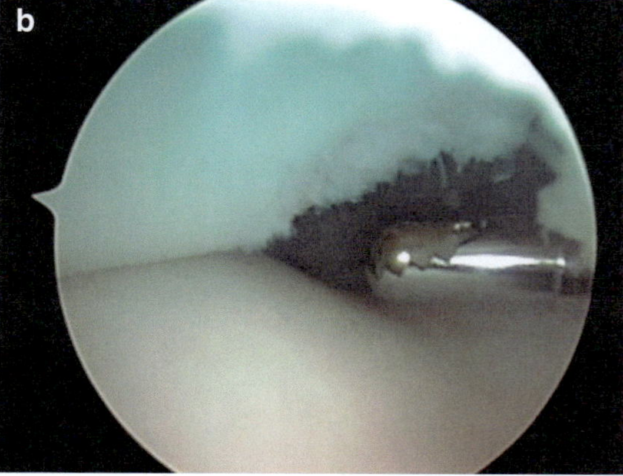

Fig. 10.2 (**a**) Shows diagnostic arthroscopy. (**b**) Distal medial patellar articular lesions

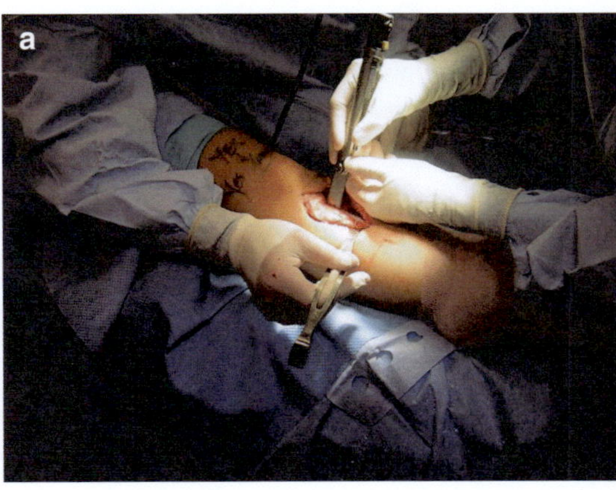

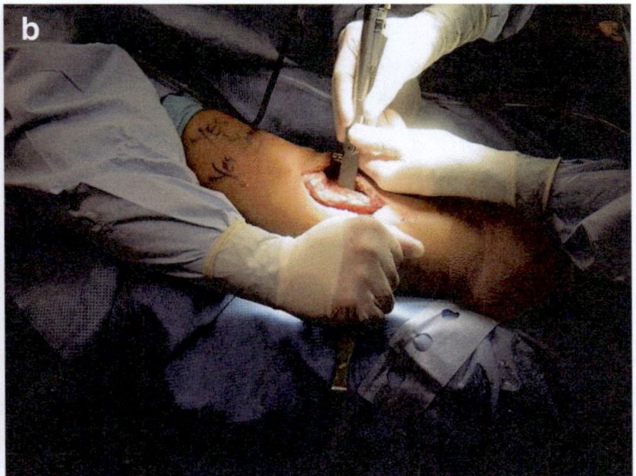

Fig. 10.3 (**a**) shows anteromedialization of the tibial tuberosity. (**b**) shows anteromedialization of the tibial tuberosity. Note the oblique orientation of the osteotome

5. OAT (osteochondral autograft transfer) or ACI (autologous chondrocyte implantation) resurfacing and MPFC reconstruction.
6. Isolated anteromedial TTO.
7. Anteromedial TTO with MPFC reconstruction.
8. Anteromedial TTO with ACI.
9. Trochleoplasty alone.
10. Trochleoplasty with MPFC reconstruction.
11. Trochleoplasty with ACI and anteromedial TTO.

10.6 Summary of Above Finding

1. *Torn MPFL.*
2. *Distal medial patellar articular lesion.*
3. *Lateral trochlear articular lesion.*
4. *Lateralization of the patella.*
5. *TT-TG 18 mm.*
6. *CD index 1.2.*
7. *Trochlear dysplasia Dejour type B.*

10.7 Definitive Management "1": Abdelkafy and Fulkerson

We choose option number 7: Anteromedial TTO with MPFC reconstruction (anteromedialization plus MPFL reconstruction) (Fig. 10.3a, b).

10.8 Definitive management "2": Kuroda

I choose option number 2: Isolated MPFC reconstruction [MPFL or MQTFL].

10.9 Discussion of definitive management "1": Abdelkafy and Fulkerson

Medialization of the tibial tuberosity moves the contact pressures of patella medially. So, in lateral patellar tracking, this improves the overall congruity and thus balances the overall contact pressures while decreasing lateral instability. However, in normal patellar tracking or in over-medialization, it will interrupt a congruous joint and thus increase the overall contact pressure focal loading and eventual iatrogenic medial overload with increase in medial forces in the medial compartment. On the other hand, anteriorization of the tibial tuberosity decreases joint reaction forces. A 10–15 mm anteriorization decreases PF stresses by 20%. While 20 mm reduces the patellofemoral compressive forces by approximately 50%. Most patients require no more than 10 mm of anteriorization, which is readily attainable with the anteromedialization (AMZ) procedure.

Combination of anteriorization & medialization (AMZ) in patients with lateral position of tibial tubercle (TT-TG > 20 mm) + PF instability + lat-

eral chondral or distal chondral lesions decreases lateral facet pressure and also decreases overall patellofemoral contact stresses by shifting contact area proximally & medially & improves PF tracking. AMZ tips-up the inferior patella, which unloads the distal patellar chondral lesion.

Once the AMZ has been completed, the surgeon should determine if restoration of MPFC [8] *is necessary by MPFL or MQTFL reconstruction. ACI is not usually necessary after unloading and rebalancing of articular tracking. Trochleoplasty is not generally necessary as the AMZ procedure gets the patella to the trochlea earlier in flexion so that the patella may engage the deeper trochlea sooner.*

Isolated MPFC reconstruction would likely prevent redislocation and might be the procedure of choice for many surgeons since it is less surgery. However, the question of whether to add AMZ is largely based on the surgeon's assessment of the articular lesions and whether unloading of these will add benefit to the outcome. It is a reasonable option given the elevated TT-TG of 18 (less than 20 mm). Examination of the other knee and establishing baseline tracking pattern would help in this decision as well, as a J-sign or defined lateral tracking of the other knee would increase the need for AMZ. In cases of this nature, it may be wise to obtain consent for MPFC reconstruction and AMZ also so that the surgeon will have the advantage of intra-operative assessment before deciding which way to go with the surgery.

10.10 Discussion of definitive management "2": Kuroda

Previous studies have shown that young patients are at high risk for having recurrent dislocation compared with adult patients [9, 10]. *Particularly, younger patients with other risk factors such as trochlear dysplasia, patellar alta, and increased TT-TG have significantly high odds ratios of having redislocation after the first-time dislocation* [11, 12] *as seen in this patient. In the preoperative assessments for the presented patient, the risk/predisposing factors need to be considered are the Dejour B proximal trochlea and the TT-TG distance of 18 mm. Other osseous abnormalities including mal-rotational and valgus limb alignment were not available.*

Considering the surgical invasiveness, MPFL reconstruction would be the most reasonable choice. Although the shape of the trochlea is flat in the proximal part, the distal part of the sulcus in flexion is maintained. Therefore, the degree of the dysplasia seems to be not severe if MPFL reconstruction is performed. In addition, we obtained favorable outcomes even in patients with increased TT-TG of >20 mm after isolated MFPL reconstruction [13]. *Therefore, TT-TG of 18 mm would not be an indication for tibial tuberosity transfer when performing MPFL reconstruction. At least for the current patient. Therefore, I recommend isolated MPFL reconstruction. During surgery, patellar tracking and patellar tightness should be checked and additional lateral release may be needed.*

Fact Box 1

Combination of anteriorization & medialization (AMZ) in patients with lateral position of tibial tubercle (TT-TG > 20 mm) + PF instability + lateral chondral or distal chondral lesions decrease lateral facet pressure and decrease overall patellofemoral contact stresses by shifting contact area proximally & medially & improve PF tracking and pain.

Fact Box 2

AMZ tips-up the inferior patella, which in turn unloads the distal patellar chondral lesion.

Fact Box 3

A 10–15 mm anteriorization decreases PF stresses by 20%. While 20 mm reduces the patellofemoral compressive forces by approximately 50%.

Fact Box 4

Isolated MPFL reconstruction is a reasonable option when the elevated TT-TG remains below 20 mm, no or mild trochlear dysplasia, normal limb alignment, normal patellar height and no lateral retinacular tightness.

Take Home Messages

1. Acute traumatic patellar dislocation has a higher recurrence rate after conservative treatment and more than half of the patients continue to have pain and fail to return to the pre-dislocation activities.
2. Radiographs, CT, and MRI are mandatory for the pre-operative assessment of every patellar instability case.
3. It is important to accurately diagnose articular cartilage injuries of the patella and trochlea as this will have a great impact on the plan of management as well as the final outcome.
4. AMZ in patients with lateral chondral or distal chondral lesions decreases lateral facet pressure and decreases overall patellofemoral contact stresses by shifting contact area proximally & medially. AMZ tips-up the inferior patella, which in turn unloads the distal patellar chondral lesion.
5. Isolated MPFL reconstruction is a reasonable option in TT-TG below 20 mm, no or mild trochlear dysplasia, normal limb alignment, normal patellar height and no lateral retinacular tightness.

References

1. Fithian DC, Paxton EW, Stone ML, Silva P, Davis DK, Elias DA, et al. Epidemiology and natural history of acute patellar dislocation. Am J Sports Med. 2004;32:1114–21.
2. Hawkins RJ, Bell RH, Anisette G. Acute patellar dislocations. The natural history. Am J Sports Med. 1986;14:117–20.
3. Atkin DM, Fithian DC, Marangi KS, Stone ML, Dobson BE, Mendelsohn C. Characteristics of patients with primary acute lateral patellar dislocation and their recovery within the first 6 months of injury. Am J Sports Med. 2000;28:472–9.
4. Kolowich PA, Paulos LE, Rosenberg TD. Lateral release of the patella: indications and contraindications. Am J Sports Med. 1990;18:359–65.
5. Zaffagnini S, Dejour D, Arendt E. Patellofemoral pain, instability and arthritis: clinical presentation, imaging, and treatment. Berlin: Springer; 2010.
6. Tjoumakaris F, Forsythe B, Bradley JP. Patellofemoral instability in athletes: treatment via modified fulkerson osteotomy and lateral release. Am J Sports Med. 2010;38(5):992–9.
7. Endo Y, Shubin Stein BE, Potter HG. Radiologic assessment of patellofemoral pain in the athlete. Sports Health. 2011;3(2):195–210.
8. Tanaka MJ, Tompkins MA, Fulkerson JP. Radiographic landmarks for the anterior attachment of the medial patellofemoral complex. Arthroscopy. 2019;35(4):1141–6. https://doi.org/10.1016/j.arthro.2018.08.052. Epub 2019 Jan 3.
9. Lewallen L, McIntosh A, Dahm D. First-time patellofemoral dislocation: risk factors for recurrent instability. J Knee Surg. 2015;28(4):303–9.
10. Christensen TC, Sanders TL, Pareek A, Mohan R, Dahm DL, Krych AJ. Risk factors and time to recurrent ipsilateral and contralateral patellar dislocations. Am J Sports Med. 2017;45(9):2105–10.
11. Arendt EA, Askenberger M, Agel J, Tompkins MA. Risk of redislocation after primary patellar dislocation: a clinical prediction model based on magnetic resonance imaging variables. Am J Sports Med. 2018;46(14):3385–90.
12. Sanders TL, Pareek A, Hewett TE, Stuart MJ, Dahm DL, Krych AJ. High rate of recurrent patellar dislocation in skeletally immature patients: a long-term population-based study. Knee Surg Sports Traumatol Arthrosc. 2018;26(4):1037–43.
13. Matsushita T, Kuroda R, Oka S, Matsumoto T, Takayama K, Kurosaka M. Clinical outcomes of medial patellofemoral ligament reconstruction in patients with an increased tibial tuberosity-trochlear groove distance. Knee Surg Sports Traumatol Arthrosc. 2014;22(10):2438–44.

11 Patellofemoral Case Study: Patella Instability, High TT-TG, Patella Alta + Trochlear Dysplasia in a Skeletally Mature Female

Jason L. Koh, Andrew Cosgarea, Phillippe Neyret, and Elizabeth A. Arendt

11.1 Case Presentation by Jason L. Koh

A 16-year-old girl presented with disabling bilateral patella instability to the office. Her dislocations occurred during activities of daily living. She was unable to run or jump. She was only able to descend stairs sideways. On physical examination she was a healthy-appearing young woman with a normal appearance, and was of above normal intelligence. Her thumb was able to bent to be parallel to the forearm and index finger could be extended to 80° angle with her forearm. She did not have elbow hyperextension. Knee range of motion was −3–145°.

She walked with a slightly crouched gait to limit lateral translation of the patellae.

The left knee had significant lateral translation/dislocation in extension and a large J sign, and had no endpoint to lateral translation. The patella could not be easily medially translated with the knee in extension. There was a positive apprehension sign. The tibial tubercle was very laterally displaced. The patella could be located with the knee in flexion but could easily be laterally translated with the knee flexed to 90°. There was no crepitus. The right knee had similar findings.

Previous treatment had included patella bracing and a recommendation to wait until the physes had closed.

Notably, there was a significantly positive family history. The patient's mother, aunt, and cousin all suffered from recurrent patella instability.

Lateral radiographs (Fig. 11.1a) demonstrated a crossing sign and significant patella alta with an Insall-Salvati and Caton-Deschamps ratio of 1:2. There was a proximal trochlear bump. Axial radiographs (Fig. 11.1c) demonstrated extremely dysplastic trochleas. The physes were closed.

CT scans showed significant dysplasia and patella alta (Fig. 11.2).

MRI (Fig. 11.3) was obtained and demonstrated MPFL insufficiency and significant troch-

J. L. Koh (✉)
Mark R. Neaman Family Chair of Orthopaedic Surgery, Orthopaedic & Spine Institute, Skokie, IL, USA

NorthShore University HealthSystem, Evanston, IL, USA

University of Chicago Pritzker School of Medicine, Chicago, IL, USA

Northwestern University McCormick School of Engineering, Evanston, IL, USA

A. Cosgarea
Department of Orthopaedic Surgery, The Johns Hopkins University School of Medicine, Baltimore, MD, USA
e-mail: acosgar@jhmi.edu

P. Neyret
Infirmerie Protestante, Lyon La Tour De Salvagny, France

E. A. Arendt
Department of Orthopaedic Surgery, University of Minnesota Twin Cities, Minneapolis, MN, USA
e-mail: arend001@umn.edu

© ISAKOS 2022
J. L. Koh et al. (eds.), *The Patellofemoral Joint*, https://doi.org/10.1007/978-3-030-81545-5_11

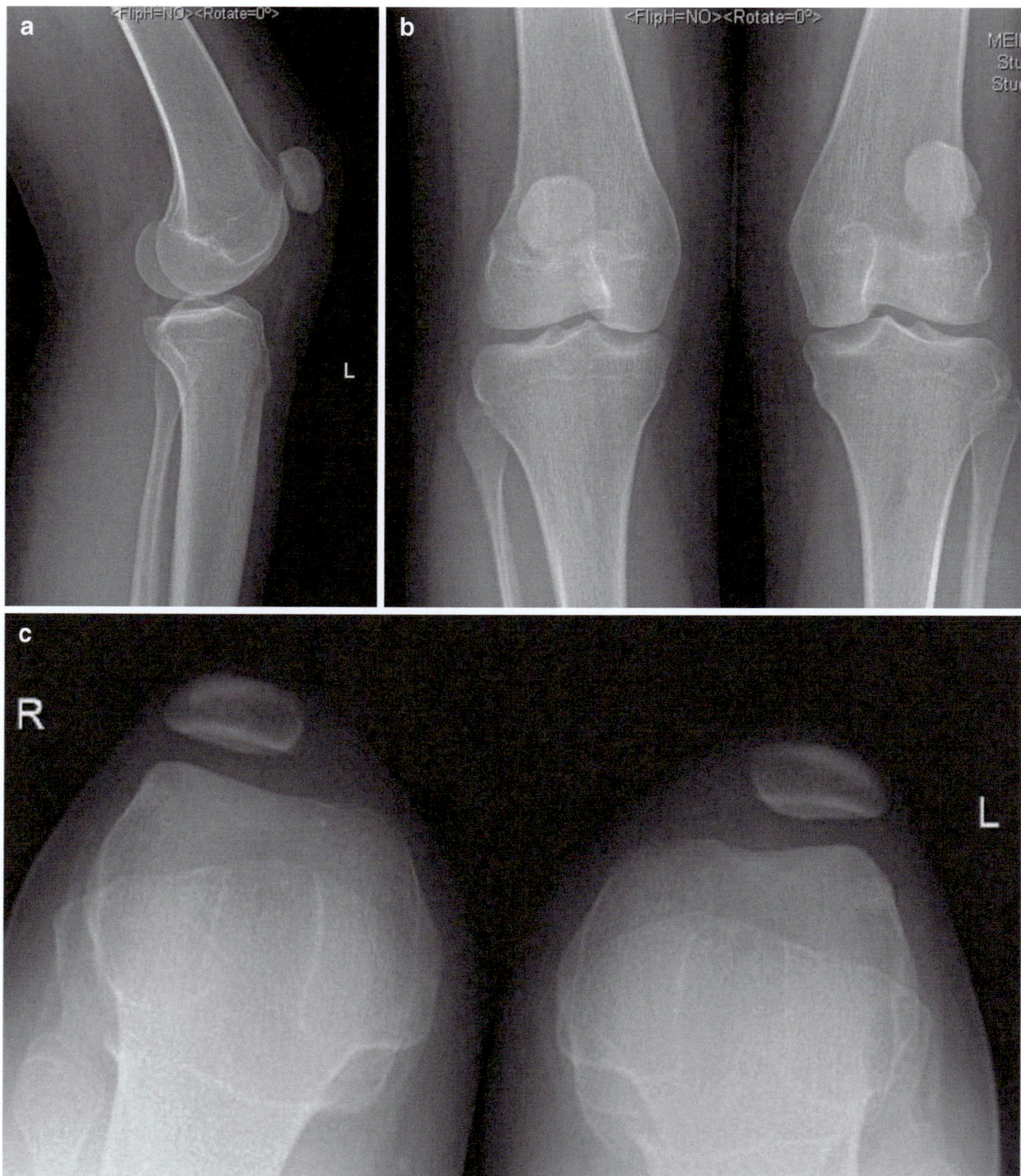

Fig. 11.1 (**a**) Lateral radiograph demonstrating patella alta, crossover sign, and proximal trochlear bump. (**b**) AP radiograph showing closed physes. (**c**) Axial radiograph showing extremely dysplastic trochleas

lear dysplasia with a bump. The tibial tubercle to trochlear groove (TT-TG) distance was 25 mm.

Given the history of disabling recurrent dislocation and failure of nonsurgical management, surgical management was proposed. Initially, treatment consisted of lateral retinacular lengthening, double limb allograft medial patellofemoral ligament reconstruction, and tibial tubercle distalization and medialization by approximately 1 cm medial and 1 cm distal [1]. The patient had improved function but continued to have a significant J sign and pain.

Given the continued discomfort, a decision was made to perform a trochleoplasty. For preop-

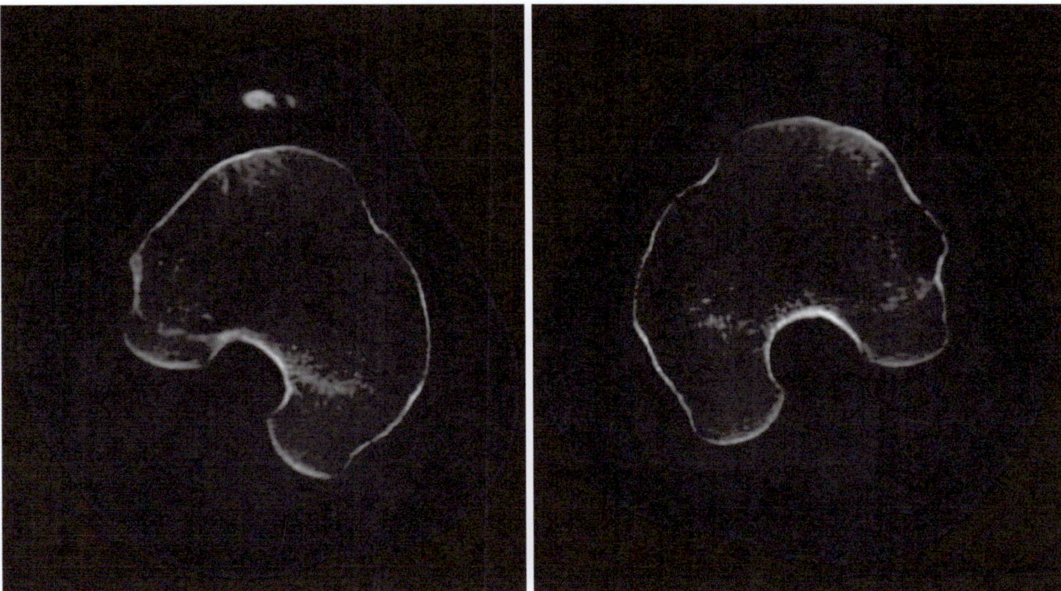

Fig. 11.2 CT Scan of bilateral knees

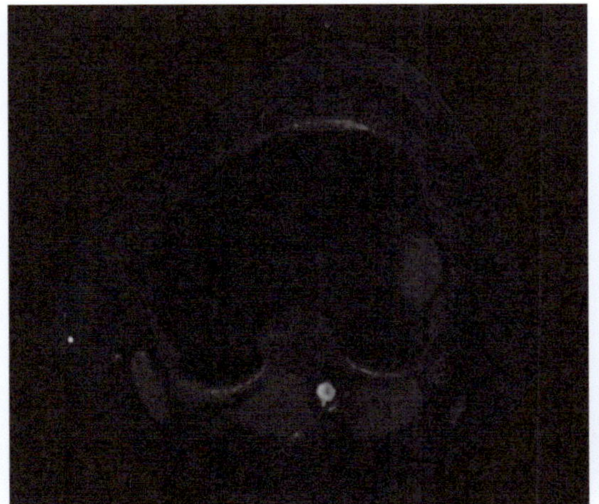

Fig. 11.3 MRI

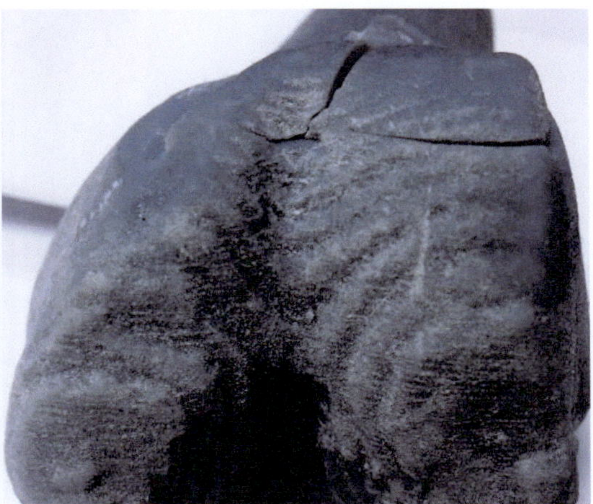

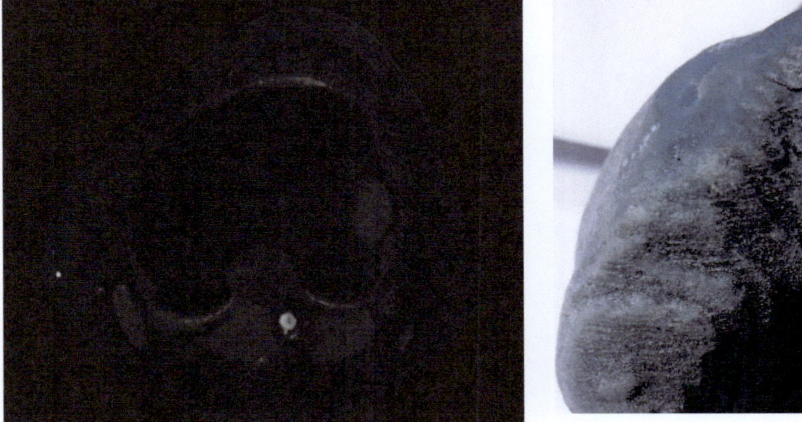

Fig. 11.4 3-D printed model of the left knee

erative planning, a CT scan was converted into a 3-D printed model (Fig. 11.4). Surgery was simulated on the model using a saw to create planned cuts and resect a segment of distal femur under the flaps. A deepened trochlear groove that matched the contour of the patella was able to be created.

The patient underwent thick flap trochleoplasty [2]. The proximal trochlea was exposed (Fig. 11.5a) and a saw was used to raise a thick osteochondral flap proximally. A saw and burr were used to remove the proximal bump and deepen the trochlear groove underneath the flap (Fig. 11.5b). The appropriate alignment of the trochlea was identified and a #10 blade was used to split the osteochondral flap. The fragments were then secured using countersunk headless metal screws.

Postoperatively immediate range of motion was initiated and partial weightbearing with the knee in extension for 4 weeks followed by progression to full weightbearing. Her postoperative course was unremarkable and she had elimination

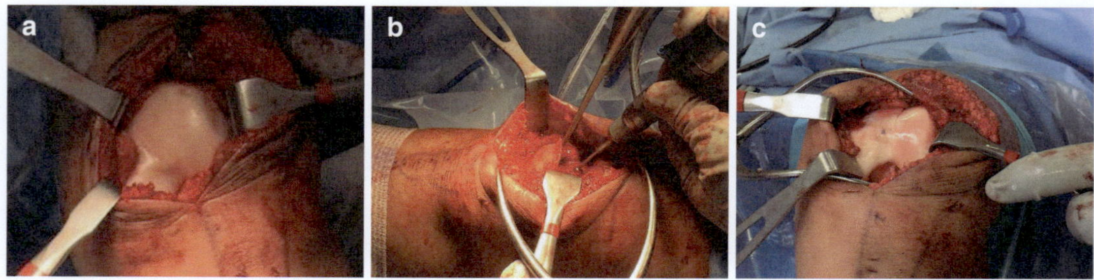

Fig. 11.5 (**a**) Exposure of the trochlea. Note significant dysplasia. (**b**) Elevation of osteochondral flap and creation of deeper groove. (**c**) Trochlear groove after reduction of osteochondral flaps

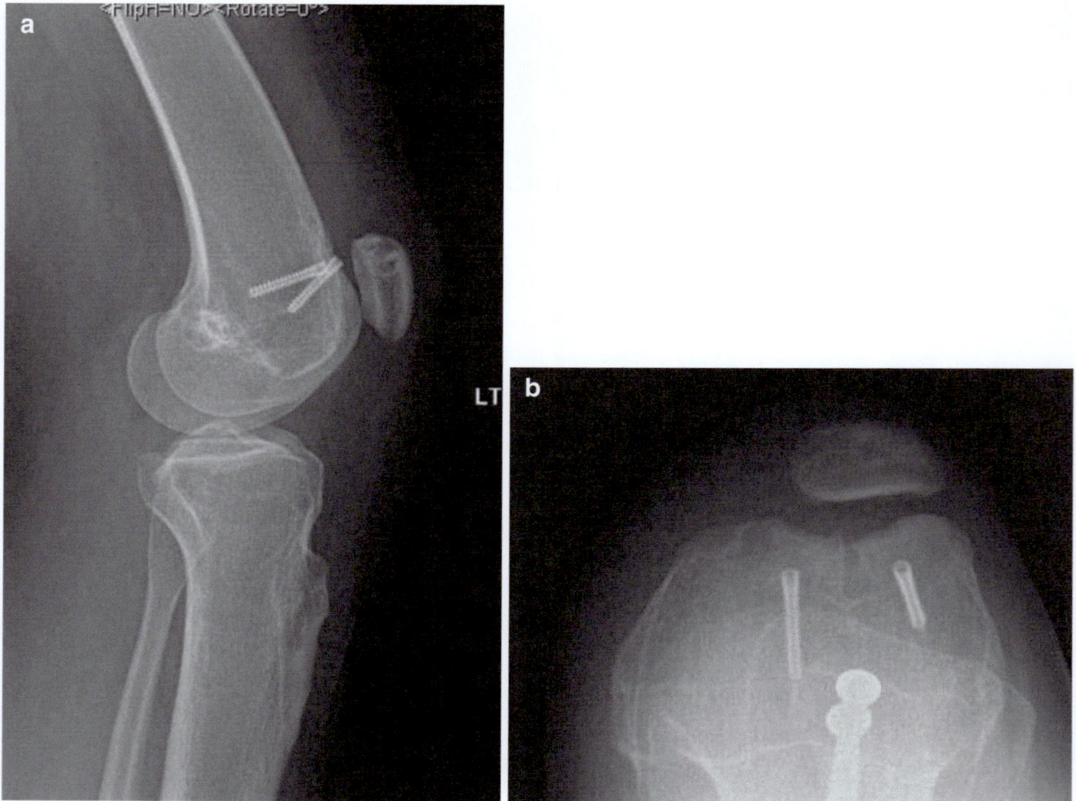

Fig. 11.6 (**a**) Lateral radiograph showing elimination of bump and deepened trochlea. (**b**) Axial radiograph showing deepened trochlea

of the J sign. Postoperative radiographs (Fig. 11.6a, b) demonstrate improved trochlea, elimination of the bump, and no crossing sign. There was some residual alta (1:1.3).

Second look arthroscopy 2 months later demonstrated healing of the osteotomy, and good congruity and fill of the central defect with fibrocartilage (Fig. 11.7).

She had no complaints and had been able to return to sports.

She had no complaints and had been able to return to sports.

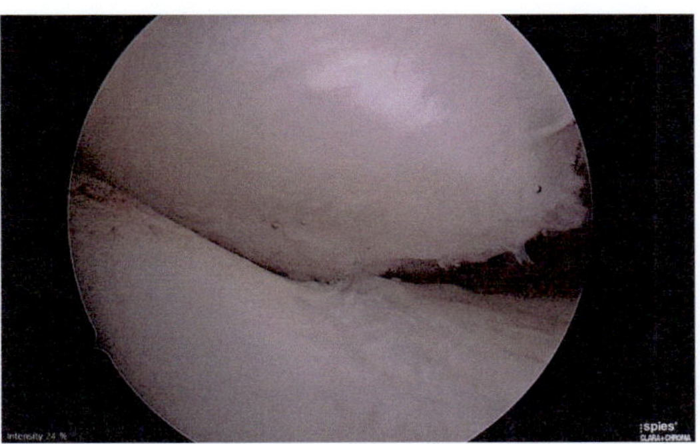

Fig. 11.7 Arthroscopic view of trochlea

11.2 Commentary by Andrew Cosgarea

After reviewing the case presentation and the images, I would have followed the same treatment algorithm, attempting first to treat her with MPFLR and TTO. My technique would have been slightly different but not substantially.

Based on her history, physical examination, and radiographic work-up, her clinical problems include:

1. malalignment (TT-TG = 25 mm)
2. patella alta (CDI = 2.0)
3. trochlear dysplasia (crossing sign, trochlear bump)
4. medial insufficiency (recurrent instability, positive apprehension sign, soft endpoint on medial translation)
5. hyperlaxity (upper extremity joint examination)

After failure of nonoperative treatment I would have recommended patellar stabilization surgery to include both a tuberosity osteotomy and MPFL reconstruction.

1. tibial tuberosity anterior medial distalizing osteotomy: I would have medialized the tuberosity 12–13 mm in order to correct the TT-TG distance to approximately 15 mm. I have found that a slight overcorrection is necessary due to slight shifting of the tibia laterally following medialization surgery. I would also distalize the tibial tuberosity the number of millimeters necessary to change the CDI from 2.0 to 1.2. That is likely approximately 20 mm, but cannot measure that distance directly on these images. Finally I would have cut the shingle with an approximately 30 degrees slope in order to compensate for the settling that occurs with screw compression, as well as the relative posteriorizing effect that occurs with shingle distalization. Although this osteotomy does not directly address the pathoanatomy associated with trochlear dysplasia, it is usually sufficient to compensate for all of the pathoanatomy and prevent further instability episodes. I would have also performed a modest lateral retinacular release at the time of the osteotomy, but I have no disagreement with a lateral retinacular lengthening.
2. after completing the TTO, I would have performed an MPFL reconstruction using single strand semitendinosus allograft. Because of her hyperlaxity, I believe allograft is preferable to autograft in her case.

If she developed recurrent instability despite normalization of her patellar height and TT-TG distance, I would have recommended trochleoplasty as a salvage procedure. I have found that most of the time a "J-sign" is minimally symptomatic or asymptomatic and does not require specific treatment, unless it is associated with instability episodes. It might have been reasonable to perform a trochleoplasty as a primary operation in her specific case, although it would have needed to be performed with concomitant tuberosity anteromedial distalization osteotomy and MPFL reconstruction.

11.3 Commentary by Phillippe Neyret

There is severe PF instability with several episodes of patella dislocations (episodic patella dislocations).

The analysis of the lateral X-Ray shows that there is a double contour sign and crossing sign with a prominence (a bump). The crossing sign is distal; it means the trochlea is flat or convex also "distally." Moreover, one can see the patella is alta. I cannot measure the Caton-Deschamps index. But one can see on the CT scan that the patella is not visible on the Roman arch level slice (patella is above) and it means the patella is alta.

The measurement of the TT-TG is not really reliable when the patella is positioned on the lateral part of the trochlea due to the increased external rotation of the knee.

TT-TG reflects not only the lateral implantation of the ATT but also the external rotation of the femorotibial joint.

In such a situation, with these morphological anatomical abnormalities I would have proposed a trochleoplasty during the first surgery.

The 3-D model is very interesting, but I believe the correction did not go far distally and the distal part of the trochlea was not corrected. We can still see the crossing sign on the post-op lateral view.

I would also distalize the tibial tubercle according to the Caton-Deschamps index. I would also medialize, but not too much if we consider the external rotation of the knee joint is reduced when the patella is in a better position, and the position of the trochlea groove is further modified by the trochleoplasty.

I suspect that a maltorsion of the lower limb or a frontal malalignment was ruled out, but this would be important to assess.

I would also assess the length of the patella tendon in order to know if a functional shortening of patella tendon is mandatory. I would recommend a "tenodesis" of the tendon into the proximal osteotomy site if the tendon is more than 52 mm in length.

11.4 Commentary by Elizabeth Arendt

It is always challenging when one tries to give clinical insight on a patient when they do not have the advantage of examination. However, in this case salient points for me are the following:

1. Patient has high grade Trochlear Dysplasia (TD) with a J-sign that is disabling. I say this based on her habitual crouched gait to keep her kneecap from dislocating in full extension, and her lack of knee confidence with descending stairs.

   Although we have not yet been able to characterize J signs with any meaningful clarity, (although several are investigating this and we are getting closer to a unified system), I classify J signs as hard or soft based on exam features.

   A ***hard J sign*** is when the patella literally jumps into the groove; typically this happens in deeper flexion. When patients are younger this can be without crepitus, but it can become associated with positive crepitus even in the young, and then it becomes less tolerable and signifies cartilage wear.

   A ***soft J sign*** is one that has a smooth transition into the groove, often without crepitus, and typically relocates in early flexion. It is this latter type of J sign that tends to be better tolerated by the patient.
2. The patient does have hyper laxity based on her upper extremity exam; she does have mild knee hyperextension but not to the extent to merit the knees being included in a hyperlaxity scale. Nonetheless this can aggravate her patella alta dynamically with knee hyperextension and quad contraction resulting in more superior migration of the patella. Since she already has a crouch gait, I do not think this is relevant for her.
3. Based on the knee exam standing images (the walls of the notch are not in an AP orientation), and to some degree the slice imaging and her history, I would be suspicious that there might be an element of limb version that is associated with this constellation of features. This should be included in the physical exam by at a minimum hip ROM. The limb version that could play an important role in a J sign is increased femoral anteversion. I would assess this with imaging (CT in my institution at this time) prior to making a concrete surgical decision.

4. I do not think that the absolute distance of the TT-TG has any relevance in this patient. Her sulcus is extremely medialized due to her TD, hence the distance is elevated due to her proximal sulcus position more than her lateral tubercle offset. Once could use TT-PCL [3], but in my hands if I make the decision to distalize, I would medialize her tubercle to reach a tubercle sulcus angle of zero, i.e., position her tubercle in the midline of the femur at 90° of knee flexion. To medialize her based on measurement distances (i.e.) trying to correct 25 mm to something <15 mm, it will likely over medialize her tibial tubercle and this could result in increasing medial tibiofemoral forces as well as medial patellofemoral forces.
5. As for patella alta, I would reduce her tibial tubercle 15 mm as maximum distance of distalization. She does have significant patella alta but she also has a fairly deep groove once the groove starts. It is not shallow throughout and therefore any degree of patella-trochlear engagement will likely be sufficient.
6. I will admit that I typically advocate for one bony procedure for patient, either a trochleoplasty or a tibial tubercle osteotomy, ideally. If I felt that I did not have the skill set to perform trochleoplasty or were reluctant to do one, I would likely do a tibial tubercle distalization of 15 mm, medialization to a tubercle sulcus angle of zero, and perform an MPFL reconstruction using allograft tissue. I would evaluate lateral tightness intra-operatively and lengthen as needed; she will likely need lateral lengthening or release.
7. If one was not to perform a trochleoplasty I would consider a grooveplasty and remove the super trochlear bump. One has to be careful because a grooveplasty does shorten the groove and this would not be advantageous in someone who will unlikely have her patella height indices completely corrected.
8. With the above thoughts in mind, for this patient, if femoral anteversion was <35°, I would advise a trochleoplasty, as well as a distal tibial tubercle osteotomy with the parameters stated above, combined with MPFL reconstruction and lateral lengthening as deemed necessary by examination on the table.

Take Home Message

Multiple factors need to be taken into consideration when attempting patella stabilization in the context of severe bony abnormalities. In some cases, significant bony work, including trochleoplasty, may be necessary or appropriate in the context of substantial trochlear dysplasia. This is a less commonly performed procedure, so the use of 3-D printing to help visualize and simulate the surgery may reduce risk of iatrogenic injury.

Fact Box

Trochlear dysplasia is a significant risk factor for recurrent instability after surgery.

Patella alta is a significant risk factor for recurrent instability after surgery.

Trochleoplasty can be a successful technique that addresses severe trochlear dysplasia.

3-D printing can assist in visualizing severe trochlear dysplasia.

References

1. Koh JL, Stewart C. Patellar instability. Orthop Clin North Am. 2015;46(1):147–57.
2. Dejour D, Saggin P. The sulcus deepening trochleoplasty-the Lyon's procedure. Int Orthop. 2010;34(2):311–6.
3. Heidenreich MJ, Camp CL, Dahm DL, Stuart MJ, Levy BA, Krych AJ. The contribution of the tibial tubercle to patellar instability: analysis of tibial tubercle-trochlear groove (TT-TG) and tibial tubercle-posterior cruciate ligament (TT-PCL) distances. Knee Surg Sports Traumatol Arthrosc. 2017;25(8):2347–51.

Patellofemoral Instability in a Young Patient with a Chondral Defect, Patella Alta and a Lateralized Tuberosity

12

Jack Farr, Jason L. Koh, Christian Lattermann, Julian Feller, and Andrew Gudeman

12.1 Clinical Vignette

Patellofemoral instability in a 23-year-old who has a large patellar chondral lesion, elevated tibial tuberosity-trochlear groove (TT-TG) distance (22), static patellar subluxation, and elevated Caton-Deschamps (1.4).

J. Farr (✉)
Indiana University School of Medicine, OrthoIndy and OrthoIndy Hospital, Greenwood, IN, USA

J. L. Koh
Mark R. Neaman Family Chair of Orthopedic Surgery, NorthShore University HealthSystem, Evanston, IL, USA

NorthShore Orthopedic & Spine Institute, Skokie, IL, USA

University of Chicago Pritzker School of Medicine, Chicago, IL, USA

Northwestern University McCormick School of Engineering, Evanston, IL, USA

C. Lattermann
Brigham and Women's Hospital Department of Orthopaedic Surgery, MassGeneralBrigham (MGB), Harvard Medical School, Boston, MA, USA

J. Feller
OrthoSport Victoria Research Unit, Melbourne, VIC, Australia

A. Gudeman
Department of Orthopaedic Surgery, Indiana University, Indianapolis, IN, USA

12.1.1 Case

History: We present a case of a 21-year-old division 1 female soccer player. She sustained a right knee injury playing hockey at home when her skate got caught. She felt her knee shift but did not feel her patella dislocate. The emergency department discharged her with crutches and knee immobilizer before being evaluated in clinic 2 days later for follow-up. She described her pain as a dull medial knee pain with intermittent sharpness with everyday activities. She denied redness, swelling, effusion, instability, or loose body sensation.

Physical examination/initial imaging: Her range of motion on the affected knee was 0/5/120 versus 0/0/150 on the contralateral knee. She had guarding on Lachman exam. She had 1+ lateral patellar displacement which was limited by apprehension and 1+ medial patellar displacement. She also had positive patellar tilt. Her hip range of motion was within normal limits. An MRI was ordered which was only notable for mild lateral facet chondrosis (Fig. 12.1).

Trial of non-operative management: The plan for her was to trial a period of formal physical therapy and bracing and return in 6 weeks for repeat evaluation. At her subsequent follow-up visit she had experienced progressive improvement in pain with physical therapy without additional episodes of instability. She had been avoiding sporting activities but wore the brace when active. At this point, she was encouraged to

© ISAKOS 2022
J. L. Koh et al. (eds.), *The Patellofemoral Joint*, https://doi.org/10.1007/978-3-030-81545-5_12

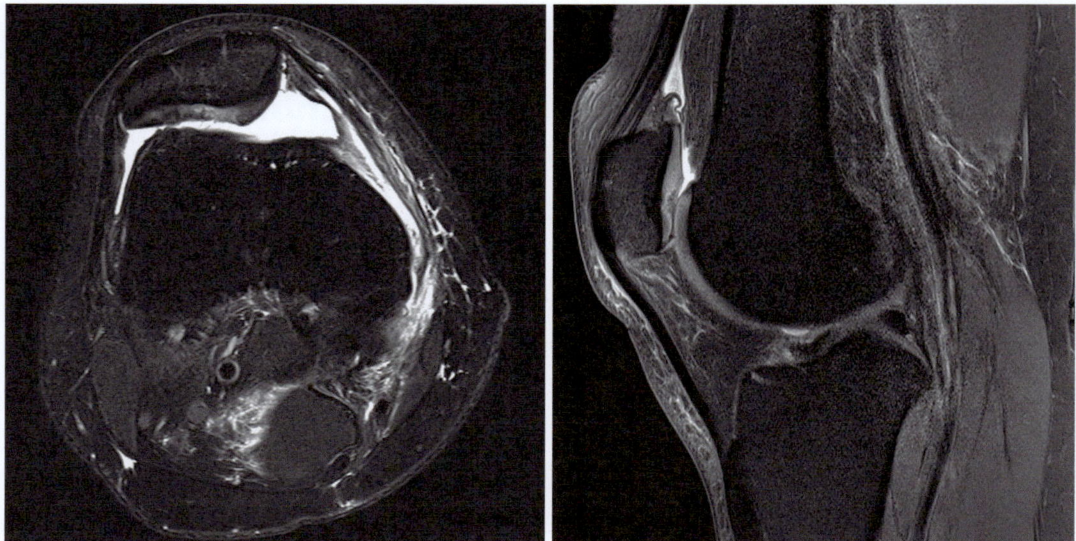

Fig. 12.1 Axial and sagittal T2-weighted images of the knee showing mild lateral facet chondrosis without loose body

continue a home exercise program and bracing during sporting events with follow-up in 6 months. She continued to have excellent progress until 4 months later when she began feeling cracking and patellar slipping with even slight pivoting. She was now a high school teacher and coaching volleyball. She had no difficulties with in line running and wore her brace while coaching and running. On exam she had anterior knee pain with range of motion from 0/0/150 bilaterally. She now had 2+ lateral displacement with apprehension and 1+ medial displacement. The decision was made to obtain further advanced imaging.

Follow-up imaging: Her new MRI showed redemonstration of lateral patellar facet chondrosis; now with a 13 × 5 mm focus of nondisplaced delamination and subchondral stress response at the patellar median ridge. There was stable mild chondrosis at the center of the trochlea, which proximally was flat (Dejour B dysplasia). There was patella alta with Caton-Deschamps (CD) ratio of 1.4 and lateral tilt. Her menisci and cruciate ligaments were intact and there was mild insertional quadriceps tendinosis without tear. Her TT-TG was 22 mm and tibial tuberosity-posterior cruciate ligament (TT-PCL) distance was 24 cm (Fig. 12.2).

Second opinion: The plan after obtaining this MRI was for diagnostic arthroscopy with subchondral bone marrow aspirate concentrate and demineralized bone matrix injection with plans for future cartilage restoration, possible MPFL reconstruction, and possible tuberosity osteotomy. However, a second opinion was encouraged. She was evaluated by Dr. FN who recommended a rotational MRI and diagnostic arthroscopy with formal MPFL reconstruction. The results of the rotational profile can be seen in Table 12.1.

Staging arthroscopy: The patient continued to have patellofemoral pain with several subluxation episodes while exercising and MRI showing lateral facet delamination. She was indicated for diagnostic arthroscopy, chondroplasty, MACI biopsy, with the plans for possible future cartilage restoration and stabilization. The findings from her arthroscopy showed lateral facet grade 3A, 3B, and 3C chondrosis 20 mm medial to lateral, 15 mm proximal to distal, and it was contained. Her menisci and cruciate ligaments were intact. Examination under anesthesia showed displacement of the patella 3 quadrants but not able to fully dislocate.

Definitive surgery: The patient returned to the OR 4 months later for MACI to the patellar defect (Fig. 12.3), anteromedialization of the tubercle in addition to distalization to normalize the CD ratio. She also was indicated for two-armed MQTFL and MPFL reconstruction and lateral lengthening.

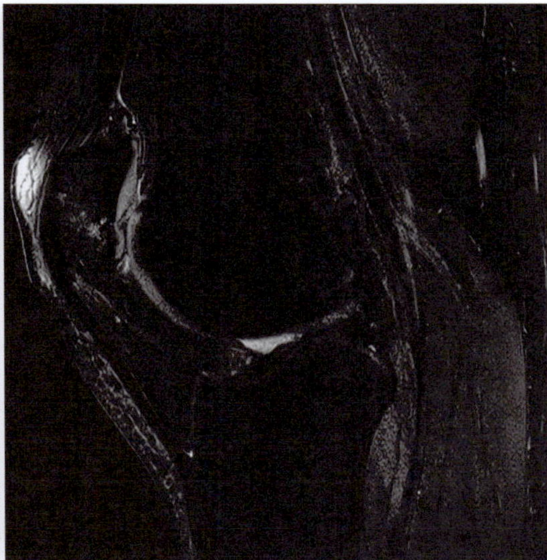

Fig. 12.2 Axial and sagittal MRI showing area of delamination and stress reaction of the patella

Table 12.1 Results of rotational MRI

Measurement	Right (affected side)	Left
Angle femoral neck axis and horizontal reference	4.5	7.5
Angle posterior condylar axis and horizontal reference	19	18.5
Angle femoral anteversion	23.5	26
Knee torsion	14.5	9.5
Lateral/external tibial torsion	28.5	17
TT-TG	22	22

Follow-up: The patient had a benign postoperative course. At her 2 week follow-up she was transitioned to weight-bearing as tolerated and started physical therapy (Fig. 12.4).

At her 6 week follow-up appointment she had full range of motion and encouraged to continue her quad strengthening with PT. At 3 months she had no pain. She had 1+ lateral and medial patellar displacement. She was cleared to perform stairs regularly but encouraged to avoid squatting (Fig. 12.5).

12.2 Commentary

12.2.1 Julian Feller

At the initial presentation, it is important to make sure a tear of the ACL is not missed. The lack of an effusion is against the diagnosis but does not completely exclude it. The presence of medial pain is consistent with an episode of patellar instability and I would check carefully for tenderness along the MPFL, particularly in the region of the medial femoral epicondyle.

This case highlights the importance of distinguishing between pain and instability. Unstable patellofemoral chondral lesions can cause both pain and a sense of instability, the latter presumably being due to pain-induced quadriceps inhibition.

These types of full-thickness chondral lesions with fluid between the articular cartilage and subchondral bone rarely heal, particularly in the patellofemoral compartment [1].

Initial non-operative management is appropriate. I am not a big user of braces. An immobilizer might be appropriate initially to control pain and prevent a fall, but I would be keen to discard it as soon as possible. If there was a question of ongoing patellar instability, a specific patellar stabilizing brace may play a role in getting the patient back to ADLs and sporting activities [2].

The representation in the subsequent follow-up visit is such that surgical intervention seems inevitable. At the point of the first surgical intervention, a decision needs to be made whether to treat the chondral lesion and stabilize the patella,

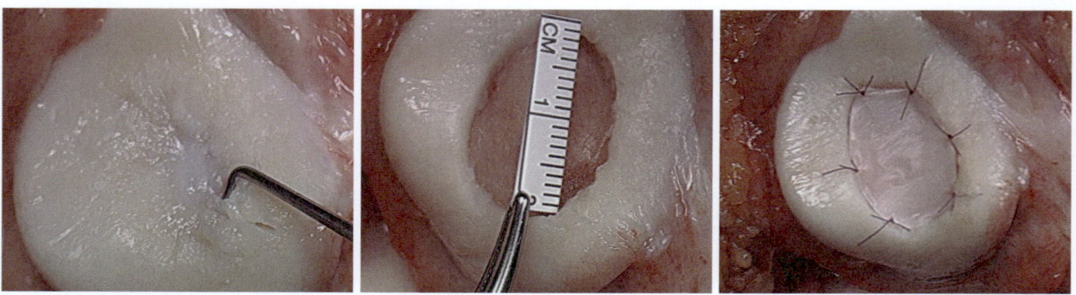

Fig. 12.3 Intra-operative images showing MACI to the patellar defect

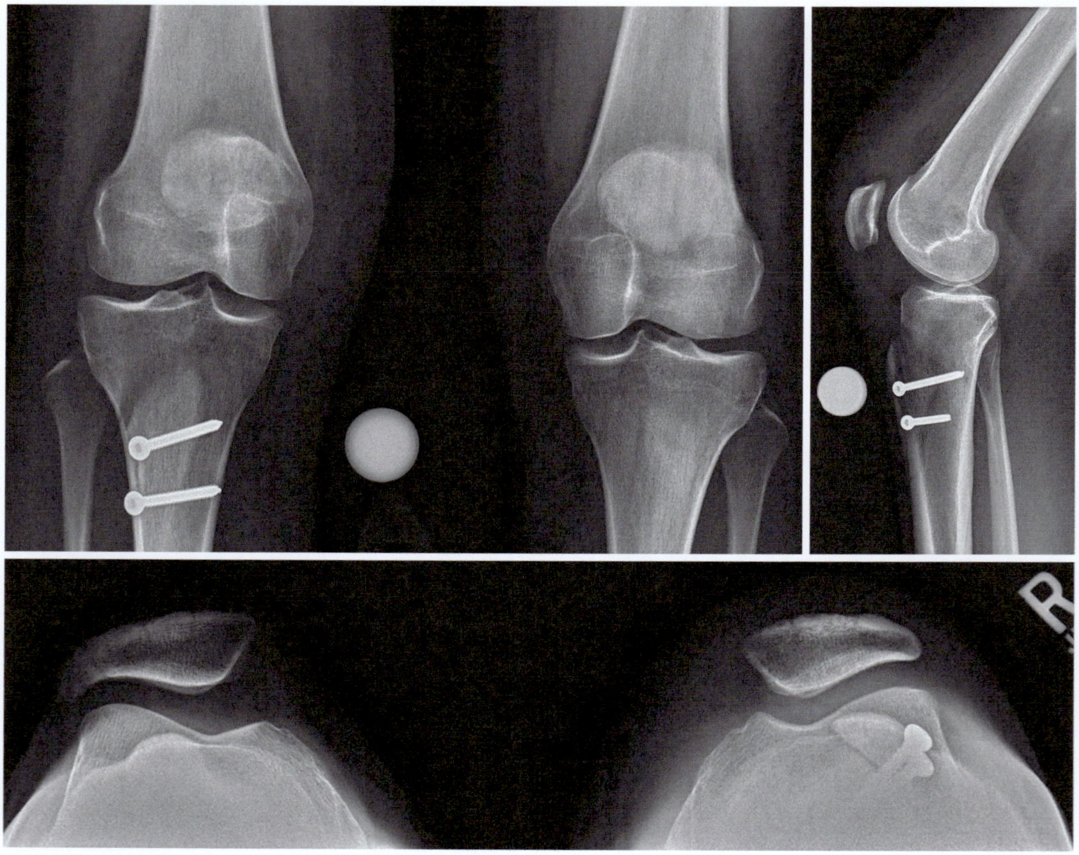

Fig. 12.4 Initial post-operative X-rays showing normalization of patella alta and improved tilt

or only deal with the chondral lesion. There is no right answer. A conservative approach would be to treat the chondral lesion only and deal with instability symptoms later, if they persist. Personally, I would address both the cartilage lesion and the patellar instability at the same sitting, but overall, my approach is more conservative than the one undertaken.

But what to do with the chondral lesion:

1. Try to get it to heal? This is unlikely to be successful.
2. Debride the unstable cartilage only? This is my preferred approach. It burns no bridges and my experience has been positive.

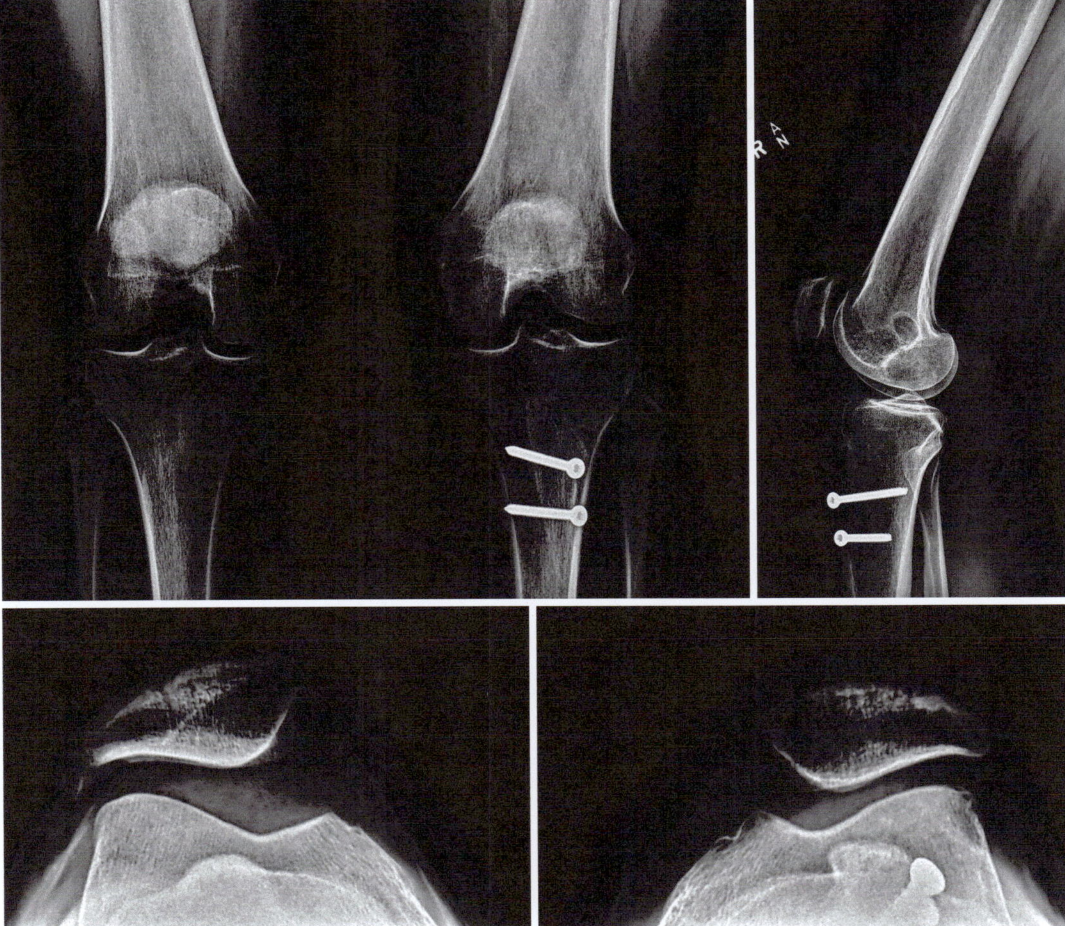

Fig. 12.5 Final post-operative X-rays showing healed osteotomy

3. Try to get the defect to fill with repair tissue? Microfracture is technically difficult in this region, due to both limited access and hard subchondral bone, which is hard to perforate. In any case, it's unlikely to result in substantial repair tissue and may compromise later ACI or MACI if these are considered an option [3].
4. ACI or MACI from the outset? In my opinion this is too aggressive as good results can be achieved with debridement alone.

A couple of technical points about chondroplasty:

1. Using a shaver with a scalloped cutting edge is less aggressive and gives a smoother margin than a traditional full radius cutter.
2. The evidence to support radiofrequency ablation, sometimes euphemistically referred to as coblation (controlled ablation), is at best limited. It is expensive compared to a shaver [4].

With regard to patellar stabilization, this raises the question of how much needs to be done. Assuming that MPFL reconstruction is the minimum, do we need to add a tibial tubercle transfer or trochleoplasty? Essentially it is a matter of defining at what threshold we would intervene. For me, the need for a bony procedure is primarily based on whether the problem is subluxation or dislocation, and whether J-tracking is present. The absolute values of radiological measurements are of secondary importance in the decision-making tree.

Subluxation, regardless of the presence or absence of J-tracking, implies a lesser severity and I would not perform an additional bony procedure. Dislocation with J-tracking is at the other end of the spectrum and in this setting, I would perform an additional bony procedure, usually a tibial tuberosity transfer. The difficult intermediate situation is dislocation without J-tracking. The decision to add an additional bony procedure then depends on the values of the radiological measurements.

In general, I think we can raise the thresholds of the values above what has generally been taught, without increasing the recurrence rate and this has been a focus of mine. The TT-TG distance is very dependent on the angle of knee flexion [5]. With the knee in extension, I am happy to accept a TT-TG distance of up to 24 mm. Using this threshold means that a medialization is only rarely required. For patellar height, it is important to recognize that the CD ratio is typically 0.1–0.2 less than the Insall-Salvati (IS) ratio. If CD is greater than 1.2 and or the IS greater than 1.4, I would not perform a distalization.

Thus, in this particular case I would not perform a tibial tuberosity transfer. The patient's subluxation rather than dislocation is the key. If it was dislocation without J-tracking, I would still use an isolated MPFLR. If it was dislocation with J-tracking, it becomes difficult. Because of the chondral lesion I would not use a trochleoplasty, but my algorithm points to the need for a bony procedure. With a CD index of 1.4, I would perform a distalization [6].

This raises the issue of the role of tibial tuberosity transfer in the setting PF chondral lesions. In the setting of patella alta, patellar chondral lesions are often located towards the inferior pole of the patella. Presumably this reflects the smaller than normal contact surface area in early knee flexion and resultant higher loads on the articular cartilage. At least theoretically, a distalization can reduce this load by increasing the contact surface area on the patella, and I have occasionally done this. The results have been satisfactory but not good enough for the individuals to have the same procedure done on the opposite knee when the same situation developed. In this particular case the patellar lesion is quite central. Unless a distalization resulted in patella infera, I would not be concerned about increasing the load on the affected area of the patella. I do not perform anteriorization due to the associated difficulties in kneeling.

12.2.2 Christian Lattermann

Regarding the initial presentation of this patient, obviously a full knee exam is needed and other injuries need to be ruled out.

A subluxation sensation or popping or sometimes even a report of "my knee cap went to the inside" is common. Interestingly, patients in those cases do not actually report seeing the knee cap on the inside, but they notice the medial femoral condyle/trochlea contour medially and incorrectly identify this as their patella "coming out on the inside."

A lack of effusion and only mild medial pain can point towards an MCL injury but that can usually be ruled out with palpation of the MPFL and the valgus stress test. The insertion of the MPFL and the MCL is close together but the palpatory "track of pain" is along the MPFL in case of an MPFL injury and along the MCL in case of an MCL injury.

A patella dislocation without effusion usually happens in habitually unstable patellofemoral joints and should give rise to a high index of suspicion for malalignment such as a patella alta or hyperlaxity. A Beighton score should be obtained in those patients.

I would always initially manage a patellofemoral instability event non-operatively unless there is clearly a loose body or a significant amount of soft tissue disruption (i.e., MPFL, MCL, and medial capsular disruption). It is important to consider analysis for risk factors of recurrence.

The RIP score should at a minimum be obtained to guide the discussion towards expectations of recurrence with non-op management [7].

12.2.2.1 Subsequent Follow-Up Visit

Patient is coping with the instability but now has pain and recurrent effusions. This is worrisome as effusions are potentially affecting additional chondral surfaces through cytokine interactions of the inflammatory synovial fluid with the chondral surfaces. The MRI representation of the chondral lesion has the character of a larger and progressive lesion than 13 × 5 mm. It appears that this lesion may extend past the ridge midline and is not simply a lateral facet lesion. The measurements of the patella alta are confusing to me. The caton is 1.4 by report, we do not have an X-ray but on the MRI it does not appear that high. The Sagittal Patella engagement (SPE) is normal (>40%). This would indicate to me that she may not functionally have a patella alta. Her TT-TG is high, her TT-PCL is borderline.

To summarize my decision-making process regarding the need to do a TTO for PF instability:

Lateralized Tubercle Assessment

1. If TT-TG and TT-PCL are both high: medialization.
2. If TT-TG and TT-PCL are both normal or high normal: no medialization.
3. If TT-TG and TT-PCL are pointing in different directions: reassess and check for rotational deformity, likely no medialization.

   In this case the TT-TG is borderline high; TT-PCL high normal

   pointing in the same direction but only borderline high. No need to medialize unless TTO done for other reasons, then I would correct to normal values and measure intraoperatively.

Alta Assessment

1. If Caton >1.4 and SPE < 30%: Patella alta, plan for distalization,
2. If Caton <1.2 and SPE > 40%: no patella Alta, no distalization.
3. If Caton >1.4, SPE > 40%: use IS to break tie, if that equivocal, no distalization.

   In this case the Caton:1.4 (still not convinced); SPE: normal:

   Not convinced that she has a patella alta, I need both to point clearly in the same direction. I would not do a distalization unless a TTO is done for other reasons. In that case, consider correction to a normal CD difference as measured in millimeters radiographically.

Addressing the Chondral Lesion

1. Is the lesion clinically symptomatic: yes, and therefore, needs to be addressed.
2. Arthroscopic debridement should always be attempted prior to a major chondral reconstruction. I would arthroscopically debride the lesion and make every effort to rehab the knee after that. The question is always if a MACI biopsy should be taken or maybe a microfracture should be done. In absence of any data suggesting successful outcomes of microfracture treatment in the patella I have stopped doing those several years ago. I always take a biopsy in patients with chondral lesions in the patella or trochlea as, in my experience, many will continue to be symptomatic [8, 9].
3. If a MACI is being done I believe that an anteriorization of the tubercle should be done. It is a very simple approach to the patellofemoral joint as well as that it provides the biomechanical benefit of reducing the mid to inferior patella engagement pressure by up to 25% [10]. Just like an HTO for TF chondral lesions, I have a very low threshold to do a TTO. If I decide to do a TTO, I would then assess the other alignment parameters and address those by correcting back to anatomic high normal values using a Fulkerson style osteotomy.
4. In this particular case after the failed scope debridement chondroplasty/rehabilitation I would plan to do a MACI and a TTO with 1 cm of anteriorization, 6–8 mm of medialization and distalization to a Caton of 1.2.
5. MPFL: this is a chronically laterally subluxed patella with a stretched MPFL and a shortened lateral retinaculum. I address the lateral retinaculum within the TTO approach and will always perform a lateral retinacular lengthening to release the lateral tightness without causing a lateral instability [11]. Regarding the MPFL, I do not believe that a redundant MPFL will shorten, hence I augment the MPFL with a two-armed MPFL reconstruction setting the length after a full range of motion cycle (with a reduced and fixed TTO in this case) and will set the length with a flexion angle of 40°.

Final Overall Approach for this Case

1. Arthroscopy and debridement, MACI biopsy and rehab and assess after 8–12 weeks. If failure, proceed to:
2. TTO: anteriorization 10 mm, medialization 8 mm, distalization to Caton of 1.2.
   MPFL reconstruction.
   Lateral retinacular lengthening.
   MACI.

12.2.3 Jason Koh

Starting with the initial presentation, despite the lack of a documented dislocation episode, the report of the knee "shifting" and causing a fall is concerning for a ligamentous injury or patella instability. The exam is notable for medial side pain, which could be related to a medial collateral ligament sprain, medial patellofemoral ligament tear, or meniscal or chondral injury. The Lachman is accompanied by guarding, so its diagnostic value is limited; however, the presence of apprehension with lateral translation of the patella is significant, and is associated in many cases with patella instability.

I would initially evaluate with X-rays but in a young athletic individual with significant guarding and apprehension after acute injury, I would typically obtain an MRI for further evaluation.

Following the MRI, which demonstrates an MPFL tear, effusion, and some chondral signal changes and patella alta, I would recommend initial nonsurgical management. The patient is at some increased risk of recurrence given the MFPL injury and alta; however, she has had only a single episode of instability, is skeletally mature, and given her upcoming graduation from college will likely no longer be playing sports at the intensity that she had previously. These factors would suggest that non-operative management would be successful.

Treatment would consist of physical therapy initially aimed at range of motion, followed by strengthening and guided therapy to avoid positions of increased risk for recurrent instability of the patella. This would include strengthening of hip external rotators and training to avoid valgus collapse of the knee in landing or pivoting. A light compression sleeve or lateral buttress brace may assist in quadriceps activation and some improvement in function, but I would not expect these to reduce the risk of dislocation [12].

At her follow-up 10 months post-injury, she is now reporting patella instability as well as pain with walking. The repeat MRI demonstrates delaminated cartilage on the lateral facet with underlying bony edema, highly elevated TT-TG, and significant patella alta.

I would be concerned about the delaminated chondral lesion; this will not heal and the adjacent cartilage may also be unstable.

At this point, given the ongoing symptoms and unstable chondral lesion, I would recommend an initial arthroscopy, debridement of the delaminated cartilage, and evaluation of patellofemoral joint laxity and tracking. As part of the discussion with the patient, I would indicate that there is a possibility that further surgery may be needed to unload the defect and restore cartilage if she remained significantly symptomatic. This initial arthroscopy would allow for appropriate staging of the chondral lesion size and location, assessment of the patella loading through a range of motion, and also help determine the amount of instability of the patella. In some cases, simple debridement of the delaminated cartilage can provide substantial symptomatic relief. In areas where there is a cystic lesion of the bone, I will also typically perform a marrow stimulation procedure using a needle or drill through an accessory portal to try to get that area to heal. The procedure would be followed by appropriate therapy.

If the patient remained symptomatic with patella instability and pain despite the arthroscopic debridement, I would recommend realignment, cartilage restoration, and MPFL reconstruction. My plan would be to perform a steeply angled tibial tubercle osteotomy to increase anteriorization and some medialization, since the lesion appears to be centrally located [13]. The amount of alta seen on the MRI does not appear to be excessive, and from our research, I would be cautious about distalization since this may increase patellofemoral contact pressures in flexion. For cartilage restoration in this age group, my preference would be for MACI since it is autologous tissue [14]. For patella stabilization, either a single or double limb MPFL reconstruction would be performed. If there was gross patella laxity found, I would perform a double limb reconstruction but with laxity, a single limb graft would be sufficient [15]. The graft could be secured with the patella located and knee flexed to 20°, with a spacer or with 1 quadrant lateral translation to avoid over tensioning. A formal lateral lengthening is typically not necessary.

Take home message: Chondral lesions of the patella are difficult to treat. These lesions rarely heal without surgery; therefore, a stabilization chondroplasty is typically required. If symptoms persist after chondroplasty and therapy, cartilage restoration and patellofemoral stabilization are usually required. Differing opinions exist on when to perform anteromedialization and/or distalization as evidenced in the above commentary. However, an MPFL reconstruction is typically indicated. Whether this reconstruction is single or double limbed and whether or not to perform lateral lengthening is also debatable.

Fact Box

1. The mainstay of treatment for a first-time incident of patellar instability is conservative management. Unstable cartilage lesions of the patella typically require chondroplasty and therapy. If pain persists, cartilage restoration and patellofemoral stabilization are typically indicated. MACI appears to do better than microfracture with larger defects in the patella.
2. Indications for tuberosity osteotomies are controversial. In general, a Caton ratio > 1.4 is an indication for distalization, as long as this distalization does not increase pressure on the chondral defect. An elevated TT-TG and/or TT-PCL necessitate a medialization although the cutoffs are controversial. MPFL reconstruction should be performed, whether this is single or double limbed and combined with lateral lengthening is debatable.

References

1. Dalal S, Setia P, Debnath A, Guro R, Kotwal R, Chandratreya A. Recurrent patellar dislocations with patellar cartilage defects: a pain in the knee? [published online ahead of print, 2021 Feb 8]. Knee. 2021;29:55–62. https://doi.org/10.1016/j.knee.2021.01.019.
2. Becher C, Schumacher T, Fleischer B, Ettinger M, Smith T, Ostermeier S. The effects of a dynamic patellar realignment brace on disease determinants for patellofemoral instability in the upright weight-bearing condition. J Orthop Surg Res. 2015;10:126. Published 2015 Aug 19. https://doi.org/10.1186/s13018-015-0265-x.
3. Schuette HB, Kraeutler MJ, Schrock JB, McCarty EC. Primary autologous chondrocyte implantation of the knee versus autologous chondrocyte implantation after failed marrow stimulation: a systematic review [published online ahead of print, 2020 Nov 6]. Am J Sports Med. 2021;49(9):2536–41. https://doi.org/10.1177/0363546520968284.
4. Kosy JD, Schranz PJ, Toms AD, Eyres KS, Mandalia VI. The use of radiofrequency energy for arthroscopic chondroplasty in the knee. Arthroscopy. 2011;27(5):695–703. https://doi.org/10.1016/j.arthro.2010.11.058.
5. Camathias C, Pagenstert G, Stutz U, Barg A, Müller-Gerbl M, Nowakowski AM. The effect of knee flexion and rotation on the tibial tuberosity-trochlear groove distance. Knee Surg Sports Traumatol Arthrosc. 2016;24(9):2811–7. https://doi.org/10.1007/s00167-015-3508-9.
6. Leite CBG, Santos TP, Giglio PN, Pécora JR, Camanho GL, Gobbi RG. Tibial tubercle osteotomy with distalization is a safe and effective procedure for patients with patella alta and patellar instability. Orthop J Sports Med. 2021;9(1):2325967120975101. Published 2021 Jan 21. https://doi.org/10.1177/2325967120975101.
7. Hevesi M, Heidenreich MJ, Camp CL, et al. The recurrent instability of the patella score: a statistically based model for prediction of long-term recurrence risk after first-time dislocation. Arthroscopy. 2019;35(2):537–43. https://doi.org/10.1016/j.arthro.2018.09.017.
8. Meyerkort D, Ebert JR, Ackland TR, Robertson WB, Fallon M, Zheng MH, Wood DJ. Matrix-induced autologous chondrocyte implantation (MACI) for chondral defects in the patellofemoral joint. Knee Surg Sports Traumatol Arthrosc. 2014;22(10):2522–30. https://doi.org/10.1007/s00167-014-3046-x.
9. Ebert JR, Fallon M, Smith A, Janes GC, Wood DJ. Prospective clinical and radiologic evaluation of patellofemoral matrix-induced autologous chondrocyte implantation. Am J Sports Med. 2015;43(6):1362–72. https://doi.org/10.1177/0363546515574063.
10. Beck PR, Thomas AL, Farr J, Lewis PB, Cole BJ. Trochlear contact pressures after anteromedialization of the tibial tubercle. Am J Sports Med. 2005;33(11):1710–5. https://doi.org/10.1177/0363546505278300.

11. Hinckel BB, Yanke AB, Lattermann C. When to add lateral soft tissue balancing? Sports Med Arthrosc Rev. 2019;27(4):e25–31. https://doi.org/10.1097/JSA.0000000000000268.
12. Dixit S, Deu RS. Nonoperative treatment of patellar instability. Sports Med Arthrosc Rev. 2017;25(2):72–7. https://doi.org/10.1097/JSA.0000000000000149.
13. Ferrari MB, Sanchez G, Kennedy NI, Sanchez A, Schantz K, Provencher MT. Osteotomy of the tibial tubercle for anteromedialization. Arthrosc Tech. 2017;6(4):e1341–6. Published 2017 Aug 21. https://doi.org/10.1016/j.eats.2017.05.012.
14. Stephen JM, Lumpaopong P, Dodds AL, Williams A, Amis AA. The effect of tibial tuberosity medialization and lateralization on patellofemoral joint kinematics, contact mechanics, and stability. Am J Sports Med. 2015;43(1):186–94. https://doi.org/10.1177/0363546514554553.
15. Kang H, Zheng R, Dai Y, Lu J, Wang F. Single- and double-bundle medial patellofemoral ligament reconstruction procedures result in similar recurrent dislocation rates and improvements in knee function: a systematic review. Knee Surg Sports Traumatol Arthrosc. 2019;27(3):827–36. https://doi.org/10.1007/s00167-018-5112-2.

Part III

Patellofemoral Pain. Chondrosis, and Arthritis: Case-Based Evaluation and Treatment

Patellofemoral Instability in the Adult Female with Recurrent Instability, Chondrosis of Patella and Trochlea

13

Kanto Nagai, Ryosuke Kuroda, Stefano Zaffagnini, and Mauro Núñez

13.1 Case Presentation

A patient was a 40 years old female who presented with recurrent patella dislocations of the left knee. Her chief complain was instability and pain of the left knee. She has had multiple prior atraumatic dislocations since she was 20 years old. At the age of 39 years old, she underwent arthroscopic lateral release at the previous hospital; however, the symptoms were not improved, and so she was referred to our hospital. Her past medical and surgical history was as follows. She had depression, which was well controlled by medications. She underwent cervical laminoplasty and anterior fusion for cervical spondylotic myelopathy at the age of 38 years, and underwent lumber spinal fenestration for lumbar spinal stenosis at the age of 39 years. No particular family history was reported.

Clinical exam reveals 5° of hyperextension and 135 degrees of flexion in both knees with slight effusion and, no swelling, or local heat. She has positive J-sign on her left knee. Her left knee is positive for patellar apprehension and has slight anterior knee pain. Her contralateral knee is also positive for patellar apprehension. The patient is 160 cm tall and weighs 62 kg with a BMI of 24.2. She does not have general joint laxity.

On radiographic assessments, plain standing antero-posterior radiograph showed no evidence of coronal knee deformity or tibiofemoral joint osteoarthritis. Plain lateral radiograph revealed thick patellar sign and the patellar width ratio is 0.67 [1] (Fig. 13.1). The axial view showed that sulcus angle is 140.0°, and congruence angle was 37.0° on her left knee at 30° of knee flexion, with joint space narrowing of the lateral facet of patellofemoral (PF) joint (Fig. 13.2). A small bony fragment was detected in the soft tissues adjacent to the medial aspect of the left patella (Fig. 13.2). On magnetic resonance imaging (MRI), the chondrosis of patella and trochlea and slight patellar subchondral edema were observed (Fig. 13.3). Computed tomography (CT) images showed that tibial tuberosity-trochlea groove (TT-TG) distance [2] was 22.4 mm, and lateral patellar tilt was 27.1° (Fig. 13.4).

The patient had symptomatic patellar instability even after conservative treatments and arthroscopic lateral release. Based on the clinical evaluations, we considered that her main symptom was due to patellar instability; thus, medial patellofemoral ligament (MPFL) reconstruction

K. Nagai · R. Kuroda (✉)
Orthopaedic Surgery, Kobe University Graduate School of Medicine, Kobe, Japan
e-mail: kurodar@med.kobe-u.ac.jp

S. Zaffagnini
Orthopaedics and Traumatology Unit—IRCCS Rizzoli Orthopaedic Institute, Bologna, Italy

DIBINEM Department, University of Bologna, Bologna, Italy

M. Núñez
Department of Orthopaedic Surgery, Hospital del Trauma, San José, Costa Rica

© ISAKOS 2022
J. L. Koh et al. (eds.), *The Patellofemoral Joint*, https://doi.org/10.1007/978-3-030-81545-5_13

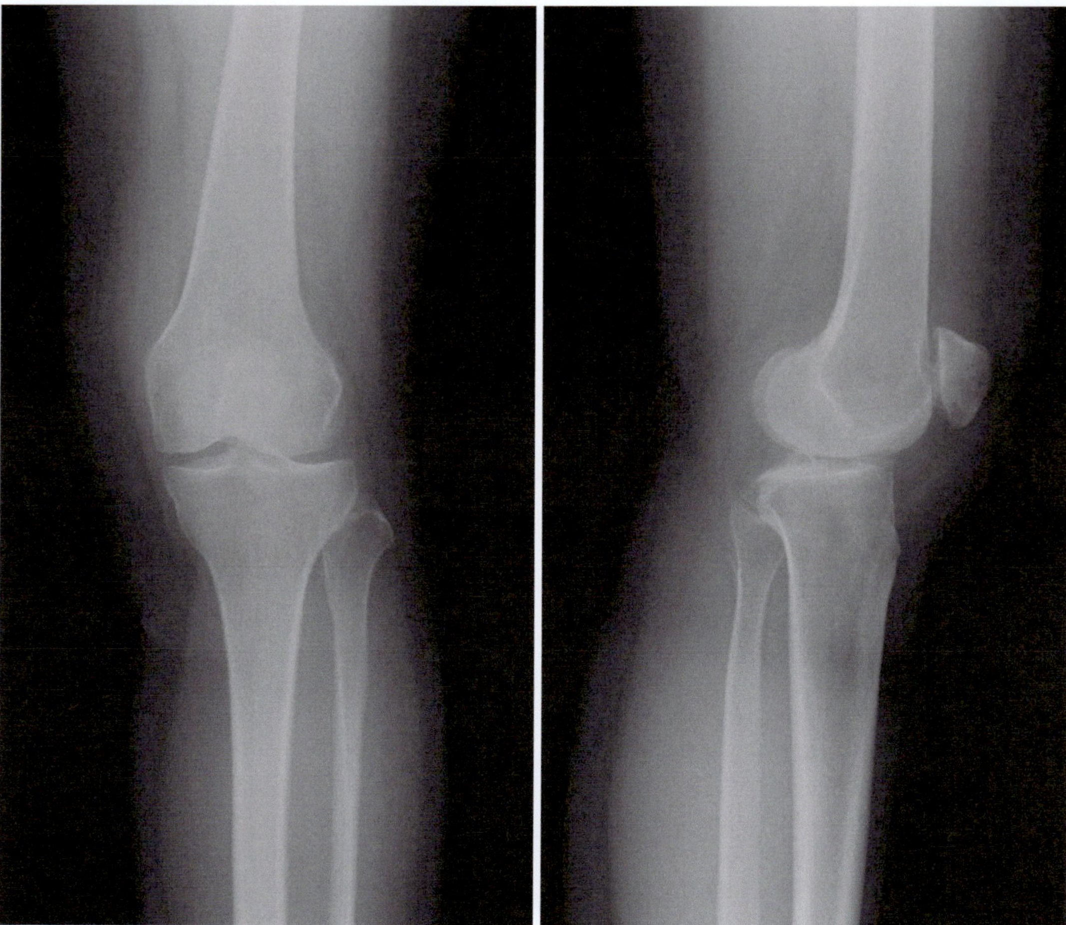

Fig. 13.1 Preoperative plain radiographs of the knee

and lateral release were performed using a doubled hamstring tendon autograft as previously described [3, 4] (Fig. 13.5). With regard to the cartilage status in PF joint, bipolar chondral lesions were observed at the lateral facet of the patella (20 mm × 10 mm) and lateral facet of the trochlea (15 mm × 10 mm) during arthroscopy, and these lesions were grade 3–4 according to International Cartilage Repair Society (ICRS) grading system [5]. Abrasion chondroplasty was simultaneously performed. Postoperative course was great without any complications. Apprehension sign has become negative, and her subjective patellar instability has been improved after the surgery; however, she still has anterior knee pain and crepitation during daily life activity at 1 year after MPFL reconstruction, and MRI shows that bipolar chondral lesions remained. Patella height is within normal range.

TT-TG distance was 22.4 mm (a white line with arrows), and lateral patellar tilt was 27.1° on her left knee when she was referred to our hospital.

13.2 Evaluation and Treatment by a Presenting Surgeon (Ryosuke Kuroda)

Although her patellar instability has been improved, the patient still has anterior knee pain and crepitation during her daily activity 1 year after MPFL reconstruction, lateral release, and abrasion chondroplasty. We have assessed that

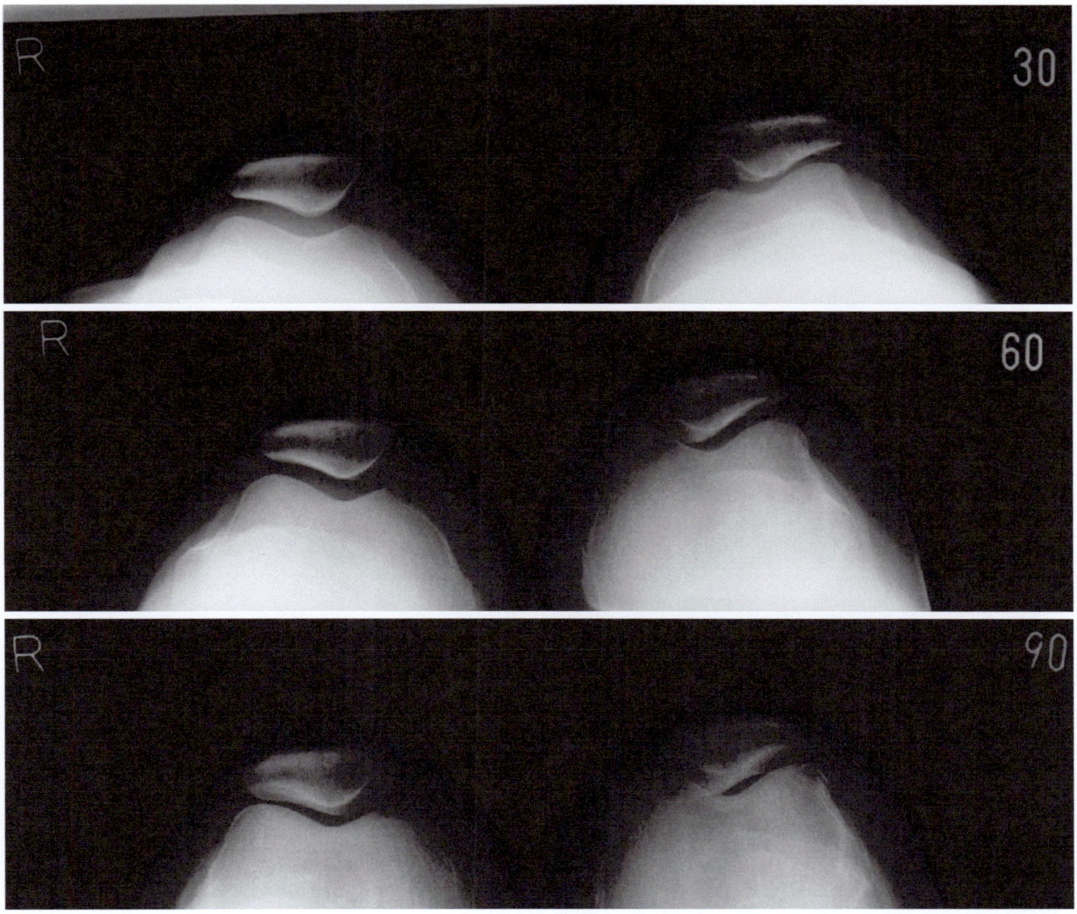

Fig. 13.2 Preoperative axial view radiographs of both knees at 30°, 60°, and 90° of knee flexion

anterior knee pain and crepitation were attributable to bipolar PF chondral lesions. She has slight large TT-TG distance (22.4 mm). We previously reported favorable clinical outcomes following only MPFL reconstruction without tibial tuberosity medialization, even in patients with greater TT-TG distance (22.7 ± 2.6 mm) [3]. That study suggested that greater than 20 mm in TT-TG distance may not be an absolute indication for medialization of the tibial tuberosity when performing MPFL reconstruction. However, the present patient still had symptomatic bipolar full-thickness chondral lesions, and slight maltracking of the patella remained, although patellar instability had been improved owing to MPFL reconstruction.

Optimal treatment of chondral lesions in PF joint remains unclear [6, 7], and bipolar/kissing lesions increase the level of difficulty; however, several recent studies reported good clinical outcomes following autologous chondrocyte implantation (ACI) in PF joint [8–11]. Ogura et al. showed high survival rate of ACI at 5 and 10 years (83% and 79%, respectively) for bipolar chondral lesions in PF joint; the best survival rates were observed among the patients who underwent ACI with concomitant tibial tubercle transfer osteotomy [11]. Thus, the present patient was indicated for ACI and anteromedial tibial tubercle transfer.

ACI was performed according to the previously reported technique which was originally developed by Ochi et al. [12] The patient underwent two-stage procedure that included cartilage harvest and subsequent implantation of autologous chondrocytes embedded in atelocollagen gel (ACC-01; Japan Tissue Engineering,

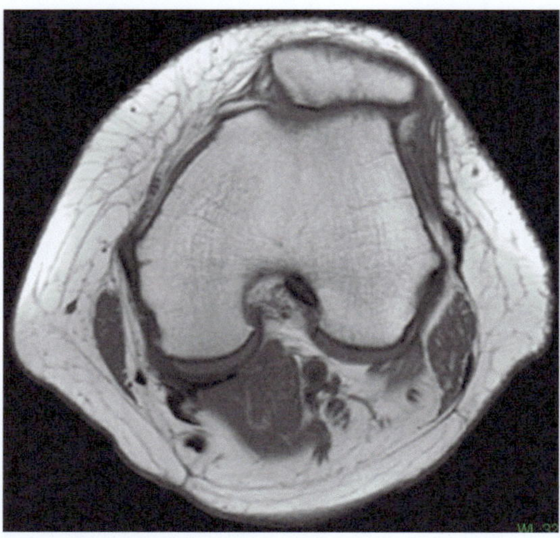
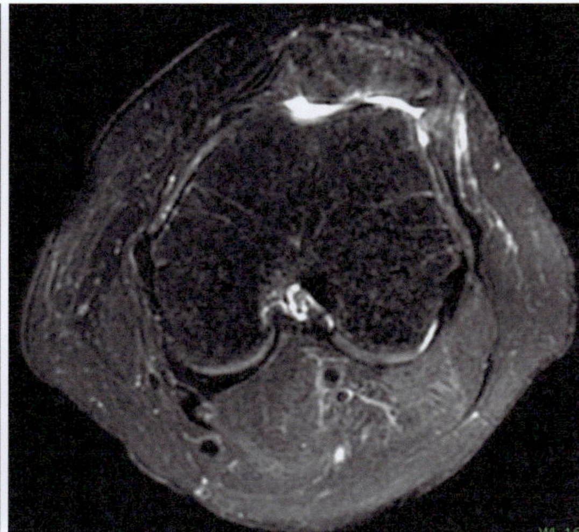

Fig. 13.3 Preoperative axial MRI

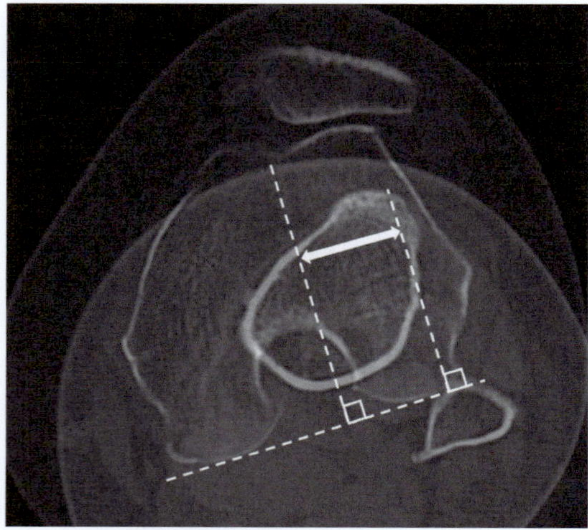

Fig. 13.4 Tibial tuberosity-trochlea groove (TT-TG) distance measurement using CT images

Gamagori, Japan) [12–14]. Anteromedial tibial tubercle transfer and ACI were performed 4 weeks after arthroscopic cartilage harvest. During the surgery, bipolar full-thickness chondral lesions were observed, and the size of the lesions were 25 mm × 15 mm at the patella and 15 mm × 10 mm at the trochlea, respectively (Fig. 13.6). The lesions were debrided as far as normal surrounding cartilage and until subchondral bone was visible. The chondrocyte-atelocollagen gels were shaped based on the shape of cartilage defects (Fig. 13.7a). The chondrocyte-atelocollagen gel was then placed in each defect, and the defects were covered by sutured periosteal flaps, which were taken from the proximal medial tibia (Fig. 13.7b, c). The anteromedial tibial tubercle transfer was performed according to the technique described by Fulkerson et al. [15, 16] The tibial tubercle was transferred about 4 mm in anteromedial direction and fixed with two cancellous screws.

13.3 Commentary and Treatment Recommendation from Dr. Stefano Zaffagnini

In the current case a 40-year-old female with recurrent patellar instability and anterior knee pain was presented. The treatment strategy should be aimed to address the instability and the chondral defect.

13.3.1 Correction of Patellofemoral Instability

The axial view of radiographic assessment showed a narrowing of the patellofemoral joint line. CT images showed a high TT-TG distance (>20 mm) and a resulting lateral patellar tilt of 27.1°.

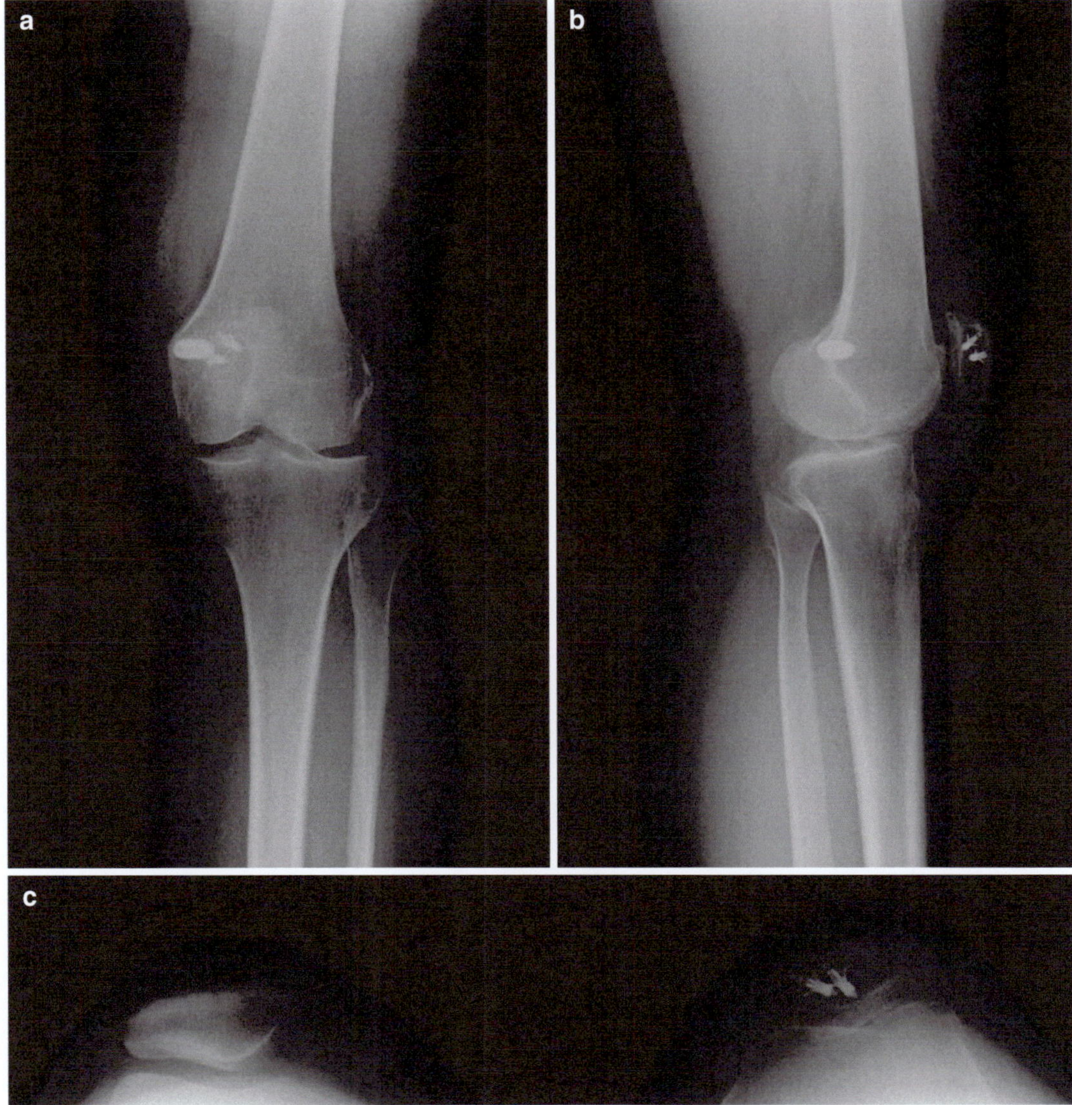

Fig. 13.5 Plain radiographs ((**a**) AP radiograph, (**b**) lateral, (**c**) sunrise view) after MPFL reconstruction and lateral release

Considering the age and the history of the patient and the anatomical risk factors for instability, I would not have chosen the MPFL reconstruction in the first surgical time, but I would have performed a tibial tubercle medialization associated with lateral release and the medial patella-tibial ligament (MPTL) reconstruction to address the patellar instability, checking the patellar tracking under dynamic evaluation and trying to avoid excessive patellar medialization and medial stress.

In this kind of case I use a direct open approach performing a straight midline skin incision from the superior patellar margin to the anterior tibial tuberosity. I perform an extensive lateral release from the tibial tubercle to the level of insertion and a dissection of the vastus medialis oblique in order to assess the medial facet of the patella. Subsequently I perform the osteotomy of the anterior tibial tubercle, in order to place the detached tibial tubercle bone plug in a more distal and medial position, fixing it with a K wire. At this

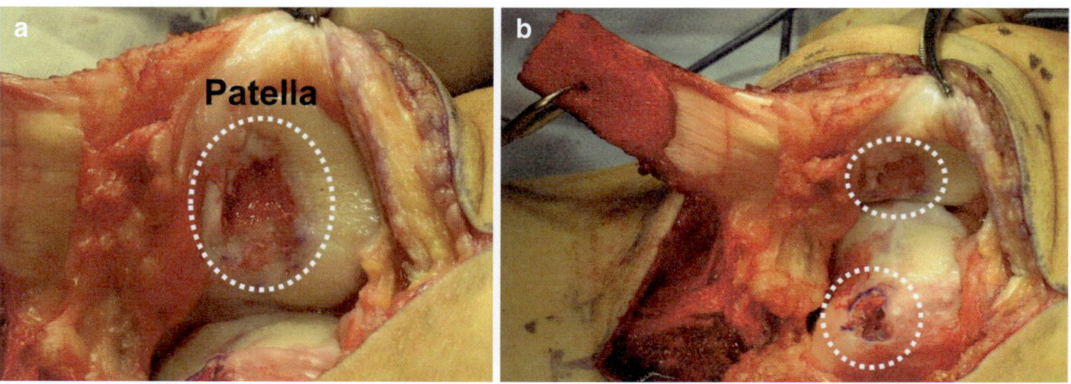

Fig. 13.6 Full-thickness chondral lesions of the patella and trochlea ((**a**) patella lesion. (**b**) Patella and trochlear lesions)

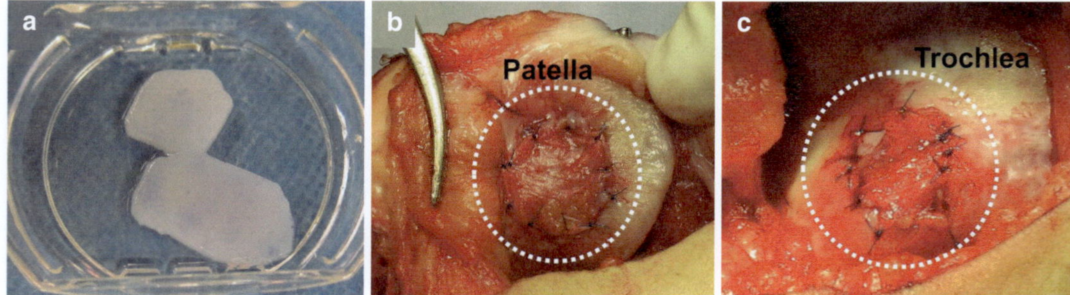

Fig. 13.7 Autologous chondrocyte implantation. (**a**) Chondrocyte-atelocollagen gels. (**b**, **c**) Atelocollagen gel was placed at the cartilage lesions of patella and trochlea, and covered by periosteal flaps

time, I do a functional evaluation of patella tracking especially in the first 30° of flexion, to check patellar stability. Finally, I would fix the tibial tubercle with 2 cortical screws. To correct the incompetent medial restraint, in association with medialization of tibial tubercle, I usually perform MPTL reconstruction. I detach the medial third of the patellar tendon distally. This ligament was then medialized and put under tension, trying to find a medial insertion location close to the anterior edge of the medial collateral ligament. The location was determined by a repeated dynamic analysis of patellar tracking that allowed us to find a reinsertion point that led to patellar stability, especially near extension without creating excessive tension in the ligament band when the knee is flexed. Finally, I fix the medial third of the patellar tendon to the tibial bone with one staple (Fig. 13.8).

Good results were provided in literature after tibial tubercle medialization associated with MPTL reconstruction in patient with recurrent patellar dislocation. In a cases series of patients affected by recurrent patellar dislocation who underwent tibial tubercle medialization associated with the detachment and medialization of the medial third of the patellar tendon no episodes of recurrent patellar dislocation since the surgical procedure and good clinical results were reported at a mean follow-up of 5 years [17].

Improving in knee function, reduction of knee pain and satisfactory radiological results were also reported in a group of 29 patients treated for patellar dislocation with MPTL reconstruction performed with medialization of the medial third of patellar tendon without associated tibial tubercle detachment or MPFL reconstruction [18].

13.3.2 Treatment of Chondral Lesion

Regarding the treatment of chondral lesion, I totally agree with the chapter presenting surgeon.

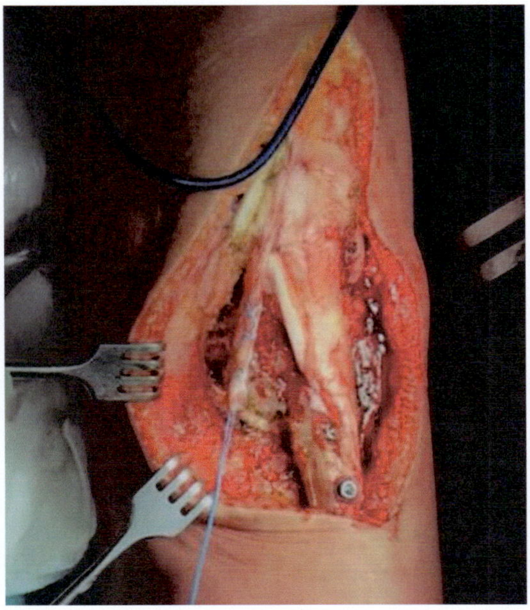

Fig. 13.8 Medial patellotibial ligament reconstruction with the medial third of patellar tendon associated with tibial tubercle medialization

In my opinion the autologous chondrocyte implantation is indicated to assess the bipolar chondral lesion of the current case.

13.4 Commentary and Treatment Recommendation from Dr. Mauro Nunez

There are several points that we must highlight in this case. First, the articular cartilage lesions are divided according to their etiology into **Focal Lesions** and **Degenerative Lesions**, the first are consequence of isolated trauma, osteochondritis dissecans, or osteonecrosis, typically they appear as focal lesions; on the other hand, degenerative lesions are diffuse and poorly demarcated, their etiology is secondary to joint instability or gross deviations from the mechanical axis [19].

In those cases, in which we find deviations of the mechanical axis and/or joint instability, we must resolve these problems before or alongside with the treatment of articular cartilage injuries. Improving patellar maltracking [20], and patellofemoral hyper pressure syndrome through a tibial tubercle transfer (anteromedialization) are crucial steps for success in this case [21].

A second factor to take into account is that the degenerative-type lesions in the patellofemoral joint commonly appear as a mirror lesion, both the articular surface of the patella and the articular portion of the trochlea are involved at the same level. It is important to address both lesions, in order to reduce the pain and other symptoms such as effusion of the joint.

As a third point, articular cartilage injuries must be resolved according to the size and the depth of the injury. There is a great variety of surgical treatment options that are usually divided into Articular Cartilage Repair procedures such as abrasion arthroplasty, subchondral perforation, and microfracture; On the other hand, we find the procedures for the Restoration of Articular Cartilage, such as osteochondral allograft or autograft transplantation, mosaicplasty, and the ACI [22].

In the case of lesions larger than 2–4 cm^2, the implantation autologous chondrocytes implantation or ACI may be a viable solution, other forms of treatment for lesions of said dimensions would be the placement of osteochondral allografts, this technique has the difficulty of requiring an exhaustive process of pairing between the receiver and possible allografts available on the market, and finally fixation can be challenging as headless screws are required.

The results of ACI are less favorable in the patellofemoral joint when compared with those obtained in the femoral condyles [23]. However, ACI has proven to be useful in improving the symptoms and pain from articular cartilage lesions at the patellofemoral level when comparing post-surgical functional scores with the previous state of surgery [24].

For less extensive lesions, the use of osteochondral autografts has been shown to be viable with variable results (Fig. 13.9), and the osteochondral autograft can be extracted from pre-established donor sites—where there is no weight load—or from the proximal fibular-tibial joint [25].

The fourth point would be the age of the patient. Many authors warn us about the age of

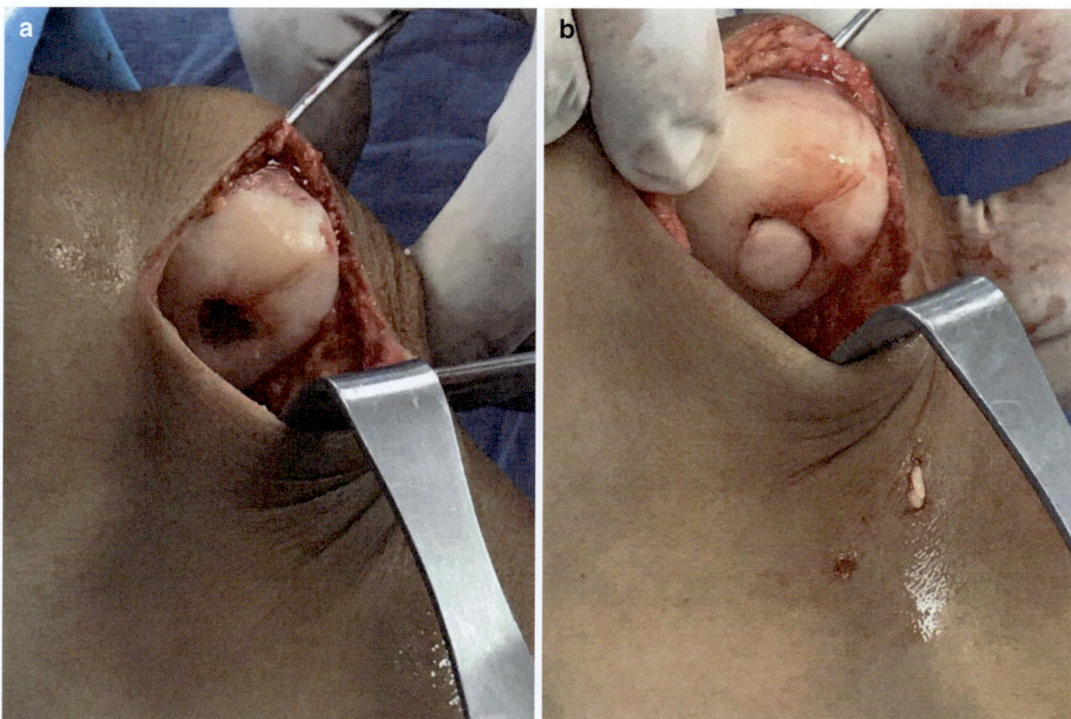

Fig. 13.9 Osteochondral autografts treatment (**a**) Phase 1, (**b**) Phase 2

the patient as a limiting factor for the restoration of articular cartilage. Fifty years seems to be the pre-established limit [26]. However, we increasingly base our decisions as clinicians on the physiological age of the patients and not on their chronological age, considering that age is a subject to be reconsidered according to the evolved surgical techniques. It is also important to take into account that the patient has a healthy lifestyle in general and that is not a sea smoker.

Finally, the patient must be able to understand what the long-term expectations are after surgery, since the return to pre-injury activity levels remains moderate and ranges from 66 to 68% of cases [27].

13.5 Postoperative Course and Outcomes (Ryosuke Kuroda)

Following the surgery, the knee was placed in a knee brace locked in extension for 2 weeks and non-weight bearing period was set for 4 weeks. Physical therapy started 3 days after the surgery. At 2 weeks after surgery, range of motion (ROM) exercise has started, and at 4 weeks, partial weight-bearing was allowed. At 7 weeks after the surgery, the patient was cleared to full weight-bearing. Postoperative course was great without any complications, and her anterior knee pain and crepitation have been resolved. Postoperative plain radiographs (Fig. 13.10) show the improvement of congruence in PF joint and joint space became slight larger compared to that at the time of the referral. At 1 year after the surgery of ACI and tibial tubercle transfer, she underwent screw removal and second-look arthroscopy; the PF chondral lesions, which underwent ACI, were well-healed (Fig. 13.11).

At her final follow-up, 6 years postoperatively, the patient was satisfied with the outcomes and still doing well with no complaints of pain or feelings of instability. There were no dislocations or subluxation episodes since her last surgery. She successfully returned to work and daily activity with no limitations in her left knee. On

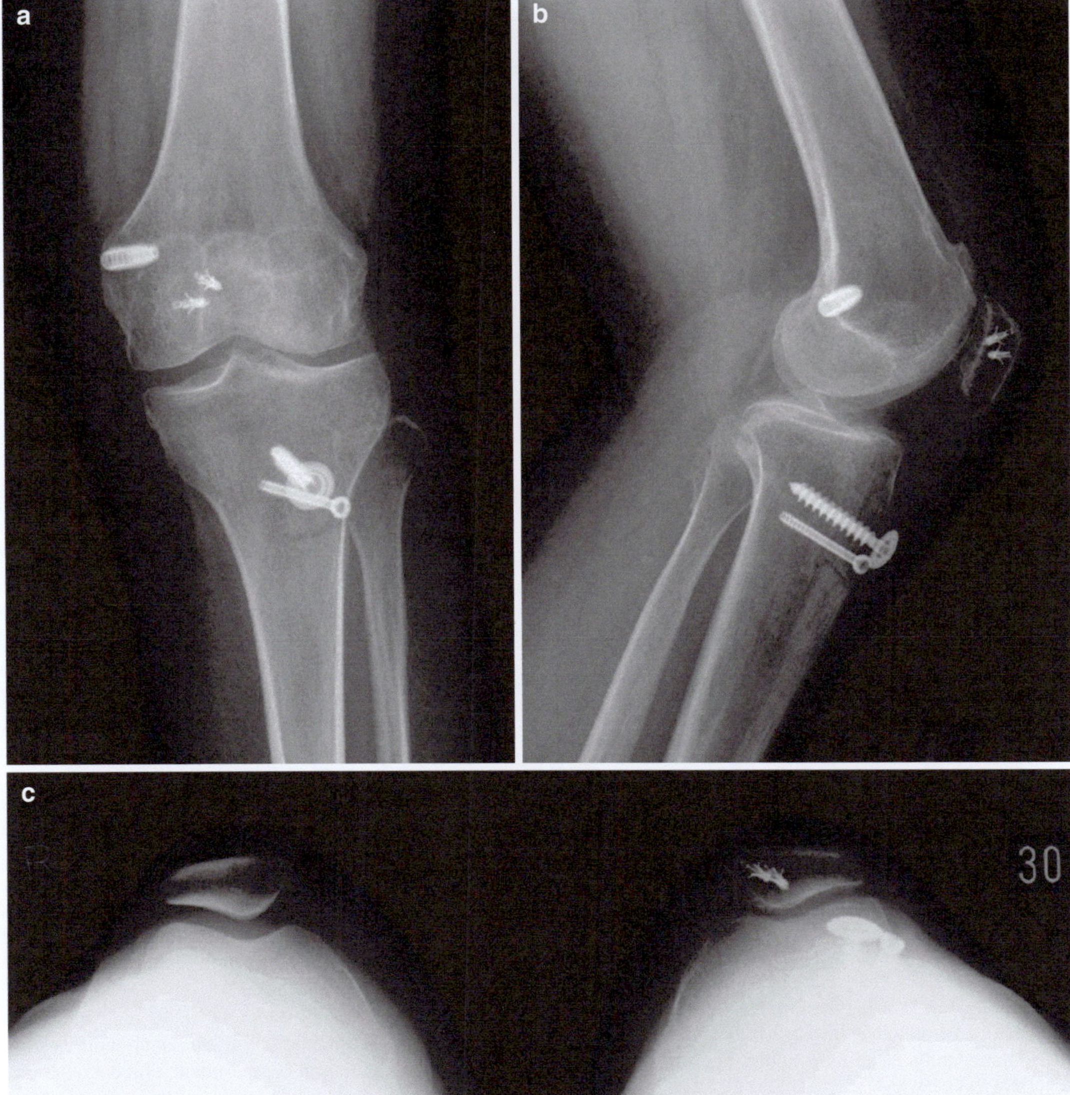

Fig. 13.10 Plain radiographs after autologous chondrocyte implantation and anteromedial tibial tubercle transfer ((**a**) AP radiograph, (**b**) lateral, (**c**) sunrise view)

clinical examination, the left knee did not have apprehension sign or crepitation. She had good MPFL endpoint, and almost full range of motion (0–125°) with no evidence of swelling or effusion.

Take Home Messages

- Patient specific risk factors for patellofemoral instability should be always fully investigated at clinical examination and imaging examinations.
- Tibial tubercle medialization procedures should be considered in a patient with objective patellar instability associated with a TT-TG distance index greater than 20 mm.
- Incompetence of medial soft tissue restraints should be treated trying to avoid excessive medial patellar constraint.

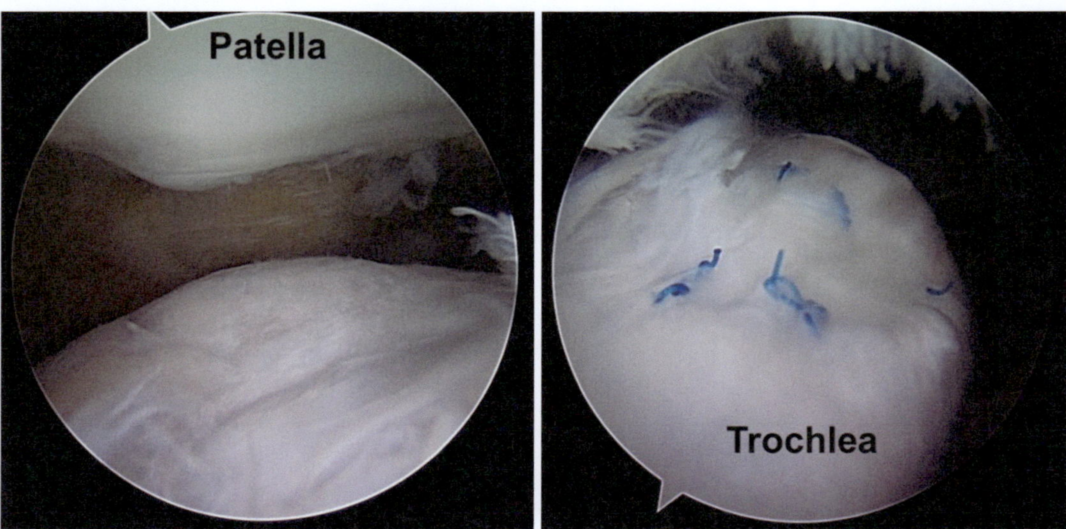

Fig. 13.11 Second-look arthroscopy 1 year after autologous chondrocyte implantation

- Autologous chondrocyte implantation represents a possible solution in treatment of chondral bipolar (kissing) lesions.

References

1. Kuroda R, Nagai K, Matsushita T, et al. A new quantitative radiographic measurement of patella for patellar instability using the lateral plain radiograph: 'patellar width ratio'. Knee Surg Sports Traumatol Arthrosc. 2017;25(1):123–8.
2. Dejour H, Walch G, Nove-Josserand L, et al. Factors of patellar instability: an anatomic radiographic study. Knee Surg Sports Traumatol Arthrosc. 1994;2(1):19–26.
3. Matsushita T, Kuroda R, Oka S, et al. Clinical outcomes of medial patellofemoral ligament reconstruction in patients with an increased tibial tuberosity-trochlear groove distance. Knee Surg Sports Traumatol Arthrosc. 2014;22(10):2438–44.
4. Matsushita T, Kuroda R, Araki D, et al. Medial patellofemoral ligament reconstruction with lateral soft tissue release in adult patients with habitual patellar dislocation. Knee Surg Sports Traumatol Arthrosc. 2013;21(3):726–30.
5. Brittberg M, Winalski CS. Evaluation of cartilage injuries and repair. J Bone Joint Surg Am. 2003;85-A Suppl(Suppl 1):58–69.
6. Mouzopoulos G, Borbon C, Siebold R. Patellar chondral defects: a review of a challenging entity. Knee Surg Sports Traumatol Arthrosc. 2011;19(12):1990–2001.
7. Noyes FR, Barber-Westin SD. Advanced patellofemoral cartilage lesions in patients younger than 50 years of age: is there an ideal operative option? Arthroscopy. 2013;29(8):1423–36.
8. Mandelbaum B, Browne JE, Fu F, et al. Treatment outcomes of autologous chondrocyte implantation for full-thickness articular cartilage defects of the trochlea. Am J Sports Med. 2007;35(6):915–21.
9. Gomoll AH, Gillogly SD, Cole BJ, et al. Autologous chondrocyte implantation in the patella: a multicenter experience. Am J Sports Med. 2014;42(5):1074–81.
10. Kon E, Filardo G, Gobbi A, et al. Long-term results after hyaluronan-based MACT for the treatment of cartilage lesions of the patellofemoral joint. Am J Sports Med. 2016;44(3):602–8.
11. Ogura T, Bryant T, Merkely G, et al. Autologous chondrocyte implantation for bipolar chondral lesions in the patellofemoral compartment: clinical outcomes at a mean 9 years' follow-up. Am J Sports Med. 2019;47(4):837–46.
12. Ochi M, Uchio Y, Kawasaki K, et al. Transplantation of cartilage-like tissue made by tissue engineering in the treatment of cartilage defects of the knee. J Bone Joint Surg Br. 2002;84(4):571–8.
13. Ochi M, Uchio Y, Tobita M, et al. Current concepts in tissue engineering technique for repair of cartilage defect. Artif Organs. 2001;25(3):172–9.
14. Tohyama H, Yasuda K, Minami A, et al. Atelocollagen-associated autologous chondrocyte implantation for the repair of chondral defects of the knee: a prospective multicenter clinical trial in Japan. J Orthop Sci. 2009;14(5):579–88.
15. Fulkerson JP, Becker GJ, Meaney JA, et al. Anteromedial tibial tubercle transfer without bone graft. Am J Sports Med. 1990;18(5):490–7.
16. Fulkerson JP. Anteromedial tibial tubercle transfer. In: Gobbi A, Espregueira-Mendes J, Nakamura N, edi-

tors. The patellofemoral joint. Springer: Berlin; 2014. p. 151–3.
17. Marcacci M, Zaffagnini S, Lo Presti M, et al. Treatment of chronic patellar dislocation with a modified Elmslie-Trillat procedure. Arch Orthop Trauma Surg. 2004;124(4):250–7.
18. Zaffagnini S, Grassi A, Marcheggiani Muccioli GM, et al. Medial patellotibial ligament (MPTL) reconstruction for patellar instability. Knee Surg Sports Traumatol Arthrosc. 2014;22(10):2491–8.
19. Willers C, Wood DJ, Zheng MH. A current review on the biology and treatment of articular cartilage defects (part I & part II). J Musculoskelet Res. 2003;7(3–4):157–81.
20. Hamby TS, Gillogly SD, Peterson L. Treatment of patellofemoral articular cartilage injuries with autologous chondrocyte implantation. Oper Tech Sports Med. 2002;10(3):129–35.
21. Falah M, Nierenberg G, Soudry M, et al. Treatment of articular cartilage lesions of the knee. Int Orthop. 2010;34(5):621–30.
22. Gudas R, Kalesinskas RJ, Kimtys V, et al. A prospective randomized clinical study of mosaic osteochondral autologous transplantation versus microfracture for the treatment of osteochondral defects in the knee joint in young athletes. Arthroscopy. 2005;21(9):1066–75.
23. D'Anchise R, Manta N, Prospero E, et al. Autologous implantation of chondrocytes on a solid collagen scaffold: clinical and histological outcomes after two years of follow-up. J Orthop Traumatol. 2005;6(1):36–43.
24. Handl M, Trc T, Hanus M, et al. [Therapy of severe chondral defects of the patella by autologous chondrocyte implantation]. Acta Chir Orthop Traumatol Cechoslov. 2006;73(6):373–379.
25. Espregueira-Mendes J, Andrade R, Monteiro A, et al. Mosaicplasty using grafts from the upper tibiofibular joint. Arthrosc Tech. 2017;6(5):e1979–87.
26. Paletta GA, Manning T, Snell E, et al. The effect of aiiograft meniscal replacement on intraarticular contact area and pressures in the human knee: a biomechanical study. Am J Sports Med. 1997;25(5):692–8.
27. Harris JD, Brophy RH, Siston RA, et al. Treatment of chondral defects in the athlete's knee. Arthroscopy. 2010;26(6):841–52.

14 Patellofemoral Pain, Chondrosis, and Arthritis in the Young to Middle-Aged Patient: A 32-Year-Old Woman with Lateral Patella and Trochlear Chondrosis

Sabrina M. Strickland, Francesca De Caro, and Robert A. Magnussen

14.1 Case

14.1.1 History

A 32-year-old female physical education teacher and competitive lifeguard presents with a 10 year history of right knee pain. She denies history of frank dislocation but reports one episode 1 year prior to presentation of subluxation. She underwent a chondroplasty 2 years prior to presentation which did not relieve her symptoms. A full course of physical therapy was completed which included quadriceps strengthening, McConnell taping, and hip abductor and core strengthening. She rates her pain as 5 with activity. She received one series of hyaluronic acid injections which resulted in temporary relief of her symptoms.

14.1.2 Physical Examination

Hip ROM 90/40/30
Knee ROM 0-135
Apprehension: Negative
J-sign: Negative
Crepitus: Positive
Tenderness to palpation lateral patellar facet
Medial patellar translation: 2 quadrants
Lateral patellar translation: 2 quadrants
Lachman: 1a

14.2 Imaging Findings (Figs. 14.1, 14.2, and 14.3)

14.3 Proposed Plan of Treatment

Arthroscopic evaluation, open lateral lengthening with lateral facetectomy, cell-based cartilage resurfacing lateral facet, osteochondral allograft lateral trochlea, anteromedializing tibial tubercle osteotomy.

14.4 Intra-operative Photos (Figs. 14.4 and 14.5)

14.5 Results

Pain score 1/10 VAS with high impact exercise. Full range of motion, no effusion. Negative apprehension. Minimal crepitus with knee extension and single leg squat (Figs. 14.6, 14.7, and 14.8).

S. M. Strickland (✉)
Hospital for Special Surgery, New York, NY, USA
e-mail: stricklands@hss.edu

F. De Caro
Department of Orthopedics, Istituto di Cura Città di Pavia, Pavia, Italy

R. A. Magnussen
Ohio State University Wexner Medical Center, Columbus, OH, USA

© ISAKOS 2022
J. L. Koh et al. (eds.), *The Patellofemoral Joint*, https://doi.org/10.1007/978-3-030-81545-5_14

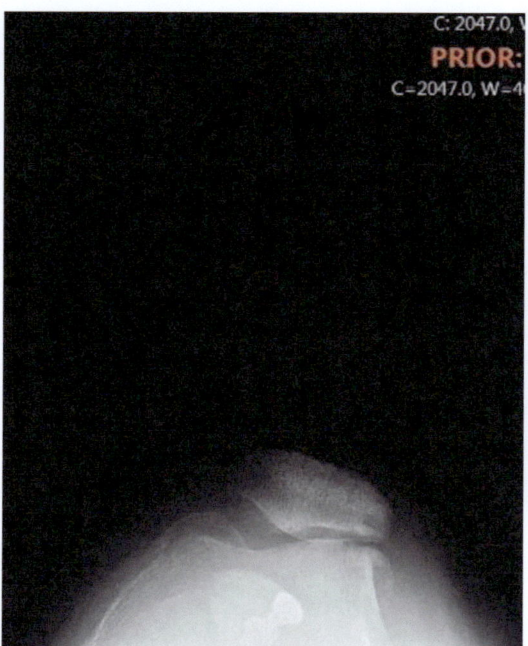

Fig. 14.1 Pre-operative X-ray, merchant view

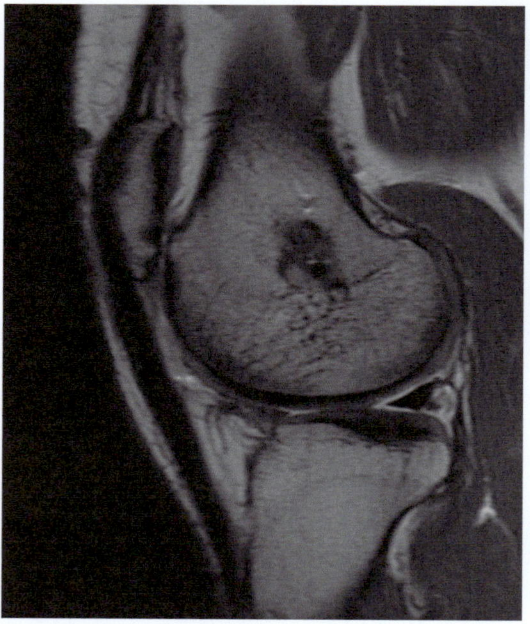

Fig. 14.3 Pre-op MRI, sagittal view. *Lateral patellar osteophyte with complete 2 cm × 2 cm chondral loss on the lateral facet, 3 cm × 1 cm chondral lesion on the lateral trochlea*

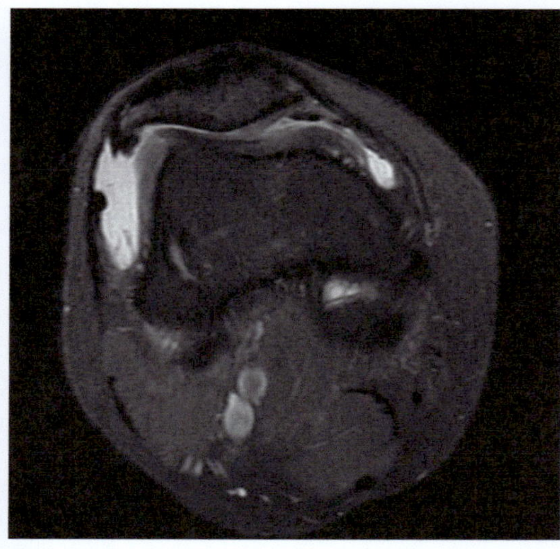

Fig. 14.2 Pre-operative MRI, axial view. TT-TG is measured 18 m and Caton-Deschamps ratio is 1.2

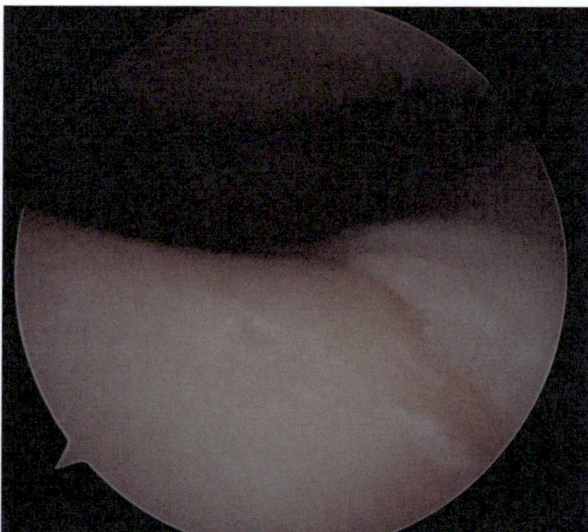

Fig. 14.4 Arthroscopic image of the lateral trochlea with complete chondral loss

14.6 Perspective: Francesca De Caro

Anterior knee pain is a very frequent disorder often associated with patellofemoral cartilage lesions. During routine arthroscopy patellofemoral cartilage defects are found in almost 44% of the cases. Moreover, in professional athletes, up to 37% of all knee cartilage lesions are located in the patellofemoral joint [1, 2]. Most of these

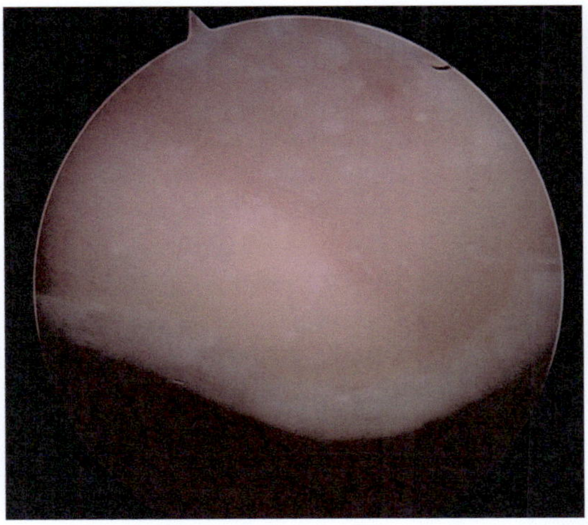

Fig. 14.5 Arthroscopic image of the inferolateral aspect of the patella demonstrating exposed bone

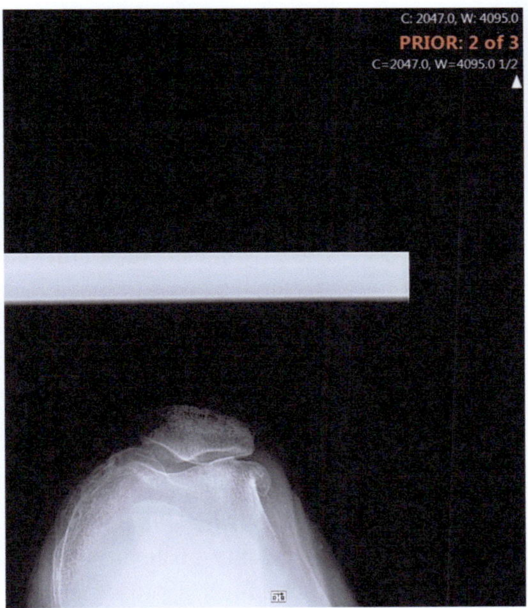

Fig. 14.6 Post-operative X-ray, merchant view

lesions are asymptomatic, but when symptomatic, patients are burdened by a debilitating disease, with high impairment and low quality of life [3]. The treatment of patellofemoral cartilage is not easy and surgeons should always consider that often these lesions are secondary to patellar instability, maltracking, malalignment, acute or repetitive traumas, and that a combined approach must be taken into consideration [4].

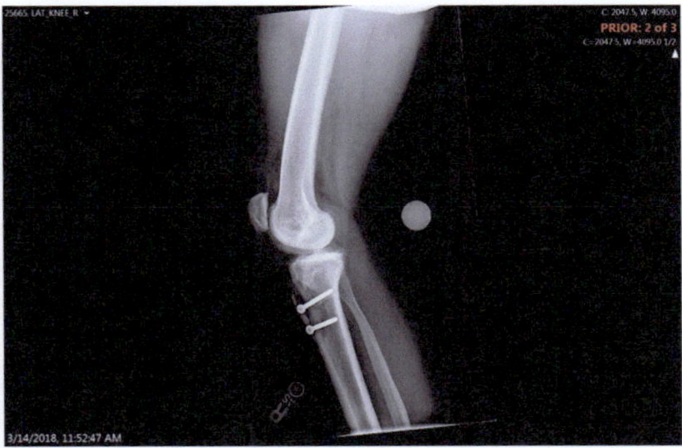

Fig. 14.7 Post-operative X-ray, lateral view

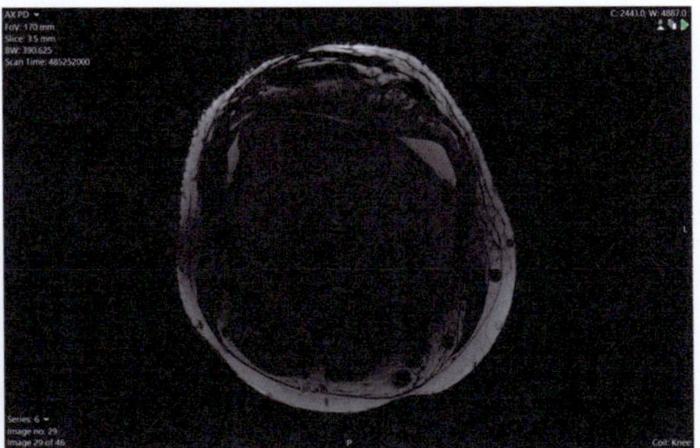

Fig. 14.8 Post-operative MRI, axial view

Considering our clinical case report, a first, non-surgical approach with a more advanced injective treatment would have been a good alternative option to hyaluronic injections. There is little evidence for the use of PRP for the treatment of osteoarthritis, moreover, a recent study by Filardo et al. reports a low rate of return to sport [5] by patients affected by early osteoarthritis and treated by multiple PRP injections. But among these emerging injective treatment options, the autologous protein solution (APS), a blood derivative that provides a milieu of bioactive factors and anti-inflammatory cytokines, seems to provide promising results in the treatment of osteoarthritis, specifically in the treatment of patellofemoral osteoarthritis, in one study there was a 30.5 point absolute improvement of KOOS pain at final follow-up [6]. In a young, active patient who already failed one surgical procedure, this single-step injective treatment could have been a valid non-surgical option instead of the hyaluronic infiltration.

Regarding surgical treatment, anteromedializing tibial tubercle osteotomy would be, in my opinion, the first step surgery.

This is the main similarity of treatment with the author.

A Fulkerson osteotomy, unloading the distal and lateral aspect of the patella can have, if you have a lesion in the lateral and distal aspect, 87% good results, reducing the need for cartilage treatment [7].

In Europe, as several regulatory burdens lead to the dismissal of ACI, advanced therapeutic options, such as the use of minimally manipulated adipose derived mesenchymal stem cells, or bone marrow derived stem cells are becoming more and more popular, even if there is a lack of high quality level studies in the literature, with only few randomized clinical studies [8]. These cells can be used by simple injections, in combination to other surgical procedures and added to a membrane for cartilage restoration. Sciarretta et al. and Gobbi et al. report good clinical results for the treatment of large patellofemoral chondral lesions both with a lipo-amic technique [9] and BMAC technique [10], reporting similar results to ACI in a randomized clinical trial for the treatment of large patellofemoral chondral lesions.

I would address the trochlear lesion of this clinical case report by a modified AMIC technique, taking in consideration that the two chondral surfaces of this joint respond to treatment in a different way, with a markedly good outcome in patients with trochlear lesions and less satisfactory results for cartilage lesions of the patella [11]. Moreover, Dhollander et al. reported good clinical outcomes after autologous matrix-induced chondrogenesis over a short-term follow-up (mean, 2 years) for the treatment of isolated patellar or trochlear cartilage defects, but no case presented bipolar defects [12].

For sure, the most frequently used treatment in Europe, in case of a lower grade of cartilage pathology, with unipolar lesions, would be the implantation of osteochondral scaffolds, an "off-the-shelf" approach with different biomaterials designed to replace the entire damaged osteochondral unit in a single-step procedure [13]. Good clinical results have been reported at mid-term follow-up for a biomimetic scaffold of type I collagen and hydroxyapatite in different concentrations to reproduce the structure and composition of the osteochondral unit [14]. A new innovative aragonite scaffold has shown an almost complete reconstruction of the osteochondral unit even in early osteoarthritis and diffuse lesions of the trochlea, but at the moment this implant is not used for patellar lesions [15].

At last, in Italy, joint replacement is becoming more and more frequent, even at younger ages. This is mainly due to the high costs of cartilage repair and reconstruction surgeries, but also because of the increasing good results of new prosthetic implants. Patellofemoral arthroplasty (PFA), as a transition operation before total knee arthroplasty, has become a more appealing option for patients and orthopedic surgeons because of an easier recovery and good survivorship [16].

I would not treat this 32-year-old patient with a PFA, as she is too young and a clear candidate for a subsequent revision surgery. In the hypothesis of not having the chance to afford a more expensive surgery, with the use of membranes and cells, I would opt for an anteromedializing osteotomy associated with an intra-articular injection of microfractured adipose tissue [17].

14.7 Perspective: Robert Magnussen

The young patient with patellofemoral articular cartilage damage can be challenging to treat and there are multiple available treatment options. The comprehensive treatment approach utilized by Dr. Strickland has addressed all of the potential pain generators in this patient and resulted in a good outcome. While this treatment worked well for this patient, it required a large and complex surgery with significant downtime for the patient as well as access to advanced cartilage repair techniques. Many patients may be unable to commit to such a large procedure and these cartilage treatments are not available worldwide. Key questions are which parts of this procedure are most important and how does one identify which patients need which procedures.

There are numerous factors in a patient's history, physical exam, and imaging that should be considered when selecting treatment. From the patient's history, it is important to identify how much of a role instability plays in the patient's complaints. While this patient has a history of subjective patellar subluxation, many patients with this presentation will have a history of one or more dislocation events in the past that have ceased as their knee becomes more osteoarthritic. Some patients with this presentation have lateral tracking but have never felt unstable. On physical examination, patellar tracking, the presence of patellar apprehension, and the location of pain are crucial to identify. This patient lacks a large J-sign and significant apprehension. Fortunately, this patient's pain is focused on the lateral aspect of the patella and is consistent with the location of chondral damage and the large osteophyte. Imaging studies of the patient demonstrate a large lateral patellar osteophyte and primarily lateral patellofemoral cartilage loss. The axial radiographs demonstrate lateral patellar subluxation.

From my perspective, the patient's pain is centered over the large lateral patellar osteophyte. I would start my treatment here and plan for a lateral patellar facetectomy via an open approach. I would perform a lateral retinacular lengthening during the approach and closure if I felt that lateral retinaculum was tight (which it nearly always is in this situation). I would judge the retinaculum to be tight if I could not evert the patellar to at least neutral from its laterally tilted position. This procedure would remove the impinging osteophyte and the lengthened retinaculum would likely allow the patella to center a bit on the trochlea. One can allow weight bearing as tolerated after the procedure, complications are rare, and no cells or grafts are required. Published series have demonstrated good results of this procedure in this situation with fairly durable results [18, 19].

The next question is whether to add an anteromedialization osteotomy to the procedure. The osteotomy would likely improve patellar tracking and serve to unload the lateral patellar and trochlear cartilage. Seminal work by Dr. Fulkerson's group has shown good outcomes of anteromedialization osteotomies in the setting of lateral patellofemoral chondral damage [7]. The downside of this additional procedure is the increased recovery time and complication risk associated with an osteotomy [20]. Factors that influence this decision for me are: (1) The patient's desires and time available for recovery, (2) a history of patellar instability, (3) patellar tracking and patellar apprehension on exam, and (4) the tibial tubercle-trochlear groove (TT-TG) distance. In a patient with a history of true patellar dislocation, persistent patellar apprehension or j-tracking, and elevated TT-TG distance, I would strongly recommend the addition of an anteromedialization osteotomy. This would reduce the risk of post-operative instability as well as gain the benefits of offloading the cartilage damage. The patient in this case does not have apprehension or a large j-sign and has never dislocated, but she does have a history of subjective patellar subluxation. She has a slightly elevated TT-TG distance on MRI, but this study is known to underestimate the TT-TG distance relative to the values obtained from a CT scan [21]. Further, recent work using osteoarthritis initiative data has demonstrated that patients with an elevated TT-TG distance are more likely to experience a worsening of the lateral patellofemoral osteoarthritis over time [22]. The bottom line for me is that this patient would likely benefit from the osteotomy and the decision to proceed would be made based on a discussion with the patient regarding risks and benefits. I personally would prefer to add the osteotomy in this case to offload the lateral cartilage and hopefully buy some additional time before additional treatments are needed.

The final question is whether it is required to address the articular cartilage damage directly as was done by Dr. Strickland or whether simply unloading the lateral patellofemoral compartment is sufficient. As mentioned above, work by Pidoriano et al. has demonstrated relatively good outcomes of unloading alone in situations with lateral patellofemoral damage [7]. However, most of the patients in that study did not have severe trochlear disease as is shown here. In the setting of more diffuse articular cartilage damage beyond the lateral side, the addition of a cartilage procedure has been shown to be advantageous [23]. In this case I would probably hold off on the cartilage procedure, but this is certainly debatable. The specific cartilage restoration procedure that is chosen is also controversial, with the approach taken by Dr. Strickland very reasonable if a cartilage restoration procedure is undertaken. The osteochondral allograft option is particularly appealing with an uncontained lesion on the trochlea such as in this patient.

In summary, I feel that the priority in treating this patient is removing the painful osteophyte and unloading the area through lateral retinacular lengthening and an anteromedialization tibial tubercle osteotomy. Direct treatment of the articular cartilage lesion could also be considered.

Take Home Message

Young to middle-aged patients with moderate patellofemoral arthritis can be treated with a combination of unloading osteotomy, ligament stabilization, and cartilage restoration surgery.

Fact Boxes

1. Bipolar lesions can be treated with a combination of surface treatment and osteochondral allograft transplantation.
2. Unloading osteotomy should be tailored to specific anatomy. Normalize patellar lateralization (increased TT-TG), Patella alta, and in some cases Genu Valgum.
3. Lateral facetectomy can be considered in cases with an overhanging lateral facet.

Useful Resources/Websites www.patellofemoralfoundation.org

www.orthoinfo.org

Eliasberg C, Diduch D, Strickland S. Failure of patellofemoral joint preservation. Operative techniques in sports medicine. 2019.

Wang D, Shubin Stein B, Strickland S. Patellofemoral issues. In: Farr J, Gomoll A, editors. Cartilage restoration: practical clinical applications, 2nd ed. Springer Science; 2018. p. 103–117.

Strickland S, Pyne A, Connors K. Non-operative treatments for patellofemoral arthritis. In: ESSKA 2nd edition Patellofemoral pain, instability, and arthritis. To be published May 2020.

References

1. Flanigan DC, Harris JD, Trinh TQ, Siston RA, Brophy RH. Prevalence of chondral defects in athletes' knees: a systematic review. Med Sci Sports Exerc. 2010;42:1795–801.
2. Andrade R, Vasta S, Papalia R, Pereira H, Oliveira JM, Reis RL, et al. Prevalence of articular cartilage lesions and surgical clinical outcomes in football (soccer) players' knees: a systematic review. Arthroscopy. 2016;32:1466–77.
3. Hinman RS, Crossley KM. Patellofemoral joint osteoarthritis: an important subgroup of knee osteoarthritis. Rheumatology. 2007;46:1057–62.
4. Andrade R, Nunes J, Hinckel BB, Gruskay J, Vasta S, Bastos R, Oliveira JM, Reis RL, Gomoll AH, Espregueira-Mendes J. Cartilage restoration of patellofemoral lesions: a systematic review. Cartilage. 2019;1947603519893076.
5. Altamura SA, Di Martino A, Andriolo L, Boffa A, Zaffagnini S, Cenacchi A, Zagarella MS, Filardo G. Platelet-rich plasma for sport-active patients with knee osteoarthritis: limited return to sport. Biomed Res Int. 2020;2020:8243865. https://doi.org/10.1155/2020/8243865. eCollection 2020.
6. Van Genechten W, Vuylstek K, Swinnen L, Martinez PR, Verdonk P. Autologous protein solution as a treatment option for symptomatic patellofemoral osteoarthritis in the middle-aged female patient: a prospective case series with one year follow-up. Knee Surg Sports Traumatol Arthrosc. 29:988–97.
7. Pidoriano AJ, Weinstein RN, Buuck DA, Fulkerson JP. Correlation of patellar articular lesions with results from anteromedial tibial tubercle transfer. Am J Sports Med. 1997;25(4):533–7.
8. Di Matteo B, Vandenbulcke F, Vitale ND, Iacono F, Ashmore K, Marcacci M, Kon E. Minimally manipulated mesenchymal stem cells for the treatment of knee osteoarthritis: a systematic review of clinical evidence. Stem Cells Int. 2019;2019:1735242. https://doi.org/10.1155/2019/1735242.

9. Sciarretta FV, Ascani C, Fossati C, Campisi S. LIPO-AMIC: technical description and eighteen pilot patients report on AMIC® technique modified by adipose tissue mesenchymal cells augmentation. GIOT. 2017;43:156–16.
10. Gobbi A, Chaurasia S, Karnatzikos G, Nakamura N. Matrix-induced autologous chondrocyte implantation versus multipotent stem cells for the treatment of large patellofemoral chondral lesions: a nonrandomized prospective trial. Cartilage. 2015;6(2):82–97.
11. Filardo G, Kon E, Andriolo L, Di Martino A, Zaffagnini S, Marcacci M. Treatment of "patellofemoral" cartilage lesions with matrix-assisted autologous chondrocyte transplantation. A comparison of patellar and trochlear lesions. Am J Sports Med. 2013;42(3):626–34. https://doi.org/10.1177/0363546513510884.
12. Dhollander A, Moens K, Van der Maas J, Verdonk P, Almqvist KF, Victor J. Treatment of patellofemoral cartilage defects in the knee by autologous matrix-induced chondrogenesis (AMIC). Knee Surg Sports Traumatol Arthrosc. 2015;23(8):2208–12. https://doi.org/10.1007/s00167-014-2999-0. Epub 2014 Apr 22.
13. Filardo G, Andriolo L, Angele P, Berruto M, Brittberg M, Condello V, Chubinskaya S, de Girolamo L, Di Martino A, Di Matteo B, Gille J, Gobbi A, Lattermann C, Nakamura N, Nehrer S, Peretti GM, Shabshin N, Verdonk P, Zaslav K, Kon E. Scaffolds for knee chondral and osteochondral defects: indications for different clinical scenarios. A consensus statement. Cartilage. 2020:1947603519894729. https://doi.org/10.1177/1947603519894729.
14. Perdisa F, Filardo G, Sessa A, Busacca M, Zaffagnini S, Marcacci M, Kon E. One-step treatment for patellar cartilage defects with a cell-free osteochondral scaffold: a prospective clinical and MRI evaluation. Am J Sports Med. 2017;45(7):1581–8. https://doi.org/10.1177/0363546517694159. Epub 2017 Mar 1.
15. Kon E, Robinson D, Verdonk P, Drobnic M, Patrascu JM, Dulic O, Gavrilovic G, Filardo GA. Novel aragonite-based scaffold for osteochondral regeneration: early experience on human implants and technical developments. Injury. 2016;47 Suppl 6:S27–32. https://doi.org/10.1016/S0020-1383(16)30836-1.
16. Strickland SM, Bird ML, Christ AB. Advances in patellofemoral arthroplasty. Curr Rev Musculoskelet Med. 2018;11:221–30.
17. Russo A, Screpis D, Di Donato SL, Bonetti S, Piovan G, Zorzi C. Autologous micro-fragmented adipose tissue for the treatment of diffuse degenerative knee osteoarthritis: an update at 3 year follow-up. J Exp Orthop. 2018;5:52.
18. Wetzels T, Bellemans J. Patellofemoral osteoarthritis treated by partial lateral facetectomy: results at long-term follow up. Knee. 2012;19(4):411–5.
19. Yercan HS, Ait Si Selmi T, Neyret P. The treatment of patellofemoral osteoarthritis with partial lateral facetectomy. Clin Orthop Relat Res. 2005;(436):14–19.
20. Payne J, Rimmke N, Schmitt LC, Flanigan DC, Magnussen RA. The incidence of complications of tibial tubercle osteotomy: a systematic review. Arthroscopy. 2015;31(9):1819–25.
21. Camp CL, Stuart MJ, Krych AJ, et al. CT and MRI measurements of tibial tubercle-trochlear groove distances are not equivalent in patients with patellar instability. Am J Sports Med. 2013;41(8):1835–40.
22. Haj-Mirzaian A, Guermazi A, Hakky M, et al. Tibial tuberosity to trochlear groove distance and its association with patellofemoral osteoarthritis-related structural damage worsening: data from the osteoarthritis initiative. Eur Radiol. 2018;28(11):4669–80.
23. Gillogly SD, Arnold RM. Autologous chondrocyte implantation and anteromedialization for isolated patellar articular cartilage lesions: 5- to 11-year follow-up. Am J Sports Med. 2014;42(4):912–20.

Patellofemoral Pain, Chondrosis, and Arthritis: A 18-Year-Old with Anterior Knee Pain and Normal Articular Cartilage

15

Sheanna Maine, Stacey Compton, and Ebenezer Francis Arthur

15.1 Case Summary

An 18-year-old Rugby sevens player "KM" who also plays touch football presents with an 8-month history of anterior knee pain (AKP) which was precipitated by a tackle in which she describes landing on her right knee. She was able to weight bear immediately after the injury but felt the knee was stiff and later that evening, she had difficulty straightening it. The stiffness resolved after 3 days. She had a similar injury 3 months later and the knee responded similarly.

KM currently experiences anterior knee pain once a week. The pain localizes to the patella and feels like an ache that is aggravated by walking, sprinting, and climbing stairs. The knee does not swell and there are no mechanical symptoms. KM has not had any rehabilitation for the knee and does feel weak.

On examination KM stands with neutral lower limb alignment and a level pelvis. She has planovalgus feet with mild atrophy of the right quadriceps and walks with a normal gait. She has full ROM in the knee with no effusion. She has slight J tracking in terminal extension and a normal Q-angle at 30 degrees of flexion. She has tenderness around the medial border of the patella with a tight iliotibial band (ITB) and reduced medial patella glide. The knee is stable with no meniscal tenderness on provocation. She has poor vastus medialis oblique (VMO) activation on straight leg raise (SLR) as well as dynamic valgus on a single leg stance (SLS). She has 7 out of 9 signs of generalized ligamentous laxity (GLL) by Beighton's criteria.

Imaging demonstrated a completely normal knee XR, with MRI findings showing no meniscal tear and normal chondral surfaces including the patellofemoral joint (PFJ).

15.2 Discussion

Anterior knee pain (AKP) is one of the most common musculoskeletal conditions seen in paediatric and young adult patients. Annual prevalence for patellofemoral pain in the general population is reported as 22.7%, with the annual prevalence in females 29.2% and males 15.5% [1]. AKP is often related to high activity level due

S. Maine (✉)
Menzies Health Institute Griffith University, Griffith, NSW, Australia

Queensland Childrens Hospital, Redcliffe Hospital, Redcliffe, QLD, Australia

Queensland Limb Reconstruction Clinic, Brisbane, QLD, Australia
e-mail: reception@maineorthopaedics.com.au

S. Compton
Sport and Exercise Medicine Physician, Brisbane, QLD, Australia
e-mail: stacey@queenslandsportsphysicians.com.au

E. F. Arthur
Makati Medical Center, Makati City, Philippines

© ISAKOS 2022
J. L. Koh et al. (eds.), *The Patellofemoral Joint*, https://doi.org/10.1007/978-3-030-81545-5_15

to sports participation. One out of every four paediatric athletes will experience AKP, with up to 70% between the age of 16 and 25 [2]. Adolescent females are 2–10 times more likely to experience AKP compared with males [3]. This condition is generally regarded as a manifestation of multiple factors rather than an isolated aetiology. The factors that influence AKP secondary to patellofemoral pain syndrome (PFPS) can be divided into biomechanical factors that relate to the function of the joint itself, as well as factors that relate to the physical demands placed on the joint when subjected to activity.

15.3 Biomechanical Factors

The movement of the PFJ results from a complex interplay between bony and soft tissue restraints. Every joint will be subtly different in its anatomy and the muscular forces that are exerted across the joint will be patient-dependent and highly variable depending on training and injury profiles. Ligamentous restraint of the PFJ is critical to maintain its stability and the spectrum of maltracking and its relationship to instability are still being defined.

Patellofemoral pain (PFP) is the most common cause of AKP. Patellar maltracking and patellofemoral malalignment are the main reasons for PFP [4]. Coronal plane malalignment in the form of genu valgum changes the relationship of the extensor mechanism to the trochlea groove and will therefore alter the force vector across the PFJ. The presence of patellar maltracking secondary to dynamic valgus alignment is more frequently observed in female athletes compared to males with causes including internal rotation of the femur, the tibia, or both [5].

Q-angle measurement is used to measure patellofemoral malalignment, with above normal values cited as a risk factor for AKP. The validity of this measurement in the paediatric population is not well researched. The most common cause of a high Q-angle is a lateralized tibial tubercle; however, Q-angle and patellar alignment can also be influenced proximally through rotation of the femur and tibia as well as medialization of the trochlea as is seen in trochlea dysplasia. Increased femoral antetorsion leading to an increased Q-angle by displacing the trochlea medially relative to the anterior superior iliac spine and tibial tuberosity [6]. PFP is also related to excessive internal torsion of the femur and lateral rotation of the patella [7].

Increased external tibial torsion may lead to displacement of the tibial tuberosity laterally and a large Q-angle. This alters patellar thus causing PFP. Increased external tibial torsion often accompanies femoral antetorsion, although it may exist separately [8]. In addition, gastrocnemius and soleus tightness reduce the amount of ankle dorsiflexion leading to excessive subtalar joint pronation and tibial internal rotation which will cause femoral internal rotation to increase the Q-angle [9].

A weak iliopsoas muscle destabilizes the pelvis. Patients tend to compensate by developing an anterior pelvic tilt with an internally rotated femur. This causes an increase in the Q-angle, leading to increased patellofemoral joint stresses [9]. External tibial torsion may also be accompanied by a pes planovalgus deformity [10].

Lateral patellar tilt angle (LPTA) is a measure of patellar inclination, and a smaller LPTA may be associated with excessive pressure on the lateral patellofemoral compartment with resultant AKP [11]. A higher sulcus angle is also related to AKP, especially at 30-degree knee flexion [12]. Excessive tightness of the lateral structures inhibits the patella from re-entering the trochlear groove when the pathologic lateral tilt is in excess of 20° when the knee is in extension [13].

The patellofemoral contact force increases with increasing knee flexion until contact occurs between the quadriceps tendon and the femoral trochlea, inducing load sharing. Patella alta causes a delay of this contact until deeper flexion. As a consequence, the maximal patellofemoral contact force and contact pressure increase significantly with increasing patellar height thus causing AKP [14].

Muscle length and contraction status of the quadriceps during dynamic and static conditions affect patellar tracking and displacement. Studies have demonstrated that lateral patellar displacement has a strong correlation with the contracted quadriceps and that a shortened and tight quadriceps plays a significant role in the development of PFP [15, 16]. Quadriceps weakness, specifically VMO weakness in comparison to the vastus lateralis (VL), can lead to lateral displacement of the patella causing the articulating pressure to be on the lateral facet [13]. ITB tightness through its relationship with the lateral retinaculum and patella will increase the lateral force vector on the patella during flexion to increase the lateral patellofemoral joint stresses [17]. Recent research indicates that the axial rotation of the quadriceps mechanism may also play a role in the biomechanics of the PFJ including contributing to a lateral force vector across the PFJ and the subsequent distribution of contact force across the articular cartilage [18].

Generalized ligamentous laxity (GLL) is said to increase the total patellar mobility which would alter patellar tracking and lead to pain symptoms [19].

15.4 Extrinsic Factors

The link between clinical overload and AKP is well established [20]. The function of the patellofemoral joint can be characterized by a load/frequency distribution that defines a range of painless loading compatible with homeostasis of the joint tissues. If excessive loading is placed across the joint, loss of tissue homeostasis can occur, resulting in pain and other dysfunction [21, 22]. This excessive loading can be a consequence of a single event (overload) or repetitive loading (overuse). An athlete who has sustained an injury must have exceeded their limits in such a way that the negative remodelling of the injured structure predominates over the repair process due to the stress placed on the structure [23]. Excessive body load, change in training habits, including increased training, and poor equipment and training technique, can lead to PFP, by overloading the patellofemoral joint and exceeding functional adaptive structural responses [24].

Patient's salient features

Positives	Negatives
18 years old	No swelling/effusion
Female	No mechanical symptoms
Active in sports	Neutral lower limb alignment
Recurrent right knee injury	Level pelvis
Recurrent right AKP	Normal gait
Right knee feels weak	Full knee ROM
Planovalgus feet	Normal Q-angle
Mild atrophy of right quadriceps	Stable knee
Slight J tracking	No meniscal signs
Tenderness on medial border of the patella	
Tight ITB	
Reduced medial patellar glide	
Poor VMO activation on SLR	
Dynamic valgus on SLS	
Ligamentous laxity on Beighton scale	

Our patient is predisposed to having AKP given her age, gender, and level of activity. She exhibits numerous signs and symptoms which would lead one to deduce that patellar maltracking is the source of her AKP. She has weak quadriceps muscles as evidenced by the knee feeling weak, the mild atrophy, and the poor VMO activation on SLR. Although her lower limb alignment is in neutral, she does have a dynamic valgus on SLS which predisposes to patellar maltracking. She has a normal Q-angle and a level pelvis but she does have planovalgus feet which may be accompanied by a certain level of external tibial torsion. She has evidence of tight lateral knee structures with the reduced medial patellar glide and tight ITB. The positive J-sign correlates with poor patellar tracking. Ligamentous laxity on the Beighton scale also predisposes to altered patellar tracking.

This patient was managed using a rehabilitation program that centred around improving the biomechanics around her PFJ. She was prescribed orthotics to assist in support of her medial arch while training, as well as gluteus medius activation and control to improve the dynamic valgus. She used an ITB roller in combination with massage to assist in release of the lateral structures and also worked on recovering her VMO strength and activation patterns. Her knee pain resolved after 6 months.

References

1. Smith BE, et al. Incidence and prevalence of patellofemoral pain: a systematic review and meta-analysis. PLoS One. 2018;13(1):e0190892.
2. Slotkin S, et al. Anterior knee pain in children and adolescents: overview and management. J Knee Surg. 2018;31(05):392–8.
3. Myer GD, et al. The incidence and potential pathomechanics of patellofemoral pain in female athletes. Clin Biomech. 2010;25(7):700–7.
4. Erkocak OF, et al. Lower extremity rotational deformities and patellofemoral alignment parameters in patients with anterior knee pain. Knee Surg Sports Traumatol Arthrosc. 2016;24(9):3011–20.
5. Ford KR, et al. Gender differences in the kinematics of unanticipated cutting in young athletes. Med Sci Sports Exerc. 2005;37(1):124–9.
6. Dejour H, et al. Factors of patellar instability: an anatomic radiographic study. Knee Surg Sports Traumatol Arthrosc. 1994;2(1):19–26.
7. Souza RB, et al. Femur rotation and patellofemoral joint kinematics: a weight-bearing magnetic resonance imaging analysis. J Orthop Sports Phys Ther. 2010;40(5):277–85.
8. Cameron JC, Saha S. External tibial torsion: an underrecognized cause of recurrent patellar dislocation. Clin Orthop Relat Res. 1996;328:177–84.
9. Piva SR, Goodnite EA, Childs JD. Strength around the hip and flexibility of soft tissues in individuals with and without patellofemoral pain syndrome. J Orthop Sports Phys Ther. 2005;35(12):793–801.
10. Meister K, James SL. Proximal tibial derotation osteotomy for anterior knee pain in the miserably malaligned extremity. Am J Orthop (Belle Mead, NJ). 1995;24(2):149.
11. Harilainen A, et al. Patellofemoral relationships and cartilage breakdown. Knee Surg Sports Traumatol Arthrosc. 2005;13(2):142–4.
12. Powers CM, Shellock FG, Pfaff M. Quantification of patellar tracking using kinematic MRI. J Magn Reson Imaging. 1998;8(3):724–32.
13. Amis AA. Current concepts on anatomy and biomechanics of patellar stability. Sports Med Arthrosc Rev. 2007;15(2):48–56.
14. Luyckx T, et al. Is there a biomechanical explanation for anterior knee pain in patients with patella Alta? Influence of patellar height on patellofemoral contact force, contact area and contact pressure. J Bone Joint Surg. 2009;91(3):344–50.
15. Guzzanti V, et al. Patellofemoral malalignment in adolescents: computerized tomographic assessment with or without quadriceps contraction. Am J Sports Med. 1994;22(1):55–60.
16. Witvrouw E, et al. Intrinsic risk factors for the development of anterior knee pain in an athletic population: a two-year prospective study. Am J Sports Med. 2000;28(4):480–9.
17. Winslow J, Yoder E. Patellofemoral pain in female ballet dancers: correlation with iliotibial band tightness and tibial external rotation. J Orthop Sports Phys Ther. 1995;22(1):18–21.
18. Maine ST, et al. Rotational malalignment of the knee extensor mechanism defining rotation of the quadriceps and its role in the spectrum of patellofemoral joint instability. JB JS Open Access. 2019;4(4):e0020. https://doi.org/10.2106/JBJS.OA.19.00020.
19. Rawi ZA, Nessan AH. Joint hypermobility in patients with chondromalacia patella. Br J Rheumatol. 1997;36:1324–7.
20. Fairbank JC, et al. Mechanical factors in the incidence of knee pain in adolescents and young adults. J Bone Joint Surg. 1984;66(5):685–93.
21. Dye SF. Therapeutic implications of a tissue homeostasis approach to patellofemoral pain. Sports Med Arthrosc Rev. 2001;9(4):306–11.
22. Witvrouw E, Van Tiggelen D, Willems T. Risk factors and prevention of anterior knee pain. In: Anterior knee pain and patellar instability. London: Springer; 2006. p. 135–45.
23. Hreljac A. Impact and overuse injuries in runners. Med Sci Sports Exerc. 2004;36(5):845–9.
24. Tumia N, Maffulli N. Patellofemoral pain in female athletes. Sports Med Arthrosc Rev. 2002;10(1):69–75.

A 24-Year-Old Female with Anterior Knee Pain, Normal Looking Cartilage, and Mild Malalignment

16

Alexandra H. Aitchison, John P. Fulkerson, Marc Tompkins, and Daniel W. Green

16.1 Introduction

Chronic anterior knee pain in adolescents and young adults with normal appearing cartilage may require arthroscopic probing to identify softening with related subchondral bone irritation from mildly abnormal tracking and resulting patella focal overload. Pain from such lesions can be debilitating and resistant to non-operative measures.

Typically, the patient develops anterior knee pain related to vigorous activities that is treated with non-operative measures including NSAIDs, physical therapy, taping, bracing, reassurance, rest, and weight loss when appropriate. Imaging may be normal or show mild tilt. When non-operative measures fail, an arthroscopy may be done and the surgeon may note softened articular cartilage, on the patella distally and laterally, by probing otherwise normal looking cartilage. When does the orthopedic surgeon intervene and what are the options?

A. H. Aitchison · D. W. Green
Department of Orthopedic Surgery, Hospital for Special Surgery, New York, NY, USA

J. P. Fulkerson (✉)
Department of Orthopedic Surgery, Yale University, New Haven, CT, USA

M. Tompkins
University of Minnesota Medical Center, Minneapolis, MN, USA

16.2 Case Presentation

A 24-year-old female presents to the office with a 6-year history of anterior knee pain that started during high school sports. She initially had relief after resting and diminishing physical activities, but continued to have intermittent, activity related pain which is worse now over the last 6 months. Stairs are painful, particularly descending stairs. Running is no longer possible. She experiences some mild swelling with increased activity as well as exacerbation of pain. She has tried several courses of physical therapy including core stabilization exercises, with minimal benefit. Braces and taping help a little, but nonsteroidal anti-inflammatories have not helped. One corticosteroid injection did nothing. She is desperate now and feels this pain is ruining her life, leading to some depression. At times she has to stop and rest with normal daily activities and avoids long walks, particularly with inclines.

Work up reveals gait with evidence of slight femoral anteversion and a tendency for the femur to roll inwards slightly more than normal when she walks. She does have a mildly painful click upon compression of the patella in early flexion of the knee. Radiographs reveal a Dejour A trochlea on lateral view and a Caton-Deschamps ratio of 1.23, very mild tilt on 30-degree knee flexion Merchant view and normal AP radiographs. An MRI is obtained and the T2 fat suppressed sagittal cuts suggest increased fluid in the distal patella articular cartilage with a TT-TG of 16 mm. The

© ISAKOS 2022
J. L. Koh et al. (eds.), *The Patellofemoral Joint*, https://doi.org/10.1007/978-3-030-81545-5_16

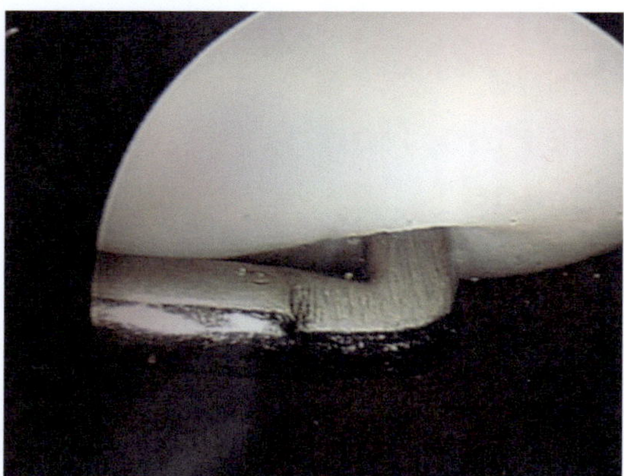

Fig. 16.1 Intraoperative arthroscopy image showing softened distal/lateral patellar articular cartilage with no breakdown or fraying

patient underwent diagnostic arthroscopy which revealed normal looking but softened, spongy distal/lateral patella articular cartilage with no breakdown (Fig. 16.1). The trochlea was normal and only mild synovial inflammation was present.

16.3 Evaluation and Treatment by Presenting Physician (Dr. John Fulkerson)

This pattern of anterior knee pain and associated chondral softening sometimes arises in situations that create focal overload on the patella. Factors such as rotational malalignment (sometimes with tilt that can be amenable to lateral release or lengthening), excessive lateral tracking vector, obesity, hyperintense activity, local trauma, or trochlear dysplasia might contribute. Not infrequently, several of these factors combine to create a point of focal cartilage failure.

In these cases, one must always be aware of underlying rheumatologic conditions and consult with a rheumatologist, particularly when recurrent swelling or other joint problems are noted.

Sometimes, catastrophizing occurs in young people who are limited in activities because of pain [1]. This psychological problem can lead to premature surgery in a patient who has an overabundance of pain and inability to cope, so psychological counseling may help in this setting.

Control of identified alignment, strength, function, and body mass disorders requires full non-operative management. Weight loss, optimal core stability, activity management, and bracing are imperative to controlling pain when indicated.

Nonetheless, some patients are severely disabled and have received all possible non-operative management, sometimes two or three courses of management, so surgical intervention may become indicated. Particularly in cases where lateral tilt and lateral overload are established, a lateral release or lengthening may be appropriate. In more extreme situations, when distal and/or lateral objective grade 1–2 cartilage breakdown is established in association lateral tilt or lateral tracking, anteromedial tibial tubercle transfer could be performed to eliminate the lateral overload, progressive cartilage breakdown, and pain.

16.4 Commentary and Treatment Recommendation from Dr. Marc Tompkins

Anterior knee pain is almost always initially treated with detailed and prolonged therapy with a patellofemoral therapist who can use multiple non-operative treatment approaches including taping, blood flow restriction therapy, etc. In the presented case it is interesting that the patient did get better with taping, which suggests that it is lateral patellofemoral overload which is at the root of the problem here. It is important to rule out other potential causes of anterior knee pain, such as patellar or quadriceps tendinitis, prepatellar bursitis, or even rare things such as a symptomatic plica or fat pad impingement. Dr. Fulkerson nicely outlined additional non-operative considerations that are paramount to managing a case such as this.

If pain persists despite involved conservative management, then it is reasonable to think about surgery, especially when there are conservative treatments that provide temporary relief suggesting that pain relief is achievable. In this case, since taping provided relief, provided that it was McConnell taping pulling the patella medially, then it is reasonable to consider an isolated lateral retinacular lengthening. This is one of the only situations in which an isolated lateral retinacular lengthening is reasonable to consider as a treatment approach. Factors that would suggest against this procedure would be minimal tilt on 30-degree knee flexion axial radiograph view, signifying that the lateral retinaculum is not necessarily too tight. A surgical decision for an isolated lateral retinacular lengthening would also be based on exam and whether the patella could be tilted to a neutral position. If the patella could not be "untilted" manually, it is reasonable to consider isolated lateral retinacular lengthening.

In this case, since the patellar problem appears to be lateral, one might consider an anteromedialization procedure. The TT-TG, however, is only 16, suggesting that the tubercle is not overly lateralized. In such a case, the surgeon would need to be careful to not overmedialize the tubercle to avoid overloading the medial tibial femoral joint. The surgeon may consider evaluating the tubercle sulcus angle intra-operatively to help prevent overmedialization. If anteromedialization is undertaken to offload cartilage, it is important to remember that it is lateral or distal cartilage lesions that are best unloaded with surgery.

Additionally, it would be valuable to get standing long leg films to assess for angular deformity in the coronal plane, and CT scan (or EOS) to assess for excessive femoral version or tibial torsion in the axial plane. If either of those factors is playing a role, it is reasonable to consider correcting the deformity with an osteotomy. Treatment could include either varus producing distal femoral osteotomy, a derotational osteotomy, or a distal femoral osteotomy that allows for both (if that was needed).

16.5 Commentary and Treatment Recommendation from Dr. Daniel Green

Step one for me in the setting of chronic anterior knee pain is to question the diagnosis. Perhaps this is a case of patellofemoral instability that is not clearly recognized by the patient. I have seen a few patients that have lateral dislocation of the patella in extension that do not verbalize that their patella is dislocating. Another type of patellar instability that can occasionally present is patients with such a severe J-sign that they refuse to actively extend their knee from a flexed position. These patients might not report having frank dislocations. I would also recommend looking closely for other MRI signs of patella dislocation sequelae such as a damaged or torn MPFL, nonarticular medial patella avulsion injuries, or cartilage wear at the lateral trochlea. Any of these findings along with a positive patella apprehension test would lead me to recommend a patella stabilizing approach for those who had failed physical therapy.

Other etiologies of intractable anterior knee pain with chondromalacia could be inflammatory in nature. In these potential patients, I would recommend a rheumatology consult to determine if there is an underlying rheumatological condition. This patient reports occasional swelling which leads me to question if there could be a case of mild Lyme disease contributing to the symptoms.

Chondral injury is another potential cause of anterior knee pain that should be considered. In this case, a traumatic isolated injury to the apex of the patella could have occurred leading to a chondral injury that perhaps could heal or perhaps could deteriorate into an OCD type lesion.

Additionally, this patient reports that she is so frustrated that she is "desperate," and the pain is "ruining her life leading to depression." Evaluating health care providers should carefully explore these feelings and their etiology prior to proceeding to surgery. Consultation with a mental health provider may unveil other sources of stress or mental health issues that may benefit from treatment, leading to a more successful conservative treatment of their anterior knee pain.

Finally, before proceeding to a surgical discussion, the provider should review the specific physical therapy program that has failed to manage the patient's symptoms in the past. Consideration could be given to a referral to a physical therapist with a special interest and expertise in patellofemoral pathology as a final attempt to solve the patient's problem without surgical intervention.

Take Home Message
Focal patellofemoral overload, which can lead to chondral softening, can be an elusive cause of pain. When all non-operative measures fail, surgical intervention may become appropriate, rarely. Specifically, objective patella tilt associated with lateral patella softening and pain may be treated successfully by lateral release or lengthening to unload the articular lesion. In more extreme cases in which lateral and distal patella overload are causing persistent focal overload and chronic pain, a moderately steep anteromedial tibial tubercle transfer is very effective to unload the softened, overloaded distal and lateral patella, but one must always be careful not to overmedialize. In more rare cases, the overload is due to axial or coronal plane malalignment and can be treated with an osteotomy to correct the plane of malalignment.

Fact Box Softened Articular Cartilage in the Setting of Chronic Anterior Knee Pain [2–7]

- Cartilage may look normal in some patients with chronic anterior knee pain, so always probe it. It is not uncommon to find softened articular cartilage on the distal and lateral patella of patients with mild malalignment.
- Softened articular cartilage related to chronic focal overload can lead to chronic anterior knee pain because of subchondral bone overload.
- Pain related to this particular lesion can be debilitating yet often responds to reducing load on the lesion by rest, bracing, taping, appropriate weight loss, NSAIDs and modification of activities.
- When non-operative measures fail, surgical intervention may be appropriate to address objective findings. The goal of surgical treatment is typically to remove inflammatory tissue, and reduce load on a pain generating articular lesion as directed by objective findings such as tilting of the patella or lateral tracking.

Resources and Websites Middleton KK, Gruber S, Shubin Stein BE. Why and where to move the tibial tubercle: indications and techniques for tibial tubercle osteotomy. Sports Med Arthrosc Rev. 2019;27(4):154–160.

Hinckel BB, Yanke AB, Lattermann C. When to add lateral soft tissue balancing? Sports Med Arthrosc Rev. 2019;27(4):e25–e31.

Fulkerson JP. Anteromedialization of the tibial tuberosity for patellofemoral malalignment. Clin Orthop Relat Res. 1983;177:176–181.

References

1. Sanchis-Alfonso V, Coloma-Saiz J, Herrero-Herrero M, Prades-Piñón J, Ramírez-Fuentes C. Evaluation of anterior knee pain patient: clinical and radiological assessment including psychological factors. Ann Jt. 2018.
2. Petersen W, Ellermann A, Gösele-Koppenburg A, Best R, Rembitzki IV, Brüggemann GP, et al. Patellofemoral pain syndrome [Internet]. In: Knee surgery, sports traumatology, arthroscopy, vol. 22. Springer; 2014 [cited 2020 Jul 31], p. 2264–74. https://pubmed.ncbi.nlm.nih.gov/24221245/.
3. Rothermich MA, Glaviano NR, Li J, Hart JM. Patellofemoral pain. Epidemiology, pathophysiology, and treatment options [Internet]. In: Clinics in sports medicine, vol. 34. W.B. Saunders; 2015 [cited 2020 Jul 31]. p. 313–27. https://pubmed.ncbi.nlm.nih.gov/25818716/.
4. Clinical rehabilitation of anterior knee pain: current concepts—PubMed [Internet]. [cited 2020 Jul 31]. https://pubmed.ncbi.nlm.nih.gov/28437492/.
5. Sanchis-Alfonso V, Dye SF. How to deal with anterior knee pain in the active young patient. Sports Health [Internet]. 2017 [cited 2020 Jul 31];9(4):346–51. https://pubmed.ncbi.nlm.nih.gov/27920260/.
6. Werner S. Anterior knee pain: an update of physical therapy [Internet]. In: Knee surgery, sports traumatology, arthroscopy, vol. 22. Springer; 2014 [cited 2020 Jul 31]. p. 2286–94. https://pubmed.ncbi.nlm.nih.gov/24997734/.
7. Slotkin S, Thome A, Ricketts C, Georgiadis A, Cruz AI, Seeley M. Anterior knee pain in children and adolescents: overview and management [Internet]. Journal of Knee Surgery, vol. 31. Georg Thieme Verlag; 2018 [cited 2020 Jul 31]. p. 392–8. https://pubmed.ncbi.nlm.nih.gov/29490405/.

Patellofemoral Pain, Chondrosis, and Arthritis: A 23-Year-Old with Patellofemoral Pain and Maltorsion of the Lower Limbs: The Place of Torsional Osteotomies

17

Magaly Iñiguez C, Phillippe Neyret, Sheanna Maine, and Shital N. Parikh

17.1 Chapter Outline

Clinical case: This chapter presents a clinical case of a young woman with chronic patellofemoral pain and failed orthopedic treatment. Torsional abnormalities were identified in the physical examination, and radiologic findings confirmed these findings. Increased femoral anteversion creates internal rotation of the distal femur, which results in increased lateral patellofemoral pressure and lateral shift, similarly, increased external tibial torsion of the proximal tibia leads to an excess pulling on the medial soft tissues, augmented patellar shift, and lateral patellofemoral contact pressure. This scenario may result in chronic pain, patellar instability, chondromalacia, and it can evolve to osteoarthritis.

- Surgical treatment principles should be aimed to correct at the level of maltorsion with a torsional osteotomy. When femur and tibia are compromised, two levels of osteotomy should be performed.
- Specifically, in this case we propose an external torsional femoral osteotomy and an internal tibial rotational osteotomy.

M. Iñiguez C
Clínica Las Condes, Hospital Sótero del Río, Santiago, Chile
e-mail: miniguez@clinicalascondes.cl

P. Neyret (✉)
Infirmerie protestante, Lyon, France

Reem Hospital, Abu Dhabi, UAE

S. Maine
Department of Orthopaedics, Children's Health Queensland Hospital and Health Service, Brisbane, QLD, Australia

Griffith Centre of Biomedical and Rehabilitation Engineering, Griffith University, Gold Coast, QLD, Australia

S. N. Parikh
Cincinnati Children's Hospital Medical Center, University of Cincinnati College of Medicine, Cincinnati, OH, USA
e-mail: Shital.Parikh@cchmc.org

17.2 Case Presentation

A 23-year-old woman presented with 6 years of anterior knee pain that had been increasing in the last year. She had pain with daily living activities, especially with stairs, while in seated position, and with squatting. There was no history of patellofemoral instability.

Physiotherapy and non-operative orthopedic treatment were performed for a long time, but limited benefit and clinical response were obtained, and she had persistent pain.

© ISAKOS 2022
J. L. Koh et al. (eds.), *The Patellofemoral Joint*, https://doi.org/10.1007/978-3-030-81545-5_17

17.2.1 Clinical Assessment

In the standing position with her feet parallel, her patella was noted to be pointing inward. This sign is also called squinting patella (Fig. 17.1a). In the same position, when feet were outwardly rotated, her patellae were pointing anteriorly (Fig. 17.1b). Hip range of motion was assessed. Internal rotation of the hip with the patient in prone position was 60° (Fig. 17.2).

In a seated or supine position with knee flexed, the feet were also outwardly rotated.

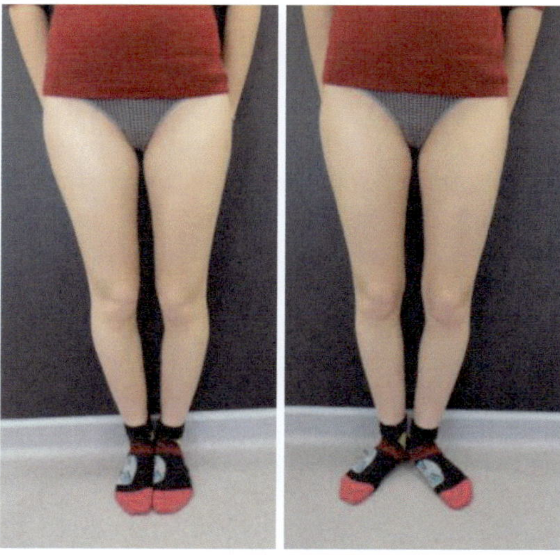

Fig. 17.1 (**a**) Standing position with parallel feet shows "squinting patella," (**b**) Standing position, patella is anterior with feet laterally rotated

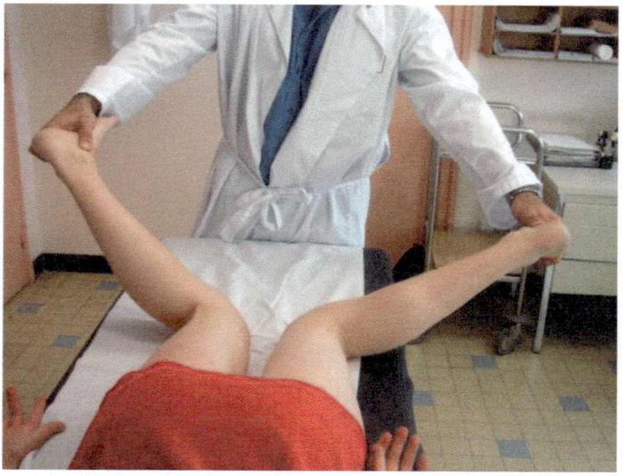

Fig. 17.2 Increased internal rotation of the hip

Gait analysis shows a neutral progression foot angle.

17.3 Radiologic Study

Knee X Rays: No crossing sign, patellar tilt 12° in the right knee, 8° at the left one. Long leg film shows medial pointing patella.

Preoperative CT Scan Values:

- Tibial tubercle-trochlear groove (TT-TG) 13.1 mm Right knee, 10.6 mm left knee
- Femoral anteversion: 24° right, 21.3° left. (Lerat Method)
- Tibial external rotation: 55.7° right, 45.7° left. (Lerat Method)

MRI: Chondromalacia of the lateral facet of patella.

17.4 Introduction

Torsional malalignment represents a complex problem; this condition usually consists of a combination of femoral anteversion (also known as femoral torsion or antetorsion) and tibial lateral torsion in variable amounts. The axis of knee motion is medially rotated. This leads to patellofemoral malalignment and can contribute to the pathogenesis of overload, chondromalacia, patellar subluxation, and even dislocation. The role of torsional deformities is less well understood; however, there is growing evidence that increased femoral antetorsion and external tibial torsion promotes abnormal lateral pressures in the patella.

Malalignment of the knee has been classified into three types depending on the affected plane [1].

1. Coronal (frontal)
2. Sagittal
3. Horizontal (transverse)

Torsional problems affect the horizontal or axial plane. These limb abnormalities can be found in different levels depending on the etiology, but the most frequent is at the proximal level.

Despite this classification, it is important to clarify that horizontal abnormalities can influence coronal and sagittal plans.

- Increased femoral neck anteversion at the hip creates an internal rotation of the distal femur, resulting in increased lateral patellofemoral pressure and lateral shift of the patella [2–5].
- Increased tibial external torsion creates a lateralization of the distal part of the extensor mechanism, increasing patellofemoral pressures in the lateral facet against the lateral trochlea and an excessive force pulling on the medial soft tissues [2–5].

17.5 Clinical Presentation

Typically, patients with significant torsional abnormalities describe patellofemoral complaints since childhood or adolescence. The main symptoms are pain and patellofemoral instability, but due to the global disturbance of the lower extremity, symptoms in the hip and feet are also common. In this group it is also not unusual to have a history of failed surgeries.

The physical examination is very important when we are assessing a patient with patellofemoral complaints.

In the standing position it is necessary to evaluate coronal, sagittal, and axial malalignments. In the last group where torsional problems take place, increased femoral anteversion causes the limb to rotate inwards, with the patella pointing inwardly. This appearance has been called "squinting patella." Compensatory external tibial torsion results in the feet pointing anteriorly while the patellae remain squinting, giving the appearance of a normal foot progression angle (Fig. 17.1a). If the patellae are oriented to point anteriorly, the feet are outwardly rotated (Fig. 17.1b). This condition where both femoral anteversion and external tibial torsion are increased is described as "Miserable malalignment syndrome."

Gait is also important; based on Staheli's work, the position of the feet will depend on which abnormality is predominant. The normal foot progression angle is between 10 and 15°. This angle can be increased when tibial external torsion is predominant; a negative progression foot angle is seen when excessive femoral anteversion is the predominant maltorsion [6].

Assessment of hip range of motion is also a key part of the physical examination. When femoral anteversion is increased, internal rotation of the hip is augmented. It is easiest to assess this in the prone position (Fig. 17.2). It is also essential to check the overall mobility of the hip, because if it is decided to correct the anteversion, the hip requires good rotation of the joint to be able to adjust.

17.6 Normal Development of Torsional Profile

The knowledge of rotational development during childhood allows us to understand associated deformity in adolescents and adults.

Population studies show that torsion of the femur and the tibia change during growth. Limb rotation starts in utero and is part of normal development. At birth, many infants have internal tibial torsion (2°–4°) and increased femoral anteversion nearly to 40° [6].

During childhood until 8–10 years old, tibial torsion increases up to 20°, and femoral anteversion decreases to 15° [6, 7].

According to Staheli [6] screening is important and that can be assessed with physical examination including measurement of foot progression angle, thigh-foot angle, and hip joint mobility, including internal and external rotation range of motion. In-toeing or out-toeing are the most common signs of torsional deformity during childhood. Patients near to 10 years old with clinical torsional abnormalities should be studied and treated, because after that age, changes in torsions are minimal.

17.7 Radiologic Study in Torsional Abnormalities

17.7.1 Femoral Torsion

The classical description of the femoral torsion angle is measured between the femoral neck axis and the posterior femoral condylar line. However, there are various published methods to measure femoral anteversion, using transverse or oblique and single or superimposed image slices. The techniques also use different anatomical landmarks for measurement. As a result, a wide range of the standard values for femoral torsion (7°–24.1° internal torsion) has been reported in the literature [8–13].

Kaiser's study comparing different techniques found significant differences depending on which method is used. A high level of reproducibility was found for all techniques, so it is possible to use different methods, but special attention should be taken to the normal values for each technique.

Lerat [13] described a line from the center of the femoral head (in one CT slice) and the base of the femoral neck in a second slice, and a second line the tangent to the posterior femoral condyles 2 cm proximal to the joint line. According to Waidelich [8], femoral anteversion was measured in the axial plane as the angle between a line connecting the center of the femoral head on one slice and the center of an ellipse around the greater trochanter on a second slice and the tangent to the posterior femoral condyles. Murphy [10] described a proximal line representing the axis of the femoral neck created by two points: the center of the femoral head taken from 1 CT slice and the center of a centroid located at the base of the femoral neck; distally, a line is placed across the posterior extension of both femoral condyles. The technique described by Yoshioka et al. is based on superimposed transverse slices [9]: the center of the femoral head on one transverse slice was connected to the center of the femoral neck at its narrowest width on another transverse slice. The same author also described an oblique measurement for the femoral neck axis.

The technique described by Hernandez et al. [11] measured on a single transverse slice the center of the femoral head connected to the center of the femoral neck on a single transverse slice, chosen in a place in which the femoral head, femoral neck, and greater trochanter were visible.

Therefore, when derotational osteotomy is being considered, it is essential to use recognized that there are different threshold values depending on the measurement technique.

The average femoral torsion (anteversion) in most studies measures between 10° and 15°. Yoshioka noted 13° average femoral torsion measuring intact dry femurs [14]. Lerat described 10.9° ± 7.1 in normal controls compared with 16.1° ± 9.4 in the patellar instability group ($p < 0.005$), and 17° ± 11.7 in the group of patients with patellar chondropathy ($p < 0.001$).

Strecker [15], using Waidelech's method, described in 505 femurs mean femoral anteversion of 24.1 ± 17.4°. No difference was found between sides.

17.7.2 Tibial Torsion

External tibial torsion (ETT) is commonly measured between the posterior tibial plateau line and the axis of the ankle malleoli [16], but similar to the evaluation of femoral torsion several measurement methods have been described [8, 13, 17, 18] and as a consequence, normal values can vary. Lerat [13] described a CT slice through the upper extremity of the tibial epiphysis (under the plateau level, beneath the articular surface), one at the level of the ankle malleoli, and also one through the foot, including the metatarsal heads and calcaneum. Normal values in this study were ETT 32.8° ± 7.8, compared with 34.1° ± 11 in patients with patellar chondropathy, and 36.1° ± 9.8° in the instability group.

Jakob [18] used proximally a trans-condylar measurement through a line in the widest part of the tibial metaphysis, and distally a juxtarticular line through the tibia and lateral malleoli. Jend [16] used the posterior condylar line proximally, and distally a line connects the midpoint of the incisura with the center of a circle created from the tibial plafond junction with the incisura. In this study, normal tibial torsion was 40° ± 9°.

Other authors have described also a posterior femoral condyle line [19], due to the variability of the measurement in proximal tibia, and distally trans-malleolar line. Reikerås [19] using these references found in normal female population, an ETT of 38° ± 9 on the right side and 37° ± 11 on the left side. In the males the values were 41° ± 6 and 40° ± 10, respectively.

Strecker [15] using Waidelich's method using proximally the posterior condylar tangent, and distally a line bisecting the ellipse formed from the incisura and the ellipse formed from the medial malleolar surface. In 504 tibiae, he described a mean external tibial torsion of 34.9 ± 15.9°, but a difference between sides was noted being 3° higher in the right side in the ETT.

Measures using CT of skeletal specimens averaged 30° by Jakob [18] and 28° by Takai [20]. Reported measures from CT of normal controls vary even more widely from 26° to 42° [17], and measures from MRI of normal controls are reported at 25°–42° [17]. These important variations make surgical recommendations challenging, yet small series indicating patient improvements after surgical correction argue the importance of tibial maltorsion correction and the need for reliable, standardized measurements, and greater validity of studies.

17.7.3 Torsional Problems Related to Patellofemoral (PF) Pain, Chondropathy, & Osteoarthritis (OA)

There are several clinical studies relating maltorsion and patellofemoral (PF) problems. [20–27].

Takai et al. [20] measured femoral and tibial torsion in patients with unicompartmental osteoarthritis of the medial, lateral, or PF joint, and described a significant correlation between increased internal femoral torsion and PF OA compared with the control group; in this study, patients with patellofemoral OA had 23° internal femoral torsion, and the control group had 9°.

Lerat et al. [13] noted a significantly increased association of internal femoral torsion with patellar chondropathy ($P < 0.001$) and patellar instability ($p < 0.005$).

Winson's [28] study described that 70% of adolescents undergoing arthroscopy for anterior knee pain had increased internal torsion of the femur, in the control group (arthroscopy due to ligament or meniscus injury) was present in only 33%.

Another study [27] reported 12 patients with chronic patellofemoral symptoms unresponsive to conservative treatment. These 12 patients had significantly increased external tibial torsion as compared to those without patellofemoral pain.

Also, multiple clinical publications show a clear improvement in clinical outcomes after treatment with surgical derotational osteotomy in patients with severe torsional problems and refractory pain [21–25].

17.8 Biomechanical Evidence

Several studies describe the biomechanical effects of torsional problems in patellofemoral contact and pressures and secondary effects on soft tissues [2–5, 29, 30].

Increased external tibial rotation produce changes in biomechanics in a rotational plane, pulling first the distal pole and then more proximally in the patella, produce increased pressure at the lateral facet and stress on medial soft tissues [5].

Lee et al. [4] also analyzed the effects of fixed femoral torsion on patellofemoral joint contact pressures. These authors reported a nonlinear increase in the patellofemoral contact pressures as a function of increasing femoral rotation. Internal rotation of the femur resulted in an increase in contact pressures on the lateral facet of the patella.

The authors concluded that femoral rotation greater than 20° induce severe alterations to the biomechanics of the patellofemoral joint.

More recent biomechanical studies confirm previous findings, especially in femoral antetorsion. Kaiser et al. [31] showed antetorsion between 10 and 20° produce changes in patellar tilt shift, and over 20° the patellofemoral biomechanics are severely affected increasing soft tissue failure.

17.8.1 The Concept of Horizontal Morphotype

Lerat introduced in 1988 two important concepts regarding torsional problems the Tibiofemoral Index (TFI), and the Sum of Femoral and Tibial Torsion (SFTT).

The TFI is the difference between tibial torsion and femoral anteversion.

The external tibial torsion is always greater than the femoral torsion.

A high TFI indicates that the external tibial torsion is the predominant problem, while a low TFI indicates that the two torsions (femoral and tibial) are more closely matched. These two torsions can be either small or large.

A high TFI appears to be more frequent in patellar pathology, and a low TFI more frequent in normal knees.

When low TFI is associated with patellar pathology, this is often related to large femoral and tibial torsions. In this morphotype, the knee is constrained between two large and inverse torsions, which cause abnormally large stresses on the patellofemoral joint. In this situation, the SFTT is high.

The Sum of Femoral and Tibial Torsion (SFTT) was found to be a relevant factor by Lerat. He found a statistically significant difference ($P < 0.005$) between the normal knees (4.5° ± 13) and those with patellar chondropathy (56.6° ± 18.2°) and also with patellar instability (59.6° ± 16.1°).

17.9 Surgical Indications

Surgery can be considered in patients who have persistent symptoms or pain despite appropriate physiotherapy and medical treatment. However, due to controversies about measurements, normal values, and evidence-based mainly on expert opinion, it is difficult to make specific indications for corrective torsional osteotomy.

According to Tietge [1], if the torsional deformity is greater than 30° or if there is significant associated genu valgus (>5°), then osteotomy is definitely beneficial. If limb torsion exceeds normal by 20°, realignment is felt to be beneficial [1].

Other authors suggest corrections depending on absolute values of the segment. Hinterwimmer [32], Biedert [33], and Imhoff [34] recommend surgical correction for femoral anteversion more than 20°.

At the tibia level, there is huge variation in the literature, and it is very difficult to find a single "normal" value.

According to Fouilleron [35] and Server [36], osteotomy is recommended for patients with clinical and radiological measurements over 30° of ETT. The mean amount of correction was 25°. Drexler [19] and Paulos [37] suggest correction in cases with over 30° based on clinical evaluation of thigh-foot angle.

Lerat and Ruguet are more conservative, and recommend tibial osteotomy when the ETT value is more than 35°. The mean amount of correction in their study was 12°, with a maximum correction of 20°.

Osteotomies of less than 10° should not routinely be attempted given the surgical morbidity of osteotomy, risk of nonunion or malunion, risk of overcorrection or undercorrection, and risk of creating abnormal varus or valgus or flexion or extension deformity [1].

It is important to note that these thresholds are mainly based on expert recommendations.

17.10 Surgical Technique

17.10.1 Femoral Derotational Osteotomy

The intertrochanteric region is selected over a supracondylar osteotomy to reduce scarring of the quadriceps near the knee joint, and to allow a longer bone length for the thigh muscles to adjust to a new direction.

A lateral trochanteric approach is used, followed by longitudinal incision of fascia lata and anterior reflection of the vastus lateralis. The metaphyseal-diaphyseal junction is identified.

A guide pin is placed in the femoral neck under fluoroscopic control. Kirschner wires can be placed proximal and distal to the proposed osteotomy site with the planned amount of rotational correction. An osteotomy is performed with a saw at the upper level of the lesser trochanter. A 95-degree condylar blade plate is positioned in the femoral neck and femoral derotation is performed.

Imhoff recommends a supracondylar femoral osteotomy for combined rotational and angular correction if the patient has an associated genu valgum or varus where the tibiofemoral angle abnormality may need to be corrected near the knee joint [34].

The goal of correction is to reach a value close to 15° of femoral anteversion, and a limb-neutral mechanical axis. In this case the rotational correction was 10°, and no frontal correction was needed.

17.10.2 Tibial Derotational Osteotomy

A lateral approach to the knee is performed, and the peroneal (external popliteal sciatic) nerve is identified and protected. A transverse osteotomy is done at the proximal tibia above the tibial tubercle. This allows for better bone healing and lowers the risk of tibial nonunion. For improved correction, it is recommended that a proximal fibular osteotomy and tibial tuberosity osteotomy are performed. The risk of neurovascular complications is higher in this particular osteotomy.

Two Kirschner wires guides are placed proximal and distal to the osteotomy site with the estimated correction angle between them.

After a fluoroscopic control of the position of the pins:

- A tibial osteotomy is performed with an oscillating saw and carefully completed with osteotomes to minimize neurovascular risk.
- The distal tibia is internally rotated to correct to the calculated angle.
- Tibial fixation is performed with 2 staples.
- Tibial tuberosity osteotomy is fixed with two 4.5 cortical screws. The position of the anterior tibial tubercle can be adjusted to achieve the desired TT-TG and patella height.

In this case, the planned amount of correction was 10° to reach a goal of less than 45 degrees of external rotation. Excessive correction was avoided due to the risk of neuropraxia.

Preoperative and postoperative long cassette X Rays are shown in Fig. 17.3.

After surgery, the patient demonstrated an improvement in terms of pain and symptoms. Full weight-bearing was allowed at the third-month postop, and normal gait was achieved. The patient was fully recovered by sixth months, with smooth patella tracking and a 0/0/130 range of motion. No other surgeries were needed.

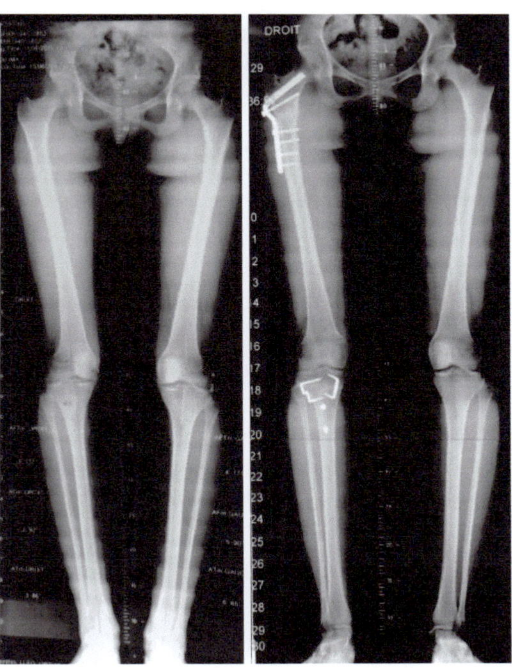

Fig. 17.3 Pre and postop X Rays

17.10.3 Commentary on Torsional Case by Sheanna Maine

With regard to managing cases of torsional malalignment, the decision to derotate the femur or tibia or both is determined by the presenting complaint being one or a combination of pain, instability or functional disability from abnormal foot progression angle. Work up includes AP and lateral long leg views to exclude deformity in other planes, as well as an MRI rotational profile with dedicated cuts through the hip, mid-thigh, knee, and ankle using a standardized protocol.

The decision to derotate is relatively easy when faced with functional disability from in-toeing. The surgery is technically straightforward and outcomes generally predictable. The decision to derotate in cases of PFP and instability is less well understood. My management of femoral anteversion as it relates to anterior knee pain/instability also involves assessment of femoral torsion as well as the quadriceps torsion angle (QTA) as this does seem to influence the lateral force vector across the patella and may influence lateral contact pressure. I am reluctant to derotate a femur too much, regardless of the femoral torsion value, if it aligns well with the quadriceps mechanism and the patella is not sitting subluxed in extension. The femoral osteotomy is easily performed proximally and can be fixed with a plate in the skeletally immature, or with a nail in the adult. The resulting position of the foot can be calculated preoperatively and will determine whether a tibial osteotomy needs to be performed.

Should it be necessary to proceed with a tibial derotation, it is important to consider the TT-TG distance, as this can indicate whether tibial alignment should be corrected using a proximal or distal osteotomy. In cases of miserable malalignment with instability, where the TT-TG distance is high, the TT-TG can be corrected using a supratubercular osteotomy as described. If the TT-TG distance is normal, then the tibial osteotomy is technically much easier performed at the supramalleolar level using a plate for fixation.

As always, the decision to operate should be based on a patient's functional assessment and their appreciation of the risk benefits profile that surgery offers. There are many patients including high functioning athletes with values outside of the "normal" range that cope very well without realignment surgery having had good rehabilitation.

17.10.4 Commentary on Torsional Chapter by Shital Parikh

The authors present an interesting case of PF pain and arthrosis in a 23-year-old female. I agree with the authors that rotational (transverse) plane malalignment, including excessive femoral anteversion and external tibial torsion, could lead to increase in lateral patellar forces. This could lead to PF pain, arthrosis, and sometimes PF instability. However, what constitutes rotational plane malalignment and the indications for surgical correction are unclear from the existing literature. As the authors have discussed, there are numerous methods to calculate femoral anteversion and external tibial torsion. Though each method has acceptable inter- and intra-observer reliability, the normative value differs by about 11° based on the method used for measurement. Thus, it is difficult to compare the literature and study the outcomes based on objective measurements of rotational profile. Also, patellofemoral pathologies are multifactorial in nature and there are several possible etiologies responsible for disruption of normal patellofemoral homeostasis and envelope of function [38]. Thus, decisions about correction of rotational malalignment are often based on individual patient, physician expertise, and preferred method of measurement of rotational profile with varying threshold values. There are physicians on either end of the spectrum: those who are committed to correction of lower limb rotational malalignment by double-level (femur and tibia) osteotomies and those who try to avoid this high-morbidity surgery unless less morbid procedures (like tibial tubercle osteotomy) have failed to compensate for rotational malalignment or in severe cases (loss of function and ambulation capacity) of rotational malalignment.

As far as patient assessment is concerned, the information presented in current case is too less to make an appropriate medical decision. Information related to history of patellar dislocation, timeline of development of symptoms, unilateral vs bilateral involvement, past treatment and predominant symptoms (pain vs instability) should be gathered. A detailed clinical evaluation should include physiologic hyperlaxity (Beighton score), standing alignment with knees extended and knees slightly flexed (to eliminate terminal tibial external rotation during screw-home mechanism), foot position (pes planus), gait analysis, patellar tracking, patellar mobility, crepitus, areas of tenderness, apprehension and muscle strength (especially gluteus maximus and quadriceps). Clinical assessment of rotational profile should include hip internal and external rotation in a prone position (Fig. 17.2), thigh-foot angle and foot progression angle. Radiographs should include AP, notch view, lateral view, and merchant view to check for joint space narrowing, osteophytes, patellar tilt, patellar subluxation, patellar height, medial patellar avulsion fracture, and trochlear dysplasia. Full-length weight-bearing radiograph would help to evaluate coronal plane alignment and limb length discrepancy, but it should be performed with the patella facing forward. If the full-length radiograph is performed with the patella facing medially, as in the presented case, an assessment of coronal plane malalignment could be spurious. MRI of the knee would help to evaluate for soft tissue structures including MPFL and lateral retinaculum and gradient-echo sequences can help to evaluate the extent of chondral lesion. CT or MRI rotational profile based on axial images of hip, knee, and ankle is used to calculate femoral anteversion and tibial torsion. Either a dedicated musculoskeletal radiologist or physician themselves should do these measurements so that consistency is maintained over period of time.

Based on the available information and for reasons mentioned below, my approach for treatment of this patient would have been different. First, rotational osteotomies of femur and tibia have shown good results in adolescent age group, preferably before the onset of PF arthritis. The mean age of patient was 14 years (range 13–18) in the classic article by Bruce and Stevens [24]. Advanced age and advanced patellofemoral chondrosis have been shown to have negative effects on outcomes. Second, prone hip internal rotation of 60°, as shown in Fig. 17.2, is within the normal hip range of motion as noted by Staheli et al. [39]. The indications for rotational osteotomies based on existing literature are summarized in Table 17.1 [24, 39, 40] I would consider correction of rotational malalignment when the hip internal rotation is beyond 75–80°. If the patient has joint hyperlaxity, the threshold would be even higher at about 80–85° or more. Third, the CT measurements of femoral anteversion (24°) do not meet the threshold criteria for correction. If normal femoral anteversion is considered to be around 13°, then the threshold to perform corrective osteotomy is about 20–30° above normal [41, 42]. Thus for this case, I would not perform femoral derotational osteotomy.

If the patient has exhausted conservative treatment (anti-inflammatories, physical therapy, bracing, cortisone injection, activity modification), then my surgical plan would include knee arthroscopy, lateral retinacular release, and anteromedialization osteotomy of tibial tubercle

Table 17.1 Indication for femoral and tibial rotational osteotomies based on physical evaluation

	Femur			Tibia	
	Hip IR[a]	Hip ER[a]	Anteversion[b]	Thigh-foot ang[a]	External tibial torsion[b]
Staheli (1989) [39]	>85°	<10°	>50°	–	>30°
Leonardi (2014) [40]	81.5° (80–85°)	27.2° (10–40°)	–	38.6° (32–45°)	–
Bruce (2004) [24]	85° (75–95°)	33° (10–60°)	35° (22–54°)	32° (25–40°)	36° (28–52°)

[a]Based on physical examination
[b]Based on cross-sectional imaging

to offload the lateral aspect of patella. If the lateral facet of patella is significantly degenerated or there is an overhanging lateral osteophyte, then lateral facetectomy would help to alleviate her pain. For minor increases in external tibial torsion, TTO should suffice. If after thorough clinical and radiographic evaluation, significant external tibial torsion is diagnosed, then I would agree with tibial rotational osteotomy. If normal tibial torsion is considered to be around 21°, the threshold for corrective osteotomy would be around 40°. In a study comparing tibial tubercle osteotomy with proximal tibial rotational osteotomy, the results favored rotational osteotomy though the article was directed towards PF instability and not PF arthrosis [37]. When performing unilateral rotational osteotomy, patient should be counseled about limb asymmetry which is inevitable and noticeable.

Take-Home Message

- Torsional abnormalities affect the femur and/or tibia in their axial plane.
- Patellofemoral symptoms are multifactorial, so a complete cause assessment should be done in this group of patients.
- Increased femoral anteversion and external tibial torsion are the main torsional abnormalities related to patellofemoral symptoms.
- Clinical and biomechanical studies confirm a relation between torsional abnormalities and patellofemoral pain and chondropathy.
- Medical treatment should be considered as primary treatment but often rehabilitation and medications have poor and limited effects. The problem is mechanical and without correction of the bone deformity one cannot expect a real improvement. But in the other hand the surgical program is complex. The surgical management consists of multiple surgeries and their potential complications.
- Differences in measurement methods and variety of "normal "values used make surgical indications more difficult to define.
- Detection and early treatment should be done during childhood and adolescence, after which the abnormalities are fixed and are more challenging to address.
- Surgery should be indicated, after a long well conducted and followed conservative management that failed. Then the surgical option is better accepted.

References

1. Tietge R. Osteotomy in the treatment of patellofemoral instability. Tech Knee Surg. 2006;5(1):2–18.
2. Van Kampen A, Huskies R. The three-dimensional tracking pattern of the human patella. J Orthop Res. 1990;8(3):372–82.
3. Hefzy MS, Jackson WT, Saddemi SR, et al. Effects of tibial rotations on patellar tracking and patello-femoral contact areas. J Biomed Eng. 1992;14(4):329–43.
4. Lee TQ, Anzel SH, Bennett KA, Pang D, Kim WC. The influence of fixed rotational deformities of the femur on the patellofemoral contact pressures in human cadaver knees. Clin Orthop. 1994:69–74.
5. Lee TQ, Morris G, Csintalan R. The influence of tibial and femoral rotation on patellofemoral contact area and pressure. J Orthop Sports Phys Ther. 2003;33(11):686–93.
6. Staheli LT, Corbett M, Wyss C, King H. Lower-extremity rotational problems in children. Normal values to guide management. J Bone Joint Surg Am. 1985;67(1):39–47.
7. Le Damany P. La torsion du tibia, normale, pathologique, experimentale. J Anat Physiol. 1909;45:598–615.
8. Waidelich HA, Strecker W, Schneider E. Computed tomographic torsion-angle and length measurement of the lower extremity. the methods, normal values and radiation load. RoFo. 1992;157(3):245–51.
9. Yoshioka Y, Cooke TD. Femoral anteversion: assessment based on function axes. J Orthop Res. 1987;5(1):86–91.
10. Murphy SB, Simon SR, Kijewski PK, Wilkinson RH, Griscom NT. Femoral anteversion. J Bone Jt Surg Am. 1987;69(8):1169–76.
11. Hernandez RJ, Tachdjian MO, Poznanski AK, Dias LS. CT determination of femoral torsion. AJR Am J Roentgenol. 1981;137(1):97–101.
12. Jarrett DY, Oliveira AM, Zou KH, Snyder BD, Kleinman PK. Axial oblique CT to assess femoral anteversion. AJR Am J Roentgenol. 2010;194(5):1230–3.
13. Lerat JL, Moyen B, Bochu M, et al. Femoropatellar pathology and rotational and torsional abnormalities of the inferior limbs: the use of CT scan. In;: Surgery and arthroscopy of the knee. 2nd congress of the European S European Society Ed. Miller & Hackenbruch: Springer; 1988.

14. Kaiser P, Attal R, Kammerer M, et al. Significant differences in femoral torsion values depending on the CT measurement technique. Arch Orthop Trauma Surg. 2016;136:1259–64. https://doi.org/10.1007/s00402-016-2536-3.
15. Strecker W, Keppler P, Gebhard F, Kinzl L. Length and torsion of the lower limb. J Bone Joint Surg Br. 1997;79-B:1019–23.
16. Jend HH, Heller M, Dallek M, Schoettle H. Measurement of tibial torsion by computer tomography. Acta Radiol Diagn (Stockh). 1981;22:271–6.
17. Tietge R. The power of transverse plane limb malalignment in the genesis of anterior knee pain—clinical relevance. Ann Jt. 2018;3:70.
18. Jakob R, Haetel M, Stüssi E. Tibial torsion calculated by computerised tomography and compared to other methods of measurement. J Bone Joint Surg Br. 1980;62-B(2):238–42.
19. Drexler M, Dwyer T, Dolkart O, Goldstein Y, Steinberg EL, Chakravertty R, Cameron JC. Tibial rotational osteotomy and distal tuberosity transfer for patella subluxation secondary to excessive external tibial torsion: surgical technique and clinical outcome. Knee Surg Sports Traumatol Arthrosc. 2014;22(11):2682–9.
20. Takai S, Sakakida K, Yamashita F, et al. Rotational alignment of the lower limb in osteoarthritis of the knee. Int Orthop. 1985;9(3):209–15.
21. Eckhoff DG, Brown AW, Kilcoyne RF, et al. Knee version associated with anterior knee pain. Clin Orthop Relat Res. 1997;339:152–5.
22. Meister K, James SL. Proximal tibial derotational osteotomy for anterior knee pain in the miserably malaligned extremity. Am J Orthop. 1995;24:149–55.
23. Delgado ED, Schoenecker PL, Rich MM, et al. Treatment of severe torsional malalignment syndrome. J Pediatr Orthop. 1996;16:484–8.
24. Bruce WD, Stevens PM. Surgical correction of miserable malalignment syndrome. J Pediatr Orthop. 2004;24:392–6.
25. Walton DM, Liu RW, Farrow LD, Thompson GH. Proximal tibial derotation osteotomy for torsion of the tibia: a review of 43 cases. J Child Orthop. 2012;6(1):81–5.
26. Cameron JC, Saha S. External tibial torsion: an underrecognized cause of recurrent patellar dislocation. Clin Orthop Relat Res. 1996;328:177–84.
27. Cooke TD, Price N, Fisher B, Hedden D. The inwardly pointing knee. An unrecognized problem of external rotational malalignment. Clin Orthop. 1990;56–60.
28. Winson, Miranda, Smith. Acta Orthop Scand. 1990;61(62 suppl):237.
29. Liska F, von Deimling C, Otto A, Willinger L, Kellner R, Imhoff AB, et al. Distal femoral torsional osteotomy increases the contact pressure of the medial patellofemoral joint in biomechanical analysis. Knee Surg Sports Traumatol Arthrosc. 2018;27(7):2328–33. https://doi.org/10.1007/s00167-018-5165-2.
30. Gresalmer R, Dejour D, Gould J. The pathophysiology of patellofemoral arthritis. Orthop Clin N Am. 2008;39:269–74.
31. Kaiser P, Schmoelz W, Schoettle P, Zwierzina M, Heinrichs C, Attal R. Increased internal femoral torsion can be regarded as a risk factor for patellar instability—a biomechanical study. Clin Biomechan. 2017;47:103.
32. Hinterwimmer S, Rosenstiel N, Lenich A, Waldt S, Imhoff AB. Femoral osteotomy for patellofemoral instability. Unfallchirurg. 2012;115(5):410–6.
33. Biedert RM. Osteotomies. Der Orthop. 2008;37(9):872, 874–876, 878–880 passim.
34. Imhoff FB, Cotic M, Liska F, Dyrna FGE, Beitzel K, Imhoff AB. Derotational osteotomy at the distal femur is effective to treat patients with patellar instability. Knee Surg Sports Traumatol Arthrosc. 2019;27:652–8.
35. Fouilleron N, Marchetti E, Autissier G, Gougeon F, Migaud H, Girard J. Proximal tibial derotation osteotomy for torsional tibial deformities generating patello-femoral disorders. Orthop Traumatol Surg Res. 2010;96:785–92.
36. Server F, Miralles RC, Garcia E, et al. Medial rotational tibial osteotomy for patellar instability secondary to lateral tibial torsion. Int Orthop. 1996;20(3):153–8.
37. Paulos L, Swanson SC, Stoddard GJ, Barber-Westin S. Surgical correction of limb malalignment for instability of the patella: a comparison of 2 techniques. Am J Sports Med. 2009;37(7):1288–300.
38. Dye SF. The knee as a biologic transmission with an envelope of function: a theory. Clin Orthop Relat Res. 1996;(325):10–8.
39. Staheli LT. Torsion—treatment indications. Clin Orthop Relat Res. 1989;(247):61–6.
40. Leonardi F, Rivera F, Zorzan A, Ali SM. Bilateral double osteotomy in severe torsional malalignment syndrome: 16 years follow-up. J Orthop Traumatol. 2014;15(2):131–6.
41. Teitge RA. Patellofemoral syndrome a paradigm for current surgical strategies. Orthop Clin North Am. 2008;39(3):287–311.
42. Duparc F, Thomine JM, Simonet J, Biga N. Femoral and tibial bone torsions associated with internal femoro-tibial osteoarthritis. Index of cumulative torsions. Rev Chir Orthop. 1992;78:430–7.

18 Anterior Knee Pain in a Patient with a Central Trochlea Defect: A 32-Year-Old Man with a Central Trochlear Defect

Jason L. Koh, Jack Farr, Yukiyoshi Toritsuka, Norimasa Nakamura, Alberto Gobbi, and Ignacio Dallo

18.1 Case 1

18.1.1 History

A 32-year-old recreational basketball player with a history of a previous allograft ACL reconstruction performed at a different institution presents with several months of anterior knee pain, a catching sensation, and difficulty with stairs, waterskiing, and basketball. There was some anterior knee pain following the ACL reconstruction but the pain progressively worsened. The knee subjectively felt stable. He underwent extensive physical therapy consisting of quadriceps and hip strengthening, McConnell taping, and other modalities, and had a cortisone injection which provided temporary relief lasting 3 weeks. His pain was at a 5–6 VAS level on a daily basis with stairs and increased to 7–8/10 with sports activity.

J. L. Koh (✉)
Mark R. Neaman Family Chair of Orthopaedic Surgery, Evanston, IL, USA

Orthopaedic & Spine Institute, Skokie, IL, USA

NorthShore University HealthSystem, Evanston, IL, USA

University of Chicago Pritzker School of Medicine, Chicago, IL, USA

Northwestern University McCormick School of Engineering, Evanston, IL, USA

J. Farr
Indiana University School of Medicine, OrthoIndy and OrthoIndy Hospital, Greenwood, IN, USA

Y. Toritsuka
Department of Orthopaedic Sports Medicine, Kansai Rosai Hospital, Amagasaki, Hyogo, Japan

N. Nakamura
Insitute for Medical Science in Sports, Osaka Health Science University, Osaka, Japan

A. Gobbi
O.A.S.I. Bioresearch Foundation, Milan, Italy

I. Dallo
Orthopaedic Arthroscopic Surgery International (O.A.S.I.) Bioresearch Foundation, Gobbi N.P.O, Milan, Italy

18.1.2 Physical Examination

The patient was 1.9 m and 100 kg. The left knee had a range of motion from −1 to 135°. There was a small effusion and patellofemoral crepitus and some pain with extension. Patella compression caused reproducible anterior knee pain. There was no medial or lateral joint line tenderness. The Lachman was 1A and the pivot shift test was normal.

18.1.3 Imaging

X-rays demonstrated good joint space preservation and a somewhat central and anterior placed ACL reconstruction.

© ISAKOS 2022
J. L. Koh et al. (eds.), *The Patellofemoral Joint*, https://doi.org/10.1007/978-3-030-81545-5_18

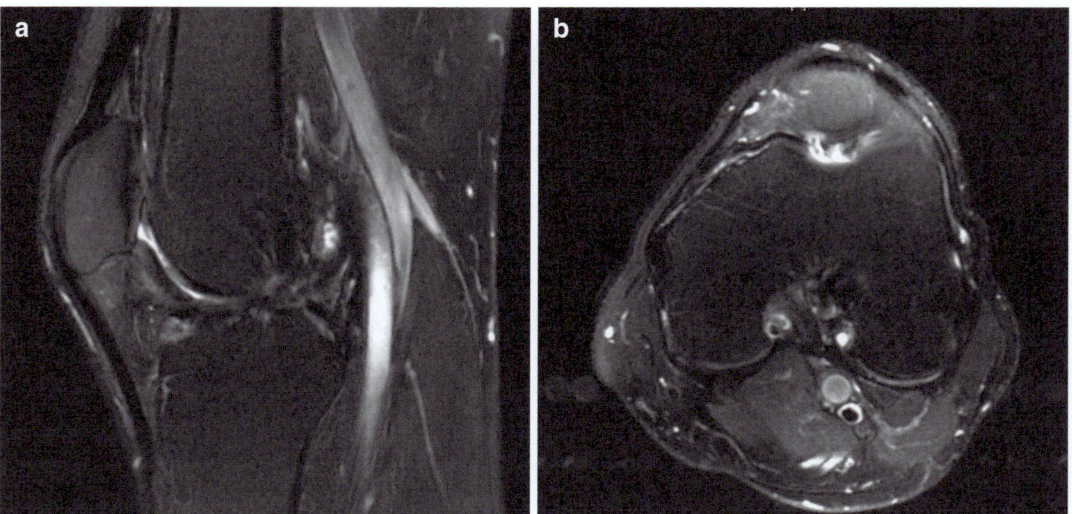

Fig. 18.1 MRI knee. (**a**) Sagittal image showing 1.2 cm full-thickness defect superior-inferior. (**b**) Axial image showing 1.4 cm defect medial-lateral

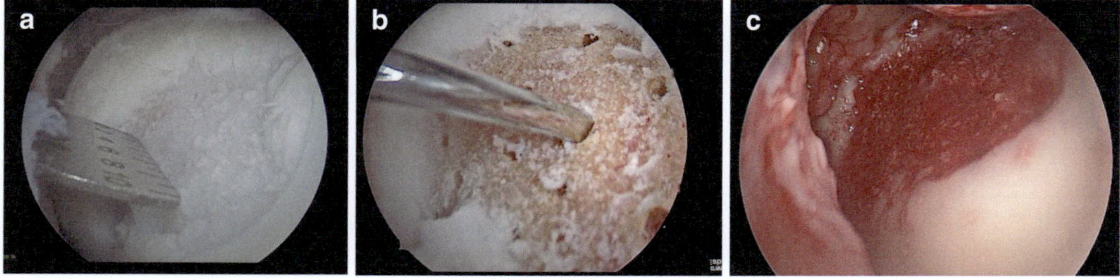

Fig. 18.2 Intra-operative photos. (**a**) Measurement of chondral defect. (**b**) Marrow stimulation with K-wire. (**c**) Defect grafted with micronized allograft cartilage mixed with autologous platelet rich plasma

MRI (Fig. 18.1) showed an intact ACL graft. There was a central articular cartilage defect measuring 1.2 cm × 1.4 cm. There was thickened anterior capsular tissue. The TTTG was 12 mm and the patella height was normal.

18.1.4 Treatment

Diagnostic arthroscopy was performed, demonstrating a full-thickness cartilage defect in the trochlea and dense anterior knee capsule limiting arthroscope mobility. The anterior interval capsule was divided, and the unstable edges of the chondral defect were debrided until the margins remained stable. The calcified cartilage layer was removed down to bare bone. The prepared defect measured approximately 12 mm superior-inferior by 16 mm mediolateral in regular shape (Fig. 18.2a). Multiple perforations were made with 0.045 K-wire (Fig. 18.2b) and then autologous platelet rich plasma was mixed with the micronized cartilage allograft. This was grafted into the defect using a cannula and then stabilized with the fibrin glue placed over the graft. The knee was placed through a range of motion and the graft remained stable within the defect (Fig. 18.2c).

18.1.5 Rehabilitation

Continuous passive motion was used for 1–2 h per session, 2–3 h per day, with the brace unlocked. Weightbearing was permitted with the knee in extension. Range of motion with weight-bearing progressed over the course of 4–6 weeks to 90°, and the brace discontinued after appropriate progress in physical therapy including the ability to straight leg raise. Progressive strengthening proceeded.

18.1.6 Results

After 10 months the patient had subjective improvement in symptoms to about 80% of normal and some pain (up to 3/10 VAS) with heavy loading and sports. He was able to perform squats and lunges without difficulty. There was trace crepitus and no effusion. Repeat MRI demonstrated near-complete defect fill (Fig. 18.3).

At 2 years follow-up the patient had minimal crepitus and no pain with daily activities. He had returned to working out at the gym and had some mild discomfort (1–2/10) with running.

18.1.7 Discussion

Patellofemoral chondral damage after ACL reconstruction is reported in up to 50% of patients [1]. Patellofemoral contact pressure is increased by 27% in the ACL reconstructed knee of elite athletes during running [2]. This may have several different causes, but one proposed mechanism is intraarticular scarring of the anterior knee joint capsule or suprapatellar pouch causing an increase in patellofemoral contact pressures and excessive load [3, 4]. Release of anterior knee capsular scarring and intraarticular adhesions has been successfully used to help address patellofemoral joint pain in the post-ACL reconstructed knee [5]. In this case, the anterior joint capsule was noted to be thickened, resulting in adhesion of the patella to the anterior tibia plateau. Capsular release resulted in improved patella mobility.

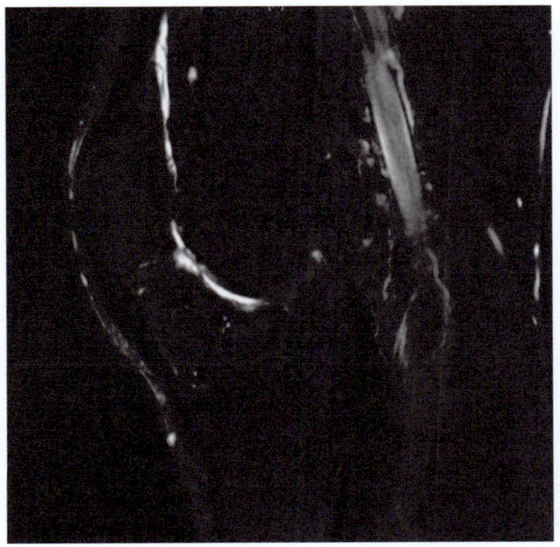

Fig. 18.3 Post-op MRI showing defect fill

Trochlear articular cartilage defects are challenging to treat given the high loads and shear forces as well as irregular shapes. Marrow stimulation remains the first line treatment for many situations involving relatively smaller defects [6]; and a recent systematic review has demonstrated good results [7]; however, some investigators have found it to be relatively less successful in the patellofemoral joint [8]. Enhancement of microfracture by using additional scaffold material may improve repair tissue formation and enhanced repair [9–11]. Smaller holes appear to have less morbidity and improved cartilage repair when compared to standard microfracture techniques [12]. Therefore, a modified and enhanced marrow access technique was chosen given the size and shape of the lesion, and the patient was able to return to normal function.

Take-Home Message

Post-surgical patellofemoral cartilage lesions maybe related to intraarticular adhesions and capsular contractures. Trochlear lesions are challenging to treat but can be successfully treated with enhanced marrow stimulation techniques.

Fact Boxes

1. Post-surgical adhesions can contribute to increased patellofemoral joint contact pressure.
2. Smaller (<2 cm^2) defects in the patellofemoral joint can be successfully treated with marrow stimulation.
3. The addition of extracellular matrix to marrow stimulation may serve as a scaffold for tissue repair.

18.2 Commentary by Jack Farr, OrthoIndy

Patellofemoral pain after ACL injury is common and future PF chondrosis is also reported [13]. These problems have a multifactorial etiology that range from the hostile chemical milieu that occurs at the time of ACL tear to altered knee biomechanics as a result of ACL deficiency. In this case, there is the additional unknown of the biomechanical consequences of a malpositioned ACL graft, yet is intact and has afforded clinical stability. It is beyond the scope of this chapter to discuss why and when to revise an ACL graft.

Independent of the ACL injury and surgery, trochlear chondral lesions are common in the same population that suffer ACL tears. That is, sporting individuals who apply high PF loads repetitively (repetitive microtrauma). Why these patents with "normal" PF morphology and tracking develop more trochlear lesions than patellar is still under debate (potential explanations include: thinner trochlear cartilage, morphology that allows concentration of shear forces in the groove that may damage the articular cartilage collagen and/or lead to central delamination near the bone cartilage interface, etc.) [14].

When assessing a trochlear lesion, it is important to measure along the contour of the base. Measuring edge to edge with a flat ruler will underestimate the dimensions of the lesion "floor." A flexible ruler will yield longer measurements. In this case, the flat ruler measured 1.2×1.4 cm^2. It is easy to envision this measuring 1.4×1.6 cm^2, which equals 2.24 cm^2. With lesions over 2 cm^2, this opens the resurfacing algorithm to cell therapy and osteochondral allograft [15].

Both autograft and allograft plugs are technically challenging to match the local topology precisely, while matching the "saddle" topology is straightforward with cultured chondrocytes on a flexible collagen patch. With larger lesions, it is appropriate to discuss the pros and cons, risk/reward ratio of adding a straight tibial tuberosity anteriorization. While there are no case matched clinical series reported to date, a biomechanical study noted a 20–30% decrease in peak trochlear loading with straight anteriorization of 15 mm [16]. Post-operatively, it is imperative that the implant is protected with a PF cell therapy protocol, which is similar to the one used with augmented marrow stimulation in this case [17].

Comments

Yukiyoshi Toritsuka
Department of Health and Sports Sciences, Mukogawa Women's University
Norimasa Nakamura
Institute for Medical Science in Sports, Osaka Health Science University

This is a case of a patient with a cartilage defect less than 2 cm^2 in the femoral groove after an ACL reconstruction, treated with a marrow stimulation procedure and grafting of the defect with micronized allograft articular cartilage extracellular matrix with autologous PRP. The clinical results of this patient seem to be rated as good to excellent.

As we have no experience to apply micronized cartilage allograft in Japan, we cannot comment on that. However, we would choose the same marrow stimulation procedure as a main surgery for this patient because the defect size was less than 2 cm^2 in the femoral groove. This is a standard application of drilling. Overall results seem to be satisfactory.

While the author did not precisely describe the potential mechanism explaining the effect of the treatment, we would think that recovery of the normal patellofemoral (PF) laxity brought by the debridement of the extensive scar tissue might have played a great role to improve the patient's symptoms. PF laxity of the patients was not described in this case, but it is reasonable to imagine that the patient had a moderate/severe arthrofibrosis around the PF joint judging from the arthroscopic findings. We would speculate that the cartilage defect might have been produced by the concentrated force on the groove generated by the forced abnormal tracking of the patella caused by the PF arthrofibrosis which limited arthroscope mobility. Thus, the PF laxity should have been evaluated both preoperatively and postoperatively in comparison with the contralateral side and if necessary, the capsular release should be additionally applied. We think another goal of the treatment is recovery of the normal PF laxity in such kind of patients. Hence, we think that rehabilitation should be focused on recovery of the PF laxity.

On the other hand, the malpositioned ACL graft may be another concern. The femoral socket seems to be anteriorly located as was mentioned and the tibial tunnel seems to be posteriorly created judging from the MR image. While the author described the Lachman test was rated as 1A and the pivot shift test was normal, that might be due to the arthrofibrosis. If so, debridement of the extensive scar tissue could give rise to potential anterior laxity become apparent. We think re-evaluation of the ACL graft should be done during this surgery. When larger anterior laxity is, revision ACL surgery could be considered because he is still 34 years old.

18.3 Patellofemoral Pain in the Patient with a Trochlear Lesion: Proposed Case 39-Year-Old Male, Basketball, with a Central Trochlear Lesion

Alberto Gobbi and Ignacio Dallo

18.3.1 Introduction

Chondral or osteochondral lesions are frequently found during knee arthroscopy; the trochlea's prevalence is 6% [18]. When we face a biological problem, we should find a biological solution. There are different ways to treat a chondral defect, but there is only one target: recreate the normal hyaline tissue and restore the natural tissue [19]. The evolution of cartilage repair technique leads to the development of new scaffolds that allowed cell proliferation. The one-step approach avoids the need for a two-step surgical procedure and reduces the operation costs by approximately five times [20].

Bone marrow aspirate concentrate (BMAC) contains bone marrow stem cells (BMSCs) and growth factors that are a promising option for cartilage repair and regeneration [21–24]. Bone marrow-derived stem cells (BMSCs) interact with a non-woven scaffold, the HYAFF 11, that supports cellular adhesion, migration, and proliferation, promoting the synthesis of extracellular matrix components under static culture conditions [25–27]. Nejadnik et al. compared the clinical outcomes of patients treated with first-generation ACI and patients treated with autologous BMSCs. Concluding that BMSCs are as effective as chondrocytes for articular cartilage repair [28].

18.3.2 History and Examination

The patient is a 39-year-old male basketball player who presented progressive worsening of left knee pain, limiting his ability to participate in competitive sport and regular exercise. The patient had undergone in the same knee a double-bundle ACL reconstruction 16 years before the present surgery. The pain was mainly localized to the patellofemoral and medial compartment, and there were no reports of instability. Physical examination demonstrated a negative Lachman test, a negative anterior drawer test, and a full range of motion.

18.3.3 Imaging

Plain radiographic examination revealed mild varus deformity and lateralization of the patella, with preserved medial compartment joint space. MRI of the knee demonstrated a large full-thickness chondral lesion in the center of the trochlea, patella, and medial femoral condyle (Fig. 18.4).

18.3.4 Treatment

We decided to treat the cartilage lesions in the trochlea, patella, and medial femoral condyle with a three-dimensional hyaluronic acid-based scaffold (Hyalofast, Anika Therapeutics, Srl, Abano Terme, Italy) plus BMAC (HA-BMAC technique). We also performed an antero-medialization (AMZ) of the tibial tubercle (TT)

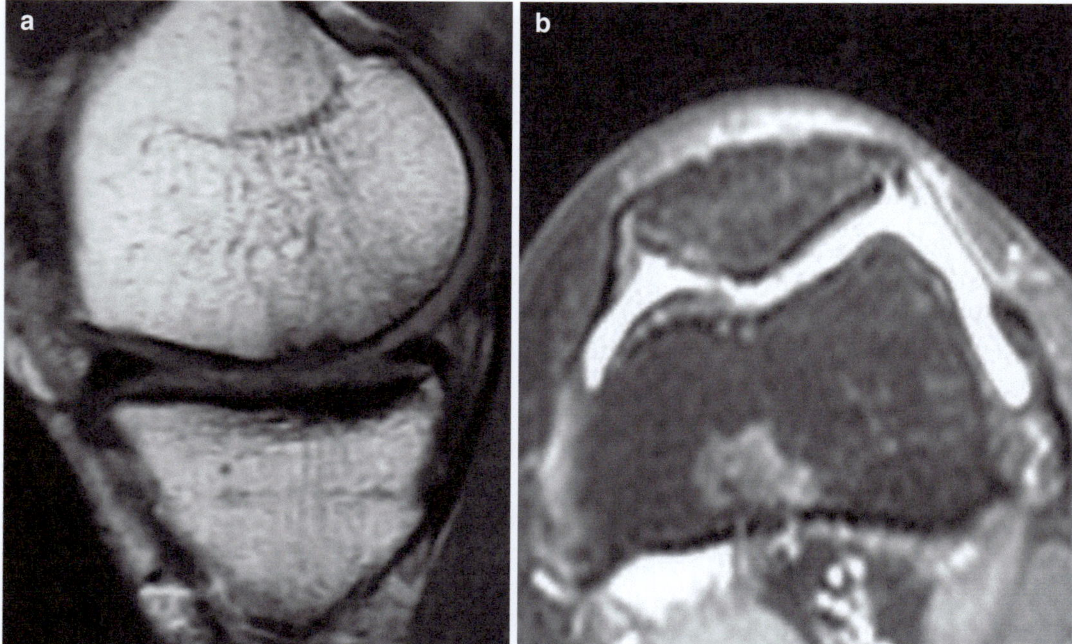

Fig. 18.4 (**a**, **b**) Preoperative sagittal and axial MRI scan through the knee showing a patellar, central trochlear, and medial femoral condyle full-thickness chondral lesion

to correct the patella malalignment and decrease the stress on the graft.

The patient was positioned supine for standard knee arthroscopy, the ipsilateral iliac crest was prepared and exposed to harvest bone marrow. Examination of the knee under anesthesia was done, cartilage lesions were identified, and the concomitant pathologies addressed during the procedure. An open technique was selected to treat this patient and the chondral lesions were prepared with vertical shoulders perpendicular to the subchondral plane. The calcified cartilage layer overlying the subchondral bone was removed (Fig. 18.5).

Sixty milliliters of bone marrow was aspirated from the ipsilateral iliac crest using a dedicated aspiration kit and was centrifuged with a commercially available system to obtain the concentrated bone marrow (Angel, ARTHREX, Naples, Florida, USA). The lesion dimensions were measured to prepare the matching implant from the three-dimensional hyaluronic acid-based scaffold. An aluminum foil model was used before, to ensure a precise fit with the lesion, then the scaffold was cut according to the aluminum foil model (Fig. 18.5b). When the scaffold was ready, BMAC was activated with batroxobin enzyme (Plateltex Act, Plateltex SRO, Bratislava, Slovakia). The activation process was necessary for BMAC to form a clot, which was then applied onto the prepared scaffold making a sticky implant that was easy to use in the lesion (Fig. 18.5c and d). The knee was then flexed and extended to check the graft stability; we did not insert a drain into the joint [29].

18.4 Rehabilitation

From 0 to 6 weeks postoperatively, there was a focus on maintaining/restoring range of motion and muscular strength, while minimizing effusion. Continuous passive motion started on postoperative day 2, 6 hours starting from 0 to 30° and gradually increasing the range of motion up to 90° in 6 weeks.

Early isometric and isotonic exercises were encouraged during the initial postoperative course. The patient started pool-based therapy and was allowed unrestricted weight bearing by 8 weeks. Active functional training begins at 9 weeks. From 3 to 8 months postoperatively, the

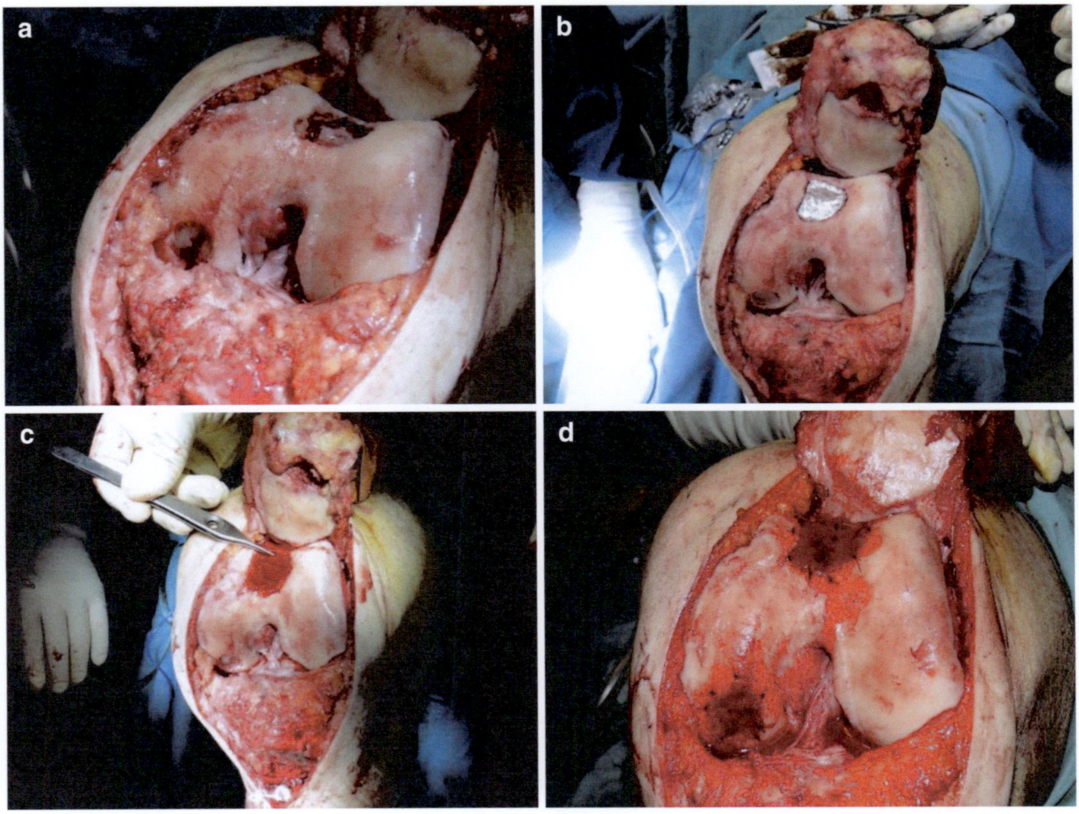

Fig. 18.5 (**a**) Intra-operative image. The trochlear and medial femoral condyle chondral defect was prepared to obtain perpendicular edges and the lesion's circumferential borders that must be perpendicular to the subchondral bone. (**b**) Intra-operative image. The lesion dimensions were measured with the aluminum template to prepare a matching implant from the three-dimensional hyaluronic acid-based scaffold. (**c** and **d**) Intra-operative image. The prepared BMAC was then placed on the Hyalofast scaffold. After a few minutes, the scaffold's activated BMAC was absorbed, creating a sticky implant easy to apply to the lesions, fibrin glue or sutures can be used to stabilize the graft

patient was progressed to straight-line running, and physical therapy was focused on strength and endurance training. Pain-free running at a moderate pace was expected by 8 months. From 8 to 10 months postoperatively, patients will focus on agility and sport-specific training, with anticipated return to sport by 10 months.

18.5 Results

The patient showed significant improvement in visual analog scale (VAS) for pain, International Knee Documentation Committee (IKDC), Knee injury and Osteoarthritis Outcome Score (KOOS), Lysholm, Tegner, and Marx scores at 6-, 12-, 24-, and 36-month follow-up ($P < 0.05$). No adverse reactions or postoperative complications were noted. IKDC objective score showed significant improvement in A and B subgroups from preoperative to final follow-up (Fig. 18.6a, b).

X-rays, and MRI, were collected at each follow-up (Fig. 18.7).

Post-treatment MRI showed complete filling of the defects, while no signs of hypertrophy were identified. Integration with adjacent cartilage was completed with restoration of the cartilage layer and subchondral bone. We also did not identify edema, cysts, or sclerosis of subchondral bone (Fig. 18.8).

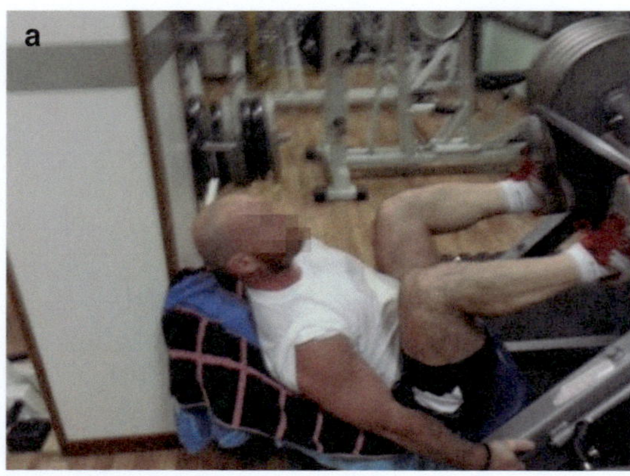

Fig. 18.6 (**a**, **b**) The patient at 4 year follow-up with a significant improvement in all clinical scores

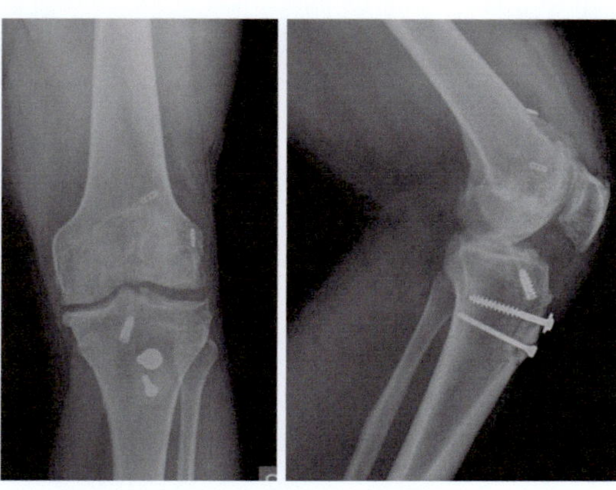

Fig. 18.7 Postoperative AP and Lateral Knee X-rays images showing the Antero-medialization (AMZ) of the tibial tubercle (TT) to optimize the patellofemoral (PF) biomechanics and maximize the result of the cartilage repair/regeneration

18.6 Discussion

The presented technique using one-step HA-BMAC implantation, in conjunction with the realignment of the extensor mechanism when needed, has shown to provide durable cartilage repair, with consistent improvements in clinical outcomes at our center [30–32]. Irrespective of cartilage repair technique, the abnormal biomechanical factors need to be evaluated for correction, to optimize the treatment of patellofemoral osteochondral lesions. Ensuring that the involved compartment is sufficiently downloaded will provide the optimal environment for cartilage repair tissue to remodel and mature. We compared patients treated with matrix-induced autologous chondrocyte implantation (MACI) with patients treated with BMSCs combined with the same scaffold and at 3 years, follow-up we did not find any significant statistical differences between the two groups, concluding that both these techniques were viable and effective [30]. It has been shown in many clinical studies that the HA-BMAC one-step technique is a viable method for the treatment of full-thickness cartilage lesions of the knee [33]. Different sizes of cartilage lesions can be treated, from small injuries to significant defects up to 22 cm^2, showing good to excellent results at 10 years follow-up [19, 34, 35].

Fact Box 1 Pearls of HA-BMAC Technique

- Complete exposure of the lesion is critical.
- Use traction methods if needed to provide a comfortable working space.
- An aluminum foil template can be used as a template to measure the defect.
- Paste the HA-based scaffold into the lesion and check the stability.
- Fibrin glue can be applied to improve implant stability.
- Cycle the knee under dry arthroscopic visualization to confirm the graft seating within the defect.
- A drain should not be inserted into the joint.

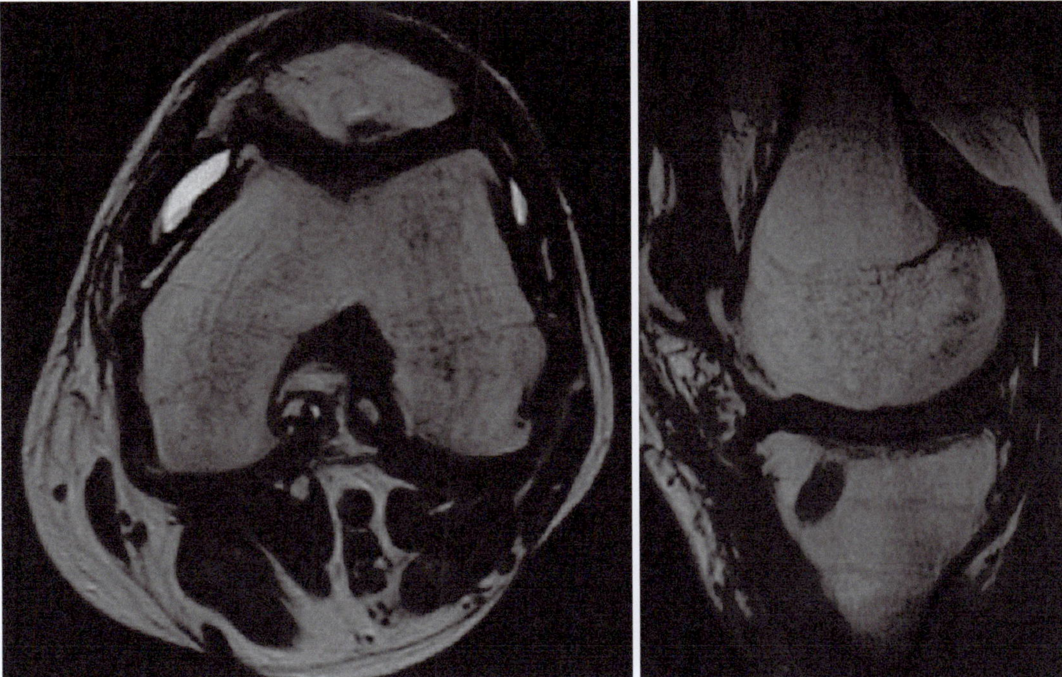

Fig. 18.8 MRI at 4 years showed good integration with the restoration of subchondral bone and complete filling of the cartilage defects

Take-Home Messages

- Single-step cartilage repair eliminates the need for a two-step procedure, reducing the patient's cost and morbidity.
- HA-BMAC procedure provides good to excellent clinical outcomes at long-term follow-up in small or large lesions, single or multiple.
- Pathological background factors such as malalignment, meniscus deficiency, or ligament laxity must be addressed to provide an optimal environment for cartilage repair.
- Results may be comparatively more successful in younger patients.
- An individualized approach based on the patient's goals and the surgeon's preferences is crucial.

References

1. Culvenor AG, Crossley KM. Patellofemoral osteoarthritis: are we missing an important source of symptoms after anterior cruciate ligament reconstruction? J Orthop Sports Phys Ther. 2016;46(4):232–4.
2. Herrington L, Alarifi S, Jones R. Patellofemoral joint loads during running at the time of return to sport in elite athletes with ACL reconstruction. Am J Sports Med. 2017;45(12):2812–6.
3. Mikula JD, Slette EL, Dahl KD, Montgomery SR, Dornan GJ, O'Brien L, et al. Intraarticular arthrofibrosis of the knee alters patellofemoral contact biomechanics. J Exp Orthop. 2017;4(1):40.
4. Ahmad CS, Kwak SD, Ateshian GA, Warden WH, Steadman JR, Mow VC. Effects of patellar tendon adhesion to the anterior tibia on knee mechanics. Am J Sports Med. 1998;26(5):715–24.
5. Steadman JR, Dragoo JL, Hines SL, Briggs KK. Arthroscopic release for symptomatic scarring of the anterior interval of the knee. Am J Sports Med. 2008;36(9):1763–9.
6. Weber AE, Locker PH, Mayer EN, Cvetanovich GL, Tilton AK, Erickson BJ, et al. Clinical outcomes after microfracture of the knee: midterm follow-up. Orthop J Sports Med. 2018;6(2):2325967117753572.
7. Orth P, Gao L, Madry H. Microfracture for cartilage repair in the knee: a systematic review of the contemporary literature. Knee Surg Sports Traumatol Arthrosc. 2020;28(3):670–706.
8. Kreuz PC, Erggelet C, Steinwachs MR, Krause SJ, Lahm A, Niemeyer P, et al. Is microfracture of chondral defects in the knee associated with different results in patients aged 40 years or younger? Arthroscopy. 2006;22(11):1180–6.
9. Gomoll AH. Microfracture and augments. J Knee Surg. 2012;25(1):9–15.

10. Drakos MC, Eble SK, Cabe TN, Patel K, Hansen OB, Sofka C, et al. Comparison of functional and radiographic outcomes of Talar osteochondral lesions repaired with micronized allogenic cartilage extracellular matrix and bone marrow aspirate concentrate vs microfracture. Foot Ankle Int. 2021:1071100720983266.
11. Strauss EJ, Barker JU, Kercher JS, Cole BJ, Mithoefer K. Augmentation strategies following the microfracture technique for repair of focal chondral defects. Cartilage. 2010;1(2):145–52.
12. Orth P, Duffner J, Zurakowski D, Cucchiarini M, Madry H. Small-diameter awls improve articular cartilage repair after microfracture treatment in a translational animal model. Am J Sports Med. 2016;44(1):209–19.
13. Culvenor AG, Collins NJ, Vicenzino B, Cook JL, Whitehead TS, Morris HG, et al. Predictors and effects of patellofemoral pain following hamstring-tendon ACL reconstruction. J Sci Med Sport. 2016;19(7):518–23.
14. Ambra LF, Hinckel BB, Arendt EA, Farr J, Gomoll AH. Anatomic risk factors for focal cartilage lesions in the patella and trochlea: a case-control study. Am J Sports Med. 2019;47(10):2444–53.
15. Krych AJ, Saris DBF, Stuart MJ, Hacken B. Cartilage injury in the knee: assessment and treatment options. J Am Acad Orthop Surg. 2020;28(22):914–22.
16. Rue JP, Colton A, Zare SM, Shewman E, Farr J, Bach BR Jr, et al. Trochlear contact pressures after straight anteriorization of the tibial tuberosity. Am J Sports Med. 2008;36(10):1953–9.
17. Nho SJ, Pensak MJ, Seigerman DA, Cole BJ. Rehabilitation after autologous chondrocyte implantation in athletes. Clin Sports Med. 2010;29(2):267–82, viii.
18. Hjelle K, Solheim E, Strand T, Muri R, Brittberg M. Articular cartilage defects in 1,000 knee arthroscopies. Arthroscopy. 2002;18(7):730–4.
19. Gobbi A, Dallo I, Kumar V. Editorial commentary: biological cartilage repair technique—an "effective, accessible, and safe" surgical solution for an old difficult biological problem. Arthroscopy. 2020;36(3):859–61.
20. Gobbi A, Lane JG, Dallo I. Editorial commentary: cartilage restoration—what is currently available? Arthroscopy. 2020;36(6):1625–8.
21. Caplan AI. Review: mesenchymal stem cells: cell-based reconstructive therapy in orthopedics. Tissue Eng. 2005;11(7–8):1198–211.
22. Caplan AI. Mesenchymal stem cells: the past, the present, the future. Cartilage. 2010;1(1):6–9.
23. Dimarino AM, Caplan AI, Bonfield TL. Mesenchymal stem cells in tissue repair. Front Immunol. 2013;4:201.
24. Huselstein C, Li Y, He X. Mesenchymal stem cells for cartilage engineering. Biomed Mater Eng. 2012;22(1–3):69–80.
25. Pasquinelli G, Orrico C, Foroni L, Bonafe F, Carboni M, Guarnieri C, et al. Mesenchymal stem cell interaction with a non-woven hyaluronan-based scaffold suitable for tissue repair. J Anat. 2008;213(5):520–30.
26. Lisignoli G, Cristino S, Piacentini A, Zini N, Noel D, Jorgensen C, et al. Chondrogenic differentiation of murine and human mesenchymal stromal cells in a hyaluronic acid scaffold: differences in gene expression and cell morphology. J Biomed Mater Res A. 2006;77(3):497–506.
27. Facchini A, Lisignoli G, Cristino S, Roseti L, De Franceschi L, Marconi E, et al. Human chondrocytes and mesenchymal stem cells grown onto engineered scaffold. Biorheology. 2006;43(3–4):471–80.
28. Nejadnik H, Hui JH, Feng Choong EP, Tai BC, Lee EH. Autologous bone marrow-derived mesenchymal stem cells versus autologous chondrocyte implantation: an observational cohort study. Am J Sports Med. 2010;38(6):1110–6.
29. Whyte GP, Gobbi A, Sadlik B. Dry arthroscopic single-stage cartilage repair of the knee using a hyaluronic acid-based scaffold with activated bone marrow-derived mesenchymal stem cells. Arthrosc Tech. 2016;5(4):e913–e8.
30. Gobbi A, Chaurasia S, Karnatzikos G, Nakamura N. Matrix-induced autologous chondrocyte implantation versus multipotent stem cells for the treatment of large patellofemoral chondral lesions: a nonrandomized prospective trial. Cartilage. 2015;6(2):82–97.
31. Gobbi A, Whyte GP. One-stage cartilage repair using a hyaluronic acid-based scaffold with activated bone marrow-derived mesenchymal stem cells compared with microfracture: five-year follow-up. Am J Sports Med. 2016;44(11):2846–54.
32. Gobbi A, Scotti C, Karnatzikos G, Mudhigere A, Castro M, Peretti GM. One-step surgery with multipotent stem cells and Hyaluronan-based scaffold for the treatment of full-thickness chondral defects of the knee in patients older than 45 years. Knee Surg Sports Traumatol Arthrosc. 2017;25(8):2494–501.
33. Gobbi A, Karnatzikos G, Sankineani SR. One-step surgery with multipotent stem cells for the treatment of large full-thickness chondral defects of the knee. Am J Sports Med. 2014;42(3):648–57.
34. Gobbi A, Karnatzikos G, Scotti C, Mahajan V, Mazzucco L, Grigolo B. One-step cartilage repair with bone marrow aspirate concentrated cells and collagen matrix in full-thickness knee cartilage lesions: results at 2-year follow-up. Cartilage. 2011;2(3):286–99.
35. Gobbi A, Whyte GP. Long-term clinical outcomes of one-stage cartilage repair in the knee with hyaluronic acid-based scaffold embedded with mesenchymal stem cells sourced from bone marrow aspirate concentrate. Am J Sports Med. 2019;47(7):1621–8.

A 35-Year-Old Woman with Painful Early Lateral Patellofemoral Degenerative Arthrosis

19

Robert A. Magnussen and John P. Fulkerson

A 35-year-old woman was active with sports including tennis and running without any difficulty until 18 months ago. She was seen then by another orthopedic surgeon who prescribed physical therapy but she got worse instead of better, noting difficulty walking hills, particularly going downhill, to the point that she stopped walking hills and was having trouble also with stairs. A lateral release was done 9 months prior to this visit and helped for 4–5 months, but now she is back to where she was, persistent anterior knee pain aggravated by long walks, hills, and stairs. She has tried bracing, taping, core stability training, topical agents, and anti-inflammatory medications with limited benefit. She is frustrated and says the pain is intolerable.

Physical examination reveals pain upon compression of the patella but particularly from 0–30 degrees of knee flexion. Squatting is particularly painful in this range of motion. She does not have an effusion. Lateral retinaculum is mildly tender. She walks with a mildly antalgic gait.

She cannot jump off of a step stool without experiencing severe pain and giving way.

Radiographs confirm a lateral tracking pattern on 45 degree knee flexion axial Merchant view (Fig. 19.1).

R. A. Magnussen (✉)
The Ohio State University, Columbus, OH, USA

J. P. Fulkerson
Yale University, New Haven, CT, USA

A precise lateral confirms lateral rotation of the patella. Other radiographs are normal. She has no evidence of patella alta.

What else is needed?
What would you do?
What are the options?

19.1 Fulkerson Perspective

This is an ideal patient for anteromedialization tibial tubercle transfer ((AMZ TTO). This situation is fairly common in my practice and so I cannot really consider seriously any other option as the patient has objective evidence of lateral facet overload and articular breakdown, A well done AMZ TTO will give her at least 90% likelihood of permanent pain relief and long term joint preservation [2]. I see no benefit to cartilage surgery in this case. 3D imaging would be helpful to better understand trochlea curvature and decide on amount of medialization. More reliable to me than TT-TG which can be variable.

19.2 Magnussen Perspective

This is a relatively common case of lateral patellofemoral overload due to maltracking that has failed conservative management. The approach outlines by Dr. Fulkerson above would be an excellent option in this setting given her young

© ISAKOS 2022
J. L. Koh et al. (eds.), *The Patellofemoral Joint*, https://doi.org/10.1007/978-3-030-81545-5_19

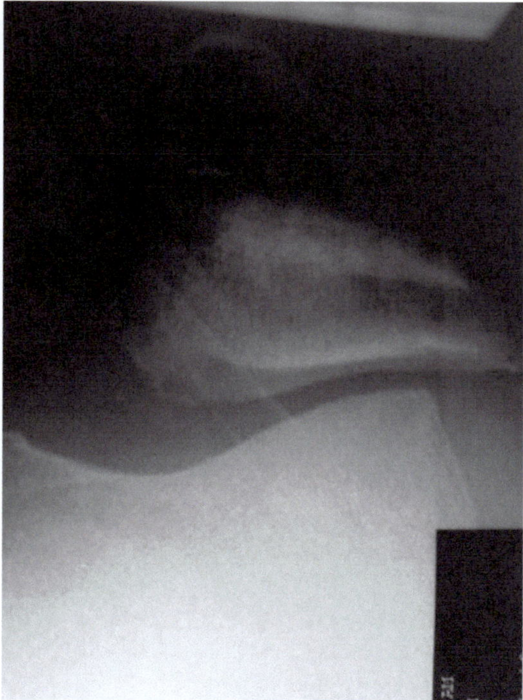

Fig. 19.1 Lateral tracking of patella

age and visible lateral patellar subluxation with articular cartilage damage laterally. While likely not critical in this case, an MRI may provide some additional useful information. First, it would allow for precise localization of the articular cartilage damage. If it is confined to the lateral and/or distal patella, the odds of a good outcome from the TTO are very high [3]. If there is damage extending on to the central, proximal, or medial patella, augmentation of the AMZ TTO with a cartilage restoration procedure may improve outcomes [1].

MRI would also allow for measurement of the tibial tubercle-trochlear groove (TT-TG) distance. While lateral subluxation in this case is clear, one must be careful not to anteromedialize the tubercle when the patella is not tracking laterally as it can potentially lead to iatrogenic medial patellar instability. A finding of a normal TT-TG distance is a warning sign that the tubercle may not be lateralized and one should be very cautious with AMZ in that situation. While we lack an MRI here, the lateral patellar tracking visible on the plain film suggests that the TT-TG distance is elevated in this patient. Further history would also be useful – especially a history of patellar instability at a younger age—as such a history would further strengthen the case to perform an AMZ osteotomy (already a very good case in this patient).

There are additional options to consider in the setting of lateral patellar overload and cartilage damage. Following failed conservative management, a lateral release is a reasonable option to consider, but as occurred in this patient, the relief is often short-lived when there is significant lateral tracking and chondral damage such as this. Lateral patellar facetectomy can also be considered with or without associated lateral release or lengthening [4]. These approaches are generally more useful in patients with less severe disease and more normal tracking. As described above, AMZ may over-medialize the tubercle in such patients. One could also consider the option of patellofemoral arthroplasty in the setting of isolated patellofemoral degenerative change; however, given the patient's young age and what appears to be isolated lateral disease, this procedure would not be indicated here.

In short, given the information presented above, I agree that an anteromedialization osteotomy is the best choice and will likely result in good pain relief in this patient.

Fact Box Indications for Anteromedialization (AMZ) Tibial Tubercle Osteotomy

- Lateral patellar tracking
  - Assessed clinically
  - Lateral patellar subluxation on axial plain films
  - Lateral curved trochlea on 3D imaging
  - Elevated TT-TG distance
- Objective evidence of lateral patellofemoral overload
  - Lateral patellofemoral joint space narrowing

  - Lateral patellofemoral articular cartilage damage and/or underlying bone edema on MRI
  - Improvement in symptoms with lateral unloaded (taping, etc.…)
  - Worsening of symptoms with activities that load the PFJ
- Articular damage localized to the lateral and/or distal patella (can consider extending the indication to more diffuse lesions with associated cartilage restoration)

References

1. Gigante A, Enea D, Greco F, et al. Distal realignment and patellar autologous chondrocyte implantation: mid-term results in a selected population. Knee Surg Sports Traumatol Arthrosc. 2009;17(1):2–10.
2. Klinge SA, Fulkerson JP. Fifteen-year minimum follow-up of anteromedial tibial tubercle transfer for lateral and/or distal patellofemoral arthrosis. Arthroscopy. 2019;35(7):2146–51.
3. Pidoriano AJ, Weinstein RN, Buuck DA, Fulkerson JP. Correlation of patellar articular lesions with results from anteromedial tibial tubercle transfer. Am J Sports Med. 1997;25(4):533–7.
4. Yercan HS, Ait Si Selmi T, Neyret P. The treatment of patellofemoral osteoarthritis with partial lateral facetectomy. Clin Orthop Relat Res. 2005;436:14–9.

Patellofemoral Chondrosis and Instability in the Middle Aged Patient

20

Seth L. Sherman, Taylor Ray, Adam Money, Stefano Zaffagnini, Mauro Núñez, and Julian Feller

20.1 Case Vignette

Patient is a 40-year-old female with right knee pain and instability. She is a factory worker but overall lives a sedentary lifestyle. She presents with recurrent low energy patella dislocations that have occurred since age 12. Treatment thus far has included NSAIDs, bracing, injections, and physical therapy, which have failed to provide continued relief. Her exam demonstrates a moderate effusion, patellar apprehension to 60 degrees of flexion, lateral retinacular tightness, positive J sign, and reproducible anterior knee pain with patellar grind. Her radiographs demonstrate patellofemoral arthrosis with lateral patellar tilt and lateral maltracking on Merchant view. Caton-Deschamps ratio is normal. There is no evidence of medial or lateral compartment arthritis and overall neutral alignment (Figs. 20.1, 20.2, and 20.3). MRI demonstrates diffuse patellofemoral chondral disease, loose body in the lateral gutter, trochlear dysplasia, and a TT-TG (Tibial Tubercle—Trochlear Groove) distance of 21 mm. Patella-Trochlear Index is normal (Fig. 20.4). An exam under anesthesia demonstrates gross lateral patellar laxity with incompetent medial patellar restraints and fixed lateral retinaculum tightness and patella tilt. Intra-operative evaluation demonstrated bipolar arthritis and confirmed the trochlear dysplasia as noted on MR (Fig. 20.5).

S. L. Sherman (✉)
Department of Orthopaedic Surgery, Stanford University School of Medicine, Redwood City, CA, USA

T. Ray · A. Money
Department of Orthopaedic Surgery, Stanford University, Redwood City, CA, USA
e-mail: raytay@stanford.edu

S. Zaffagnini
Orthopaedics and Traumatology Unit, IRCCS Rizzoli Orthopaedic Institute, Bologna, Italy

DIBINEM Department, University of Bologna, Bologna, Italy
e-mail: stefano.zaffagnini@unibo.it

M. Núñez
Orthopaedic Surgery Department, Sports and Trauma Unit, Hospital del Trauma, San José, Costa Rica

Universidad de Ciencias Médicas (UCIMED), San José, Costa Rica

J. Feller
School of Allied Health, Human Services and Sport, La Trobe University, Melbourne, VIC, Australia

OrthoSport Victoria, Epworth HealthCare, Melbourne, VIC, Australia
e-mail: jfeller@osv.com.au

20.2 Expert Opinions

20.2.1 Stefano Zaffagini, MD

In the current case a 40-year-old female factory worker with objective recurrent patellar dislocation and anterior knee pain was presented. In the management of objective patellar dislocation all the instability factors must be identified and corrected one by one [1].

© ISAKOS 2022
J. L. Koh et al. (eds.), *The Patellofemoral Joint*, https://doi.org/10.1007/978-3-030-81545-5_20

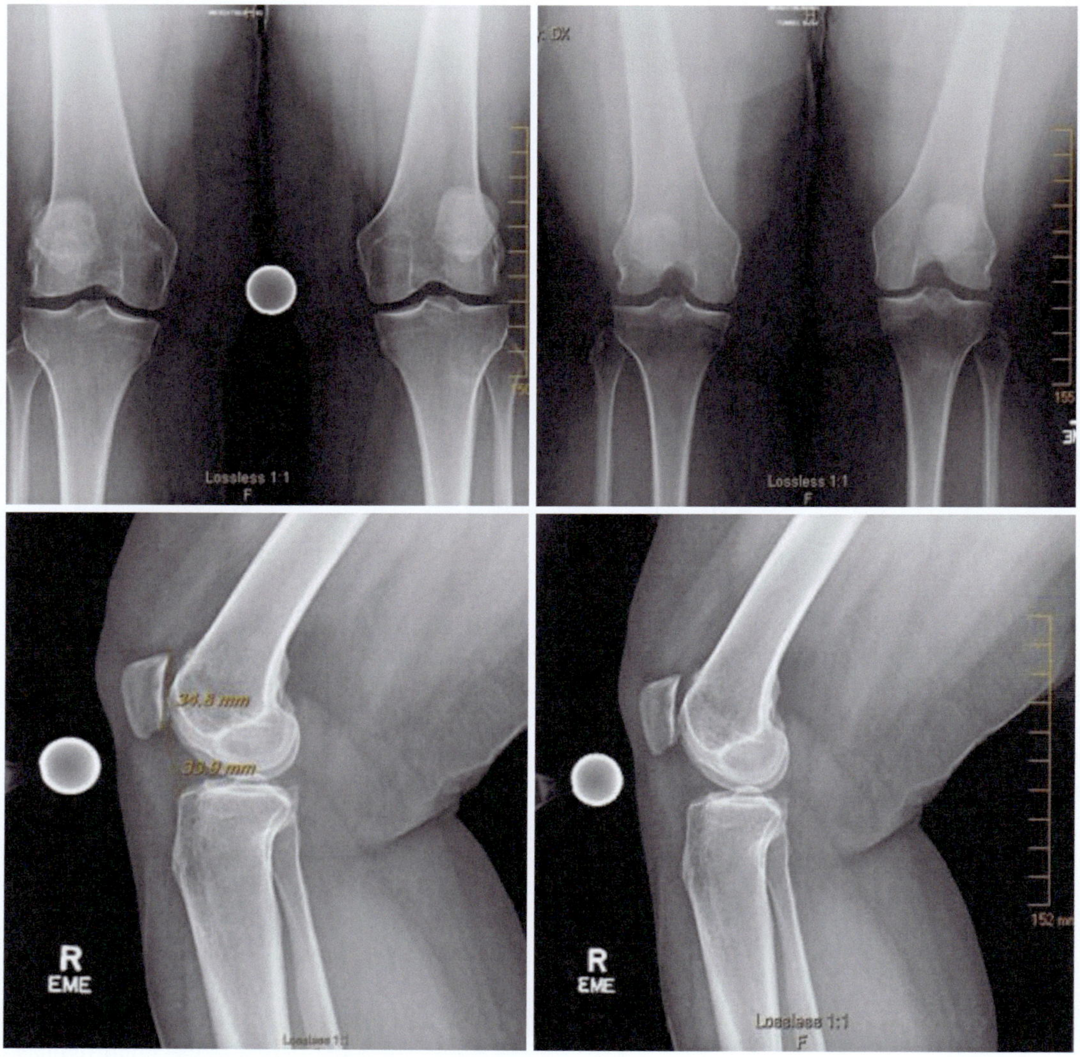

Fig. 20.1 AP, PA flexion, and lateral X-rays. Lateral X-rays demonstrate patellofemoral arthritis with supratrochlear spur formation and a Caton-Deschamps ratio of 0.97

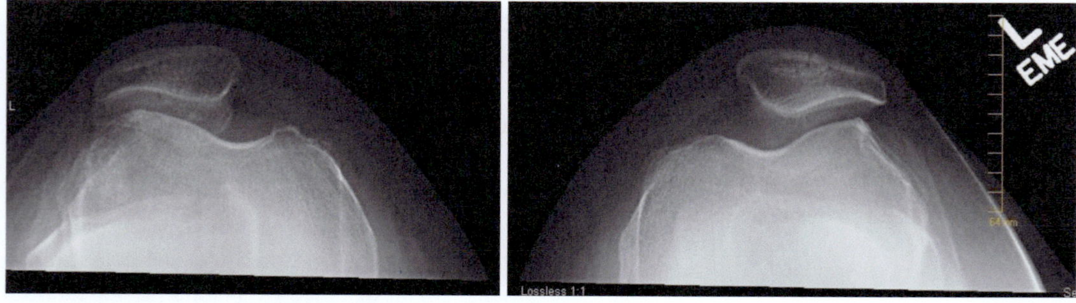

Fig. 20.2 Merchant view demonstrates joint space narrowing, osteophyte formation in the medial trochlea on right knee, lateral patellar tilt, and lateral patellar tracking

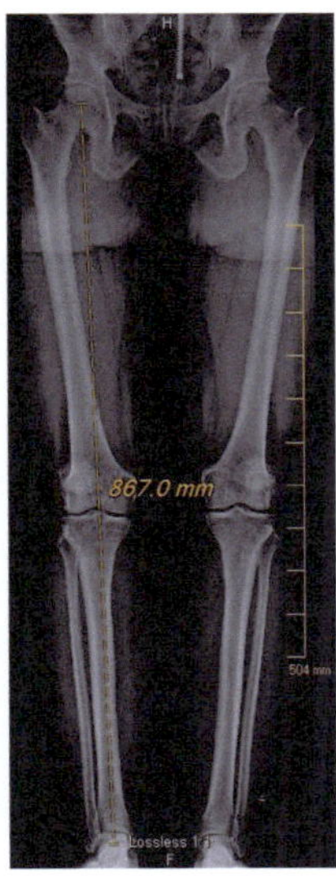

Fig. 20.3 Alignment X-rays demonstrate neutral alignment

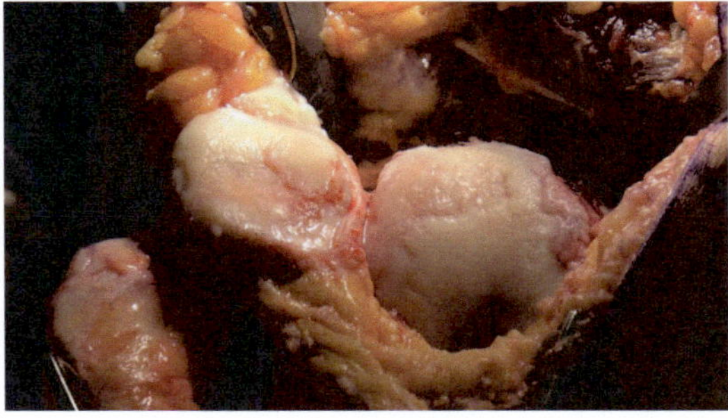

Fig. 20.5 Intra-operative photo demonstrated diffuse chondral damage of both patella and femur with confirmation of trochlear dysplasia

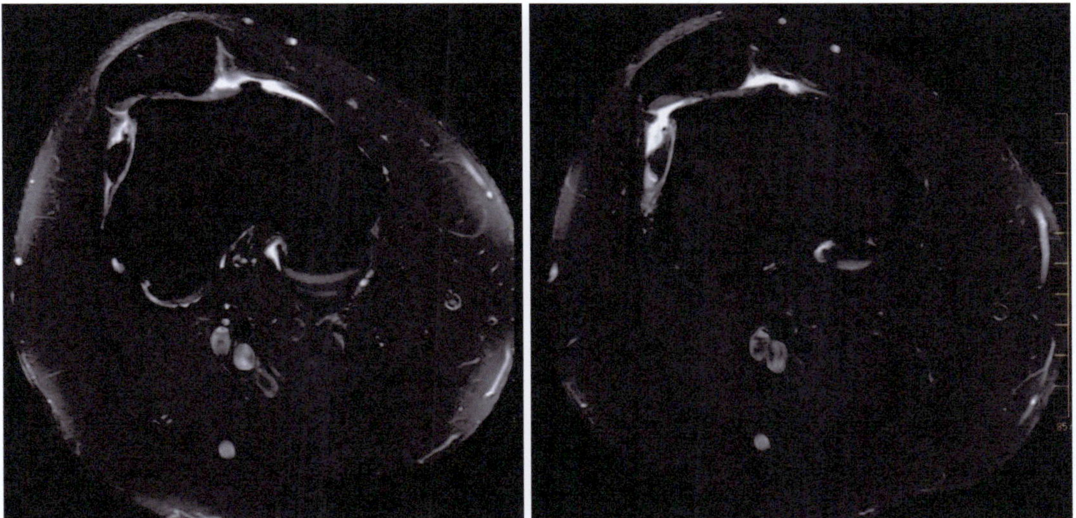

Fig. 20.4 Axial MRI images demonstrating loose body, diffuse bipolar near full thickness chondrosis, trochlear osteophytes, patulous medial capsular structures, lateral patellar tilt, and shallow trochlear dysplasia

In the case under consideration the instability factors were the abnormal axial alignment (TT-TG 21 mm), low grade trochlear dysplasia ("type c" according to Dejour classification [2]), lateral patellar retinaculum tightness, and incompetent medial soft tissue restraints. Moreover, patient presented with diffuse patellofemoral chondral disease.

In this kind of case, I use a direct open approach, performing a straight midline skin incision from the superior patellar margin to the anterior tibial tuberosity. I then proceed addressing all the instability factors of the patient. In this specific case I would not perform trochleoplasty, because current patient presented low grade trochlear dysplasia ("type C"), and a diffuse patellofemoral chondral disease ("grade IV") [2].

20.2.1.1 Lateral Patellar Retinaculum Tightness

I would perform an extensive release of the lateral retinaculum.

20.2.1.2 Abnormal Axial Alignment

I would perform an osteotomy of the anterior tibial tuberosity. Subsequently I would position the detached tibial tubercle bone plug in a more distal and medial position and fixing it with a K wire. At this time, I would do a functional evaluation of patella tracking especially in the first 30 degrees of flexion, to check patellar stability. Finally, I would fix the tibial tubercle with two cortical screws.

20.2.1.3 Incompetent Medial Soft Tissue Restraints

To correct the incompetent medial restraint, in association with medialization of tibial tubercle, I usually perform medial patella-tibial ligament (MPTL) reconstruction. I detach the medial third of the patellar tendon distally. This ligament was then medialized and put under tension, trying to find a medial insertion location close to the anterior edge of the medial collateral ligament. The location was determined by a repeated dynamic analysis of patellar tracking that allowed us to find a reinsertion point that led to patellar stability, especially near extension without creating excessive tension in the ligament band when the knee is flexed. Finally, I fix the medial third of the patellar tendon to the tibial bone with one staple [3]. Finally, I would perform an advancement plasty of vastus medialis obliquus muscle.

20.2.1.4 Patella Cartilage Lesion

After removing the articular osteochondral fragment, I would manage the patella cartilage lesion using microfracture associated with the CarGel bio-scaffold on the osteochondral defect [4].

20.2.2 Mauro Nunez, MD

In the illustrative case, we must address a five-problem list.

In regard to the axial malalignment with a high TT-TG, a TTT should be performed, the usual transfer would be an anteromedialization as Fulkerson described in 1990 [5].

However, there is a diffuse osteochondral lesion that should make us reconsider if medialization is appropriate in the light of the study published by Pidoriano and Fulkerson regarding the pressure on the patella after a TTT with pre-existing chondral lesions [6].

In the Pidoriano study, patients with pre-existing chondral lesions at the distal level of the patella (type I) and at the lateral level (type II) had 87% excellent and good results after the anteromedialization of the patella, those who had medial lesions (type III) obtained excellent and good results in 55%, while patients with proximal or diffuse lesions (type IV) had excellent and good results in only 20% of cases.

These results lead us to think that the definitive treatment of the chondral lesion should be carried out with an osteochondral allograft or evaluating techniques for restoration of articular cartilage such as ACI or MACI.

As for trochlear dysplasia, my preference would be to treat it conservatively since at 40 this patient will probably lack adequate elasticity for surgical molding of the trochlea.

If it seems pertinent, I would do a reconstruction of the MPFL that acts as a medial checkrein of the patella during flexion and extension of the knee.

Finally, perhaps one of the most controversial issues is the treatment of patellar tilt, in a survey of a group of experts in patellofemoral injuries, no consensus can be reached as when to perform a lateral retinaculum release in a patient [7].

However, the patellar inclination would seem to be one of those few indications for a lateral retinaculum release, even if it continues to be a sensitive and difficult issue to reach a consensus, but my preference in this case would be to perform a lateral retinaculum release to improve the patellar tilt.

20.2.3 Julian Feller, MD

As is so often the case in this age group, this patient highlights the importance of distinguishing between pain and instability. I would want to understand more about the pain; location, aggravating factors (up or down stairs and inclines), how much it affects her ADLs, walking distance, and sleep disturbance. I assume that swelling is a feature, given that there is a moderate effusion on examination.

The examination findings certainly suggest a significant instability—J sign and patellar apprehension to 60°. To have a J sign, I would expect either patella alta given the mild trochlear dysplasia, or a higher grade of trochlear dysplasia in the absence of patella alta. The tightness of the lateral retinaculum does not help from the diagnostic point of view, but is relevant in terms of treatment options.

A TT-TG distance of 21 mm is below my threshold for considering altering this in any intervention. In general, I think we can raise the threshold of the TT-TG value above what has generally been taught, without increasing the rate of recurrence of patellar instability after patellar stabilization. The TT-TG distance is very dependent on the angle of knee flexion and this can be variable when using MRI measurements as opposed to CT measurements. With the knee in extension, I accept a TT-TG distance of up to 24 mm.

I will assume that patellar instability is in fact the main problem. Therefore, I want to do some kind of patellar stabilization procedure. Using arthroscopy, I would have removed the loose body and lightly debrided loose and unstable articular cartilage in the patellofemoral compartment. A small technical point about chondroplasty is that using a shaver with a scalloped cutting edge is less aggressive and gives a smoother margin than a traditional full radius cutter.

Given the chondral changes, I would take a minimalist approach to patellar stabilization, for fear of aggravating the pain arising from the chondral degeneration. My overall approach to patellar stabilization is that MPFL reconstruction is the basis of any surgery and the real question comes down whether and what additional surgery is needed. Normally, the need for a bony procedure is primarily based on whether the problem is subluxation or dislocation (dislocation indicating more severe instability and therefore being a relative indication for a bony procedure), and whether J-tracking is present.

For dislocators, I would be thinking about a tibial tubercle transfer or a trochleoplasty. However, there is no patella alta, the TT-TG is below my personal threshold for intervention, the trochlear dysplasia is mild, and there is significant patellofemoral chondral damage. In addition, there is lateral retinacular tightness which may result in increased pressure in the patellofemoral compartment if an MPFL reconstruction is performed.

My approach would be a MPFL reconstruction with an autograft gracilis tendon and an arthroscopic lateral retinacular release. I think one could equally perform an open lateral retinacular lengthening in addition to the MPFL reconstruction. I would not perform a medialization of the tibial tubercle because of concerns about increasing pressure in the medial half of the patellofemoral compartment and, perhaps, creating the potential for medial patellar instability given the lateral release/retinacular lengthening. As a rule, I do not perform anteriorization/anteromedialization of the tibial tubercle based on my experience of seeing unhappy patients for a second opinion following these procedures, but that may just be local experience.

20.2.4 Author's Treatment Plan

The patient presents with multiple factors that require thorough evaluation and treatment. We must approach the patellofemoral joint as an "organ." We need to consider alignment, stability, and the health of the cartilaginous surfaces to maximize the chance of a good outcome. She has axial malalignment as demonstrated by her TT-TG >20 mm [8–10]. Her trochlear dysplasia can be classified as type C based on the Dejour classification [2]. The exam under anesthesia demonstrates an incompetent medial patellofemoral ligament (MPFL). Her imaging demonstrates patellar tilt and her exam confirms lateral retinacular tightness. Her imaging and intra-operative evaluation demonstrate grade IV diffuse chondral disease involving both the patella and the trochlea. Therefore, our extensive problem list includes: lateral retinacular tightness, incompetent medial soft tissue restraints, axial malalignment, trochlear dysplasia, and bipolar grade IV chondral disease of the patellofemoral joint.

20.2.4.1 Lateral Retinacular Tightness

Our plan was to use an open anterior midline approach similar to Dr. Zaffagnini. We would address the tightness of the lateral retinaculum through an open lateral retinaculum lengthening (Fig. 20.6). With this technique, we were able to lengthen the lateral retinaculum approximately 18 mm (Fig. 20.7). We chose a lateral retinacular lengthening, as we believe it provides reproducible results without the unpredictability or complication profile of a lateral release. The three expert opinions all agreed that some type of lateral soft tissue balancing would be helpful in this patient.

20.2.4.2 Incompetent Medial Soft Tissue Restraints

We utilized a MPFL reconstruction to address this problem. Our graft of choice is hamstring allograft placed using a combination of anatomy, fluoroscopy, and isometry to find the appropriate origin and insertion (Figs. 20.8, 20.9, 20.10, 20.11, and 20.12). While the specific type of medial soft tissue reconstruction varied per surgeon preference, addressing the incompetent medial soft tissue envelope was a critical part of

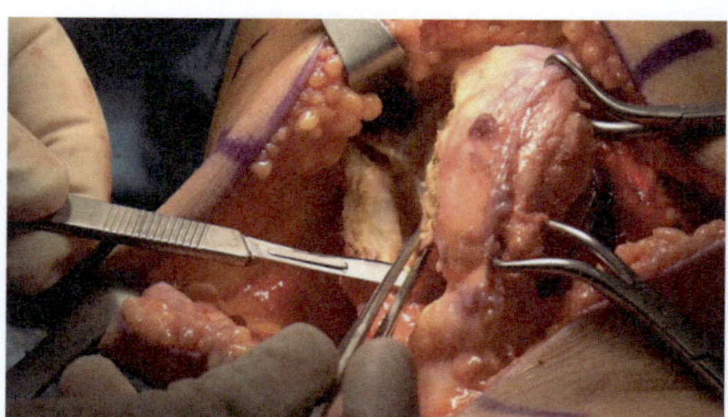

Fig. 20.6 Intra-operative photo demonstrating our open anterior approach, dissection of the lateral retinaculum, and direct lengthening

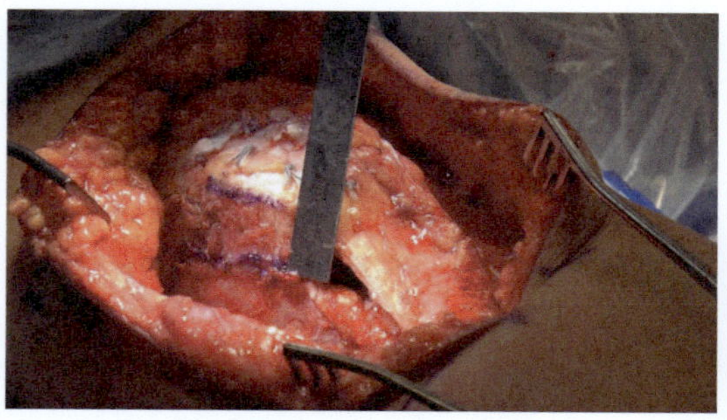

Fig. 20.7 Completed lateral retinacular lengthening demonstrating approximately 18 mm of added lateral length

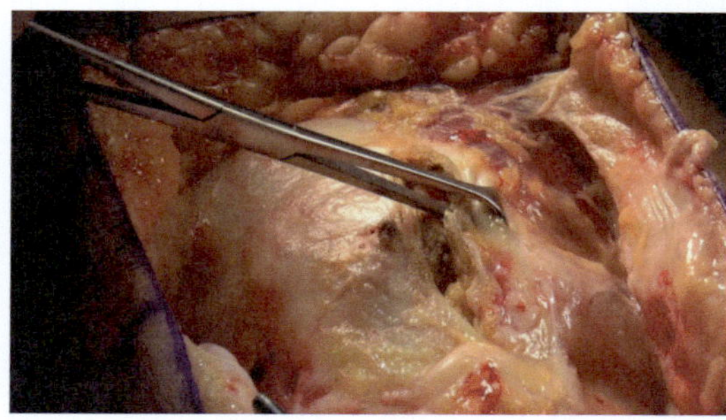

Fig. 20.8 Dissection of the medial soft tissues with isolation of the native MPFL

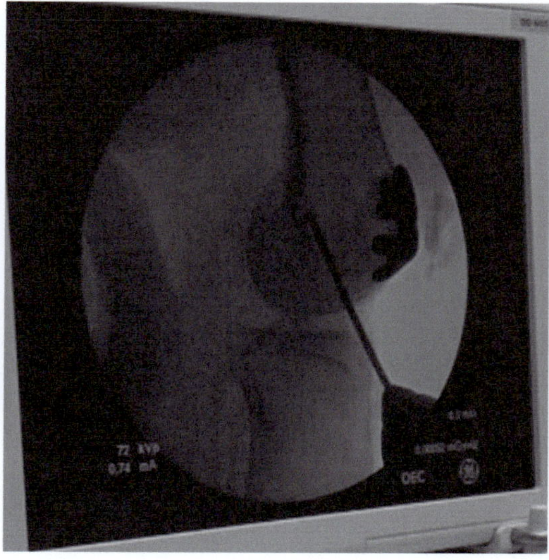

Fig. 20.9 Intra-operative fluoroscopic image demonstrating placement of the guide pin at Schottles point

Fig. 20.10 Hamstring allograft prepared on back table prior to implantation

each surgeon's plan. MPFL reconstruction is a reliable method for treating this essential lesion and should be addressed in all complex instability cases.

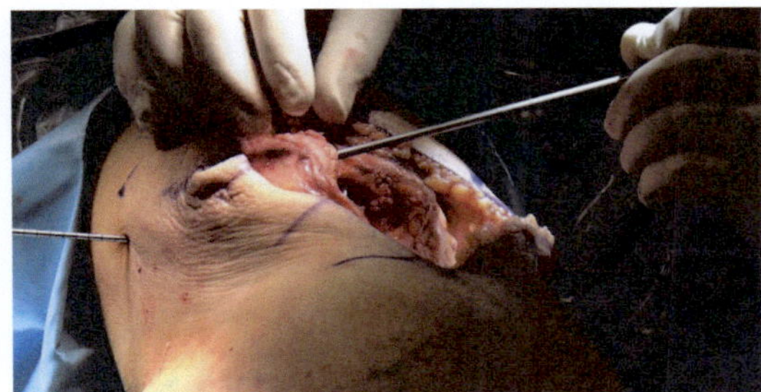

Fig. 20.11 Intra-operative photo demonstrating placement of the guide wire in the femur as well as placement of the drill holes on the medial aspect of the patella

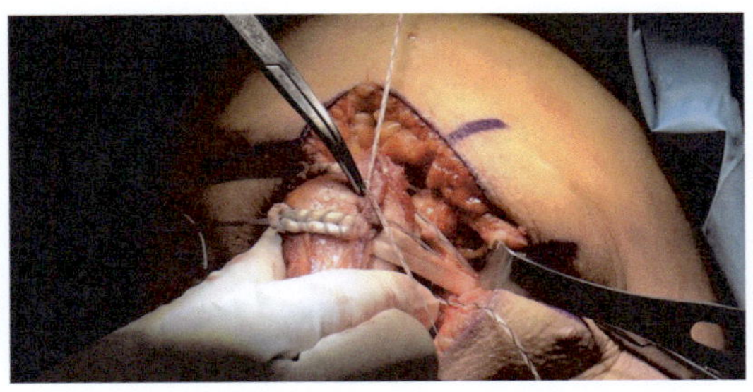

Fig. 20.12 Intra-operative photo demonstrating tensioning of the allograft for reconstruction of the MPFL

20.2.4.3 Axial Malalignment, Trochlear Dysplasia, and Bipolar Grade IV Chondral Disease of the Patellofemoral Joint

While admittedly controversial, we felt that patellofemoral arthroplasty was an elegant way to address the complex problems of axial malalignment, trochlear dysplasia, and bipolar chondrosis with a single procedure. Despite being relatively young patient for arthroplasty, she is low demand with a sedentary lifestyle. Her arthritis and malalignment make arthroplasty combined with soft tissue balancing (as above) a reasonable solution for a challenging problem (Fig. 20.13a, b). Advantages include straightforward surgical technique and relatively easy rehabilitation (WBAT, ROM as tolerated). Disadvantage includes potential progression of arthritis and conversion to TKA in the future.

Regarding axial malalignment, three separate treatment plans were outlined for this problem: TTO, benign neglect, and patellofemoral arthroplasty. The discussion by Dr. Feller regarding TT-TG cutoff values and treatment thresholds demonstrates regional treatment variations and the complexity of an ever-evolving discussion. As above, we chose arthroplasty due to our patient's age and lifestyle as well as its ability to address multiple problems with a single procedure. TTO is certainly a reasonable option; how-

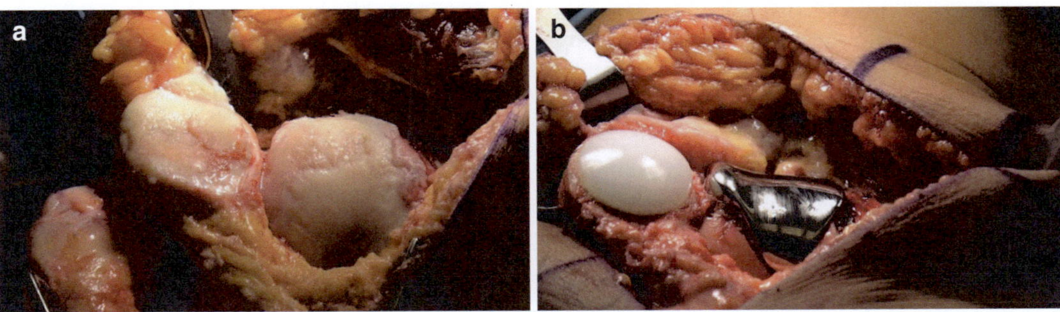

Fig. 20.13 (**a**) Intra-operative imaging demonstrating Grade IV bipolar chondral lesion of the patellofemoral joint. (**b**) Intra-operative photo demonstrating placement of the patellofemoral arthroplasty

ever, this would have to be accompanied by bipolar cartilage restoration procedure to optimize chance of a good outcome. Results of techniques such as MACI or OCA drop precipitously in patients with bipolar disease, older age, osteophytes, and frank joint space narrowing. Similarly concerning, the risk of complication is higher and rate of success is lower with osteotomy and bipolar cartilage restoration (MACI, OCA) when compared to isolated osteotomy or osteotomy combined with treatment of a solitary unipolar defect.

Regarding trochlear dysplasia, the second-generation onlay patellofemoral arthroplasty designs re-shape the proximal trochlea, improving force vectors, and providing inherent entrance stability to the patella. This case also exemplifies well known challenges with inter- and intra-observer reproducibility using the DeJour classification [11]. There was discrepancy noted between low vs. high grade dysplasia in this case which did have an influence on treatment plan. Note that no surgeon chose trochleoplasty as an option, citing either the "low grade" dysplasia, patient age, or amount of chondrosis as the reason for avoiding the procedure in this case.

20.3 Conclusion

This patellofemoral joint preservation case demonstrates that four experienced orthopedic surgeons can have very different approaches to a complex problem. The literature provides support for various treatment plans without a definitive answer as to which is ultimately correct. We believe it is imperative to approach the joint as an "organ" and to define a problem list before performing any surgical intervention. This allows the surgeon to systematically address each contributing factor to maximize the opportunity for success and to reduce the risk of surgical morbidity.

References

1. Weber AE, et al. An algorithmic approach to the management of recurrent lateral patellar dislocation. J Bone Joint Surg Am. 2016;98(5):417–27.
2. Ntagiopoulos PG, Dejour D. Current concepts on trochleoplasty procedures for the surgical treatment of trochlear dysplasia. Knee Surg Sports Traumatol Arthrosc. 2014;22(10):2531–9.
3. Zaffagnini S, et al. Medial patellotibial ligament (MPTL) reconstruction for patellar instability. Knee Surg Sports Traumatol Arthrosc. 2014;22(10):2491–8.
4. Shive MS, et al. BST-CarGel® treatment maintains cartilage repair superiority over microfracture at 5 years in a multicenter randomized controlled trial. Cartilage. 2015;6(2):62–72.
5. Fulkerson JP, Becker GJ, Meaney JA, Miranda M, Folcik MA. Anteromedial tibial tubercle transfer without bone graft. Am J Sports Med. 1990;18(5):490–7.
6. Pidoriano AJ, Weinstein RN, Buuck DA, Fulkerson JP. Correlation of patellar articular lesions with results from anteromedial tibial tubercle transfer. Am J Sports Med. 1997;25(4):533–7.
7. Fithian DC, Paxton EW, Post WR, Panni AS. Lateral retinacular release: a survey of the International Patellofemoral Study Group. Arthroscopy. 2004;20(5):463–8.
8. Dejour H, Walch G, Nove-Josserand L, Guier C. Factors of patellar instability: an anatomic radiographic study. Knee Surg Sports Traumatol Arthrosc. 1994;2(1):19–26.
9. Hinckel BB, Gobbi RG, Filho EN, et al. Are the osseous and tendinous-cartilaginous tibial tuberosity-trochlear groove distances the same on CT and MRI? Skeletal Radiol. 2015;44(8):1085–93.
10. Schoettle PB, Zanetti M, Seifert B, Pfirrmann CW, Fucentese SF, Romero J. The tibial tuberosity-trochlear groove distance; a comparative study between CT and MRI scanning. Knee Surg Sports Traumatol Arthrosc. 2006;13(1):26–31.
11. Rémy F, Chantelot C, Fontaine C, Demondion X, Migaud H, Gougeon F. Inter- and intraobserver reproducibility in radiographic diagnosis and classification of femoral trochlear dysplasia. Surg Radiol Anat. 1998;20(4):285–9.

21 Patellofemoral Pain and Arthritis in Latter Middle-Aged Patient, with Marginal Osteophytes and General Patellofemoral Chondrosis

Stefano Zaffagnini, Giacomo Dal Fabbro, Margherita Serra, and Elizabeth A. Arendt

21.1 Case Presentation

A 58-year-old female university professor presented with bilateral chronic knee pain worse on the left side. In particular, the patient complained of worsening left knee pain with stair climbing and rising from a chair, associated with recurrent knee swelling. The patient denied prior knee trauma or surgery, and prior episodes of patellar dislocation. Medical co-morbidities included a post-Hashimoto disease hypothyroidism which treated with levothyroxine. No other pathologies were reported in patient's history.

Clinical examination showed a slight swelling left knee with a physiological valgus limb alignment. Acute pain was evoked by palpation of the anterior left knee, and the patellar grind test (Clarke's sign) was positive. Full knee range of motion was present, with associated crepitance to active knee motion.

The left knee MRI showed a severe patellofemoral osteoarthritis with marginal osteophytes and general chondrosis concentrated laterally (Fig. 21.1). Moreover, further imaging including MRI examination showed no evidence of coronal knee deformity, tibiofemoral joint osteoarthritis, or tibiofemoral bony edema.

21.2 Evaluation and Treatment by Presenting Physician (Stefano Zaffagnini)

Based on the history, clinical examination, and imaging evaluation, I have assessed that patient's anterior knee pain and crepitation were attributable to isolated patellofemoral osteoarthritis.

At first, a conservative treatment strategy was chosen: a left knee intra-articular corticosteroid injection was performed, and daily ice-therapy, vastus medialis oblique muscle enforcement, and internal rotary footbeds were recommended. Moreover, the patient received recommendation to avoid bike, climbing or descending stairs, and walking on inclined terrain. One month later an intra-articular injection with hyaluronic acid with high molecular weight was performed.

S. Zaffagnini (✉)
IRCCS—Istituto Ortopedico Rizzoli—Clinica Ortopedica e Traumatologica II, Bologna, Italy

University of Bologna, Bologna, Italy
e-mail: Stefano.zaffagnini@unibo.it

G. Dal Fabbro · M. Serra
IRCCS—Istituto Ortopedico Rizzoli—Clinica Ortopedica e Traumatologica II, Bologna, Italy
e-mail: giacomo.dalfabbro@studio.unibo.it; margherita.serra2@studio.unibo.it

E. A. Arendt
Department of Orthopedic Surgery, University of Minnesota, Minneapolis, MN, USA
e-mail: arend001@umn.edu

© ISAKOS 2022
J. L. Koh et al. (eds.), *The Patellofemoral Joint*, https://doi.org/10.1007/978-3-030-81545-5_21

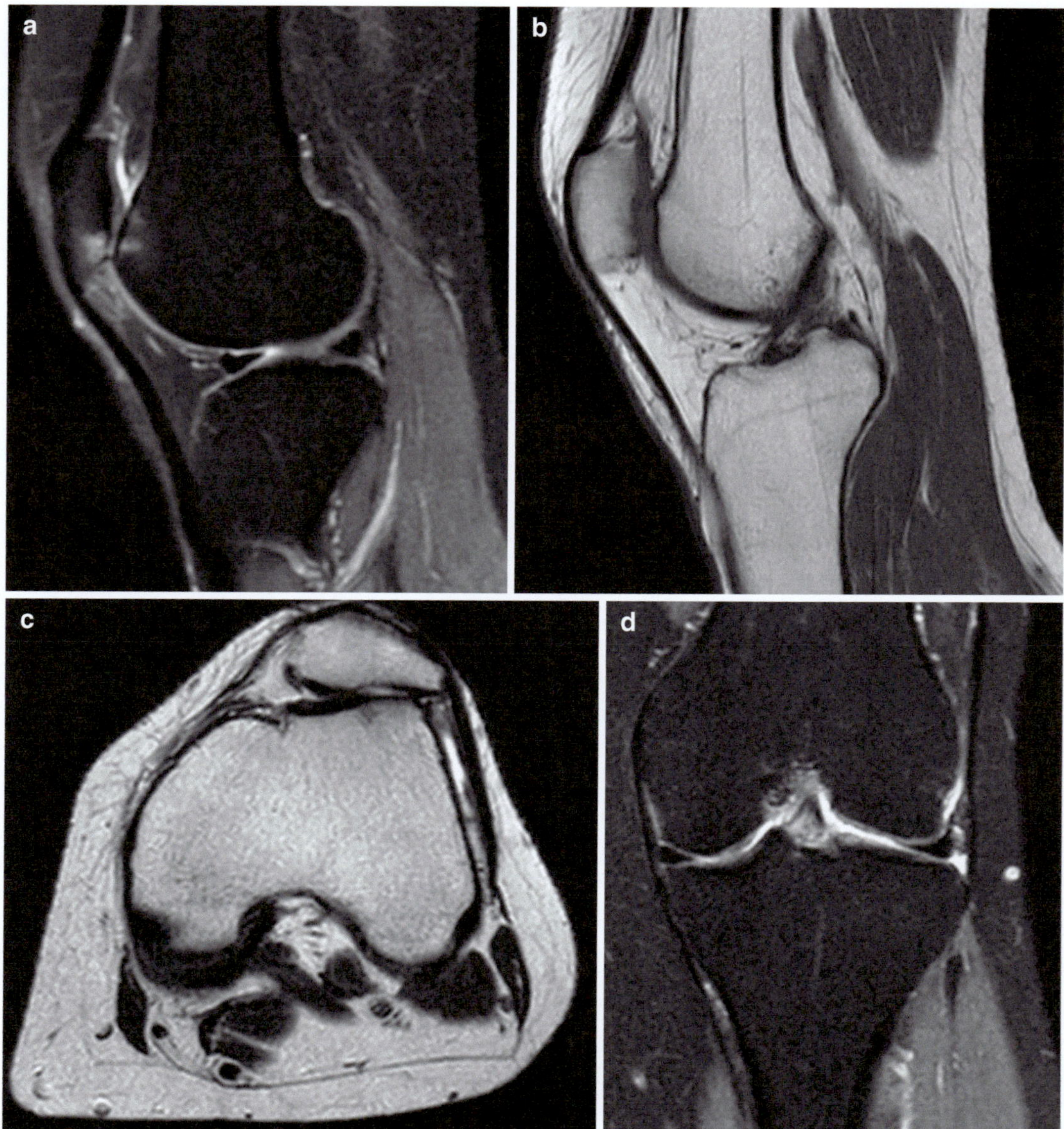

Fig. 21.1 Sagittal and axial view showed a severe patellofemoral osteoarthritis with osteophytes and chondrosis (**a–c**). Frontal view showed no evidence of tibiofemoral osteoarthritis (**d**)

However, after 6 months patients presented with persistent anterior knee pain, complaining of serious limitations during normal daily activities.

In literature, satisfactory clinical outcomes were provided after surgical treatment of isolated patellofemoral osteoarthritis with patellofemoral arthroplasty [1]. Furthermore, in the current patient, the non-operative strategies did not lead to any clinical improvement. Thus, the present patient was indicated for patellofemoral arthroplasty.

21.2.1 Surgical Treatment

The patient was positioned supine, under regional anesthesia supplemented with sedation. The entire low extremity was prepared and draped. A straight midline skin incision was carried out; subsequently parapatellar medial arthrotomy was performed. Before proceeding with the arthroplasty, the joint was carefully inspected to confirm that the tibiofemoral compartment was free of disease or degeneration

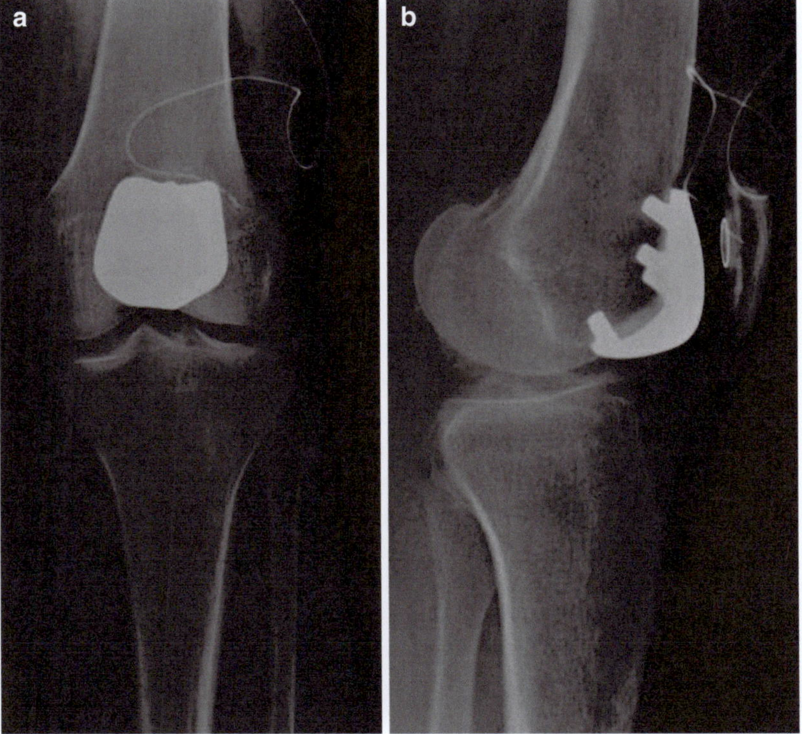

Fig. 21.2 Post-operative X-ray. Antero-posterior view (**a**); lateral view (**b**)

signs. The osteophytes bordering the intercondylar notch were removed. With the appropriate cutting guides, the femoral and patellar surface were prepared, and a patellofemoral joint prosthesis was implanted (Journey PFJ Smith&Nephew, London UK, trochlear component size x-small, patellar component symmetric 29 × 7.5 mm). The patella was resurfaced restoring the original patellar thickness and medializing the component on the native patella.

Finally, the patellar tracking during flexion and extension was assessed and provided a satisfactory result. At the end of the surgery post-operative image of the operated knee was performed (Fig. 21.2).

21.2.2 Post-operative Management

Patient started physical therapy the first day after surgery with isometric exercise to enhance quadriceps muscle strength. Progressive weight bearing using crutches was allowed during the first 15 days postoperatively. After 15 days patient was allowed to achieve full weight bearing to gradually stop using crutches as motion and strength allowed. After 21 days patients started swim, resumed all normal daily activities, and returned to work.

21.3 Commentary and Treatment Recommendation from Dr. Elizabeth Arendt

I agree that with the stated history, physical exam, and imaging, that a patellofemoral arthroplasty is a preferred surgical option. In my hands, at the age of 58, one would have to be physiologically younger than age 58 with a near pristine tibiofemoral joint to have this be a preferred option.

For surgical treatment, regional anesthesia supplemented with peri-articular multimodal pain cocktail is performed. A straight midline incision is used, with a vastus muscle splitting incision to enter the joint.

With the trial prostheses in place, a trial of capsular closure is performed to assess patella tracking through passive knee motion. Although this does not guarantee good tracking of the patella actively, one should look for patellar tracking in two key areas:

1. As the patella enters the groove in early flexion, if there is any deviation of the femoral prosthesis into a varus alignment, one can create a type of J tracking. This is unusual but clearly can be an unexpected complication.
2. When the knee goes from full flexion to extension, if there is any prominence of the metal component's distal edge, this will create a type of jumping of the patella into the groove in deep flexion. This typically can happen when you have a slight (native) femoral valgum; in order to achieve a flush surface between your lateral prosthesis and the native cartilage, one has to recess the medial side. In such a case if one leaves the medial side flush (not recessed) it may cause lateral sided prominence and a hop into the groove from deep flexion to extension.

From the enclosed lateral X-rays, one can identify mild knee hyperextension. With knee hyperextension and relative patella alta, the surgeon must be mindful not to have the patellar button exceed the length of the femoral arthroplasty flange when the knee is hyperextended. In this case the patient has normal height and there was no concern. One might consider placing the button as inferior as possible on the native patellar bone, to avoid a catch in knee hyperextension.

A drain is not used; the tourniquet is deflated after closure of the capsulotomy, and typically the patient is able to go home on the same surgical day.

Postoperatively, the patient's knee is placed in a soft compressive dressing, weight bearing as tolerated is allowed, using strength motion and pain as a guide to advance to full weight bearing. Stair climbing in a tandem fashion is discouraged until appropriate quad strength is achieved.

In cases where there has been significant bone erosion, the patient is warned that there may be a feeling of relative tightness of their knee in flexion, especially in the early rehabilitation phase, as the patella is now relocated centrally with patella height and trochlea groove restoration. Full motion is expected to be achieved post-operatively.

Realistic activity expectation is discussed, with the knowledge that quadricep muscle activity, already compromised pre-operatively, will take months to achieve greater strength than the pre-operative state.

21.4 Post-operative Follow-Up

At 1 year follow-up the patient referred to carried out daily and slight physical activities, such as trips to the hills or pilates class, with no limitations in her left knee. She was satisfied with the outcomes and doing well with no complaints of left knee pain. On clinical examination patient had not crepitation or apprehension sign, a full active and passive range of motion with no evidence of swelling or effusion. The knee X-ray examination showed the correct placement of the prosthetics components (Fig. 21.3).

21.5 Patellofemoral Arthroplasty: Current Concepts and Evidence

Patellofemoral osteoarthritis (PFOA) is described radiographically in up to 36% of the population [2] and isolated PFOA accounts for 10–24% of all patients presenting with knee pain [3]. In particular, middle-aged female and high-BMI patients tend to be more often affected by PFOA [4]. While the total knee arthroplasty represents the standard for treatment of knee OA, patellofemoral (PF) arthroplasty has emerged as an excellent option for patients presenting with isolated PFOA.

Patient selection and precise indications are crucial to obtain satisfactory clinical outcomes. The ideal patient is one with isolated anterior knee pain due to an isolated PFOA confirmed at imaging examination, felt during daily activities such as climbing stairs or rising from a chair. On physical examination is important to assess the stability of the patella as well as the knee in its entirety. Imaging exams must be used to exclude

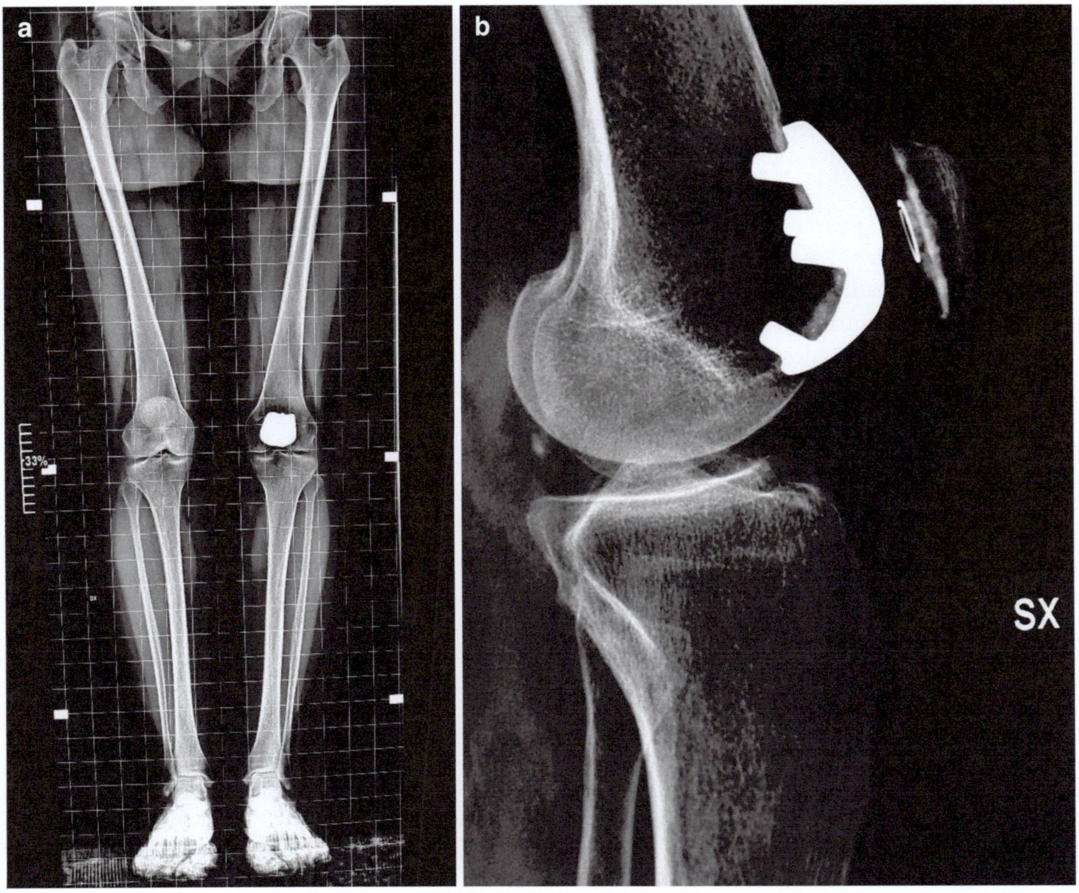

Fig. 21.3 1 year follow-up X-ray. Full standing weight bearing view (**a**); lateral view (**b**)

tibiofemoral disease and to identify existing deformity or dysplasia [5]. In these patients, when the first line conservative management fails (including treatment with anti-inflammatory drugs, injections and physical therapy), the surgical treatment is recommended. Despite, in clinical practice the PF arthroplasty is more often reserved for younger or middle-aged patients, if the tibiofemoral joint is unaffected, PF implants represents a good options regardless of patient's age. In support of this treatment choice, the revision to total knee arthroplasty after PF arthroplasty resulted relatively uncomplicated [6, 7].

However, controversy remains about the management of isolated PFOA. A systematic review provided fairly good outcomes of PF arthroplasty [1]; in another systematic review, in which the total knee replacement and the PF arthroplasty for the treatment of isolated PFOA were compared, similarity was reported between the two procedures in terms of complications and reoperation rate [8]. On the other hand, a later systematic review suggested that the reoperation rate for PF arthroplasty may be higher than the reoperation rate for total knee arthroplasty [9]. A 5- and 10-year revision rate of 9.75% and 18.70% respectively for PF arthroplasty compared with 2.16% and 3.39% for cemented total knee arthroplasty was reported in the National Registry for England, Wales, Northern Irland and Isle of Man (NJR) [10]. However, more recently published studies reported lower revision rates of PF arthroplasty in the setting of isolated patellofemoral OA, when compared to prior studies [1]; moreover a meta-analysis showed that second-generation PF implants had equivalent reoperation and revision rates, pain and mechanical complications when compared to total knee arthroplasty [11]. These findings suggest that, while the first generation PF arthroplasty offset the potential advantages of maintaining the knee's native soft tissues, the second-generation PF arthroplasty incorporated changes in implant design and instrumentation, showing promising results in the properly selected patient population. Furthermore, a recent randomized trial, in which the cost effectiveness analysis was performed, showed that patients with isolated PFOA achieved better short-term outcomes at lower costs from treatment with PF arthroplasty than from total knee arthroplasty [10].

Take Home Message

Isolated patellofemoral arthroplasty should be consider approaching latter middle-aged patient, often female, whose presenting complaint is anterior knee pain, and imaging reveals isolated PFOA. Patellofemoral arthroplasty represents an excellent treatment option, and in the management of patients with isolated patellofemoral osteoarthritis must always be considered.

References

1. van der List JP, Chawla H, Zuiderbaan HA, Pearle AD. Survivorship and functional outcomes of patellofemoral arthroplasty: a systematic review. Knee Surg Sports Traumatol Arthrosc. 2017;25(8):2622–31.
2. Davies AP, Vince AS, Shepstone L, Donell ST, Glasgow MM. The radiologic prevalence of patellofemoral osteoarthritis. Clin Orthop Relat Res. 2002;(402):206–12.
3. Duncan RC, Hay EM, Saklatvala J, Croft PR. Prevalence of radiographic osteoarthritis—it all depends on your point of view. Rheumatology (Oxford). 2006;45(6):757–60.
4. Collins NJ, Oei EHG, de Kanter JL, Vicenzino B, Crossley KM. Prevalence of radiographic and magnetic resonance imaging features of patellofemoral osteoarthritis in young and middle-aged adults with persistent patellofemoral pain. Arthritis Care Res (Hoboken). 2019;71(8):1068–73.
5. Hurwit D, Strickland S. Indications for patellofemoral arthroplasty in isolated patellofemoral arthritis. In: Dejour D, Zaffagnini S, Arendt EA, Sillanpää P, Dirisamer F, editors. Patellofemoral pain, instability, and arthritis. Berlin: Springer; 2020. p. 507–9.
6. Hutt J, Dodd M, Bourke H, Bell J. Outcomes of total knee replacement after patellofemoral arthroplasty. J Knee Surg. 2013;26(4):219–23.
7. Parratte S, Lunebourg A, Ollivier M, Abdel MP, Argenson JN. Are revisions of patellofemoral arthroplasties more like primary or revision TKAs. Clin Orthop Relat Res. 2015;473(1):213–9.
8. Vasta S, Papalia R, Zampogna B, Espregueira-Mendes J, Amendola A. Current design (onlay) PFA implants have similar complication and reoperation rates compared to those of TKA for isolated PF osteoarthritis: a systematic review with quantitative analysis. J ISAKOS. 2016;1(5):257–68.
9. Woon CYL, Christ AB, Goto R, Shanaghan K, Shubin Stein BE, Gonzalez Della Valle A. Return to the operating room after patellofemoral arthroplasty versus total knee arthroplasty for isolated patellofemoral arthritis-a systematic review. Int Orthop. 2019;43(7):1611–20.
10. Fredborg C, Odgaard A, Sørensen J. Patellofemoral arthroplasty is cheaper and more effective in the short term than total knee arthroplasty for isolated patellofemoral osteoarthritis: cost-effectiveness analysis based on a randomized trial. Bone Joint J. 2020;102-b(4):449–57.
11. Dy CJ, Franco N, Ma Y, Mazumdar M, McCarthy MM, Gonzalez Della Valle A. Complications after patello-femoral versus total knee replacement in the treatment of isolated patello-femoral osteoarthritis. A meta-analysis. Knee Surg Sports Traumatol Arthrosc. 2012;20(11):2174–90.

Part IV

Traumatic Injuries to the Patellofemoral Joint: Case-Based Evaluation and Treatment

Stellate Patella Fracture Case

22

Mauro Núñez, Rajeev Garapati,
Stefano Zaffagnini, Khalid Alkhelaifi,
and Ashraf Abdelkafy

In this chapter, several cases of stellate patella fractures are presented, along with their proposed treatment, and the rationale for the plan.

M. Núñez
Orthopaedic Surgery Department, Sports and Trauma Unit, Hospital del Trauma, San José, Costa Rica

Universidad de Ciencias Médicas (UCIMED), San José, Costa Rica

R. Garapati
Division of Orthopaedic Trauma, NorthShore University HealthSystem, Evanston, IL, USA

University of Chicago Pritzker School of Medicine, Chicago, IL, USA

S. Zaffagnini
Orthopaedics and Traumatology Unit, IRCCS Rizzoli Orthopaedic Institute, Bologna, Italy

DIBINEM Department, University of Bologna, Bologna, Italy
e-mail: Stefano.zaffagnini@unibo.it

K. Alkhelaifi
Orthopaedic Surgery Department, AspetarOrthopaedic and Sports Medicine Hospital, Al Rayyan, Qatar
e-mail: Khalid.Alkhelaifi@aspetar.com

A. Abdelkafy (✉)
Orthopaedic Surgery Department, Suez Canal University, Ismailia, Egypt
e-mail: ashraf.abdelkafy@med.suez.edu.eg

22.1 Case Presentation

22.1.1 History and Physical Examination

A 27-year-old male presented to the emergency room after a motor vehicle collision with another motorcycle. He was unable to straighten his left leg or weight bear on the affected side after a direct blow to his left knee on the motorcycle gas tank.

Upon inspection there was swelling and ecchymosis over the left knee. There was marked tenderness and crepitus at the anterior knee. The patient was unable to perform a straight leg raise and attempts at active knee extension were painful. He was neurovascularly intact distally.

22.2 Imaging Findings

X-rays (Figs. 22.1 and 22.2) demonstrated a left knee with a comminuted patella fracture with displacement and intraarticular involvement. CT scans (Figs. 22.3 and 22.4) also demonstrated the same comminuted patella fracture. A 3D CT reconstruction (Fig. 22.5) assisted in visualizing the fracture pattern.

© ISAKOS 2022
J. L. Koh et al. (eds.), *The Patellofemoral Joint*, https://doi.org/10.1007/978-3-030-81545-5_22

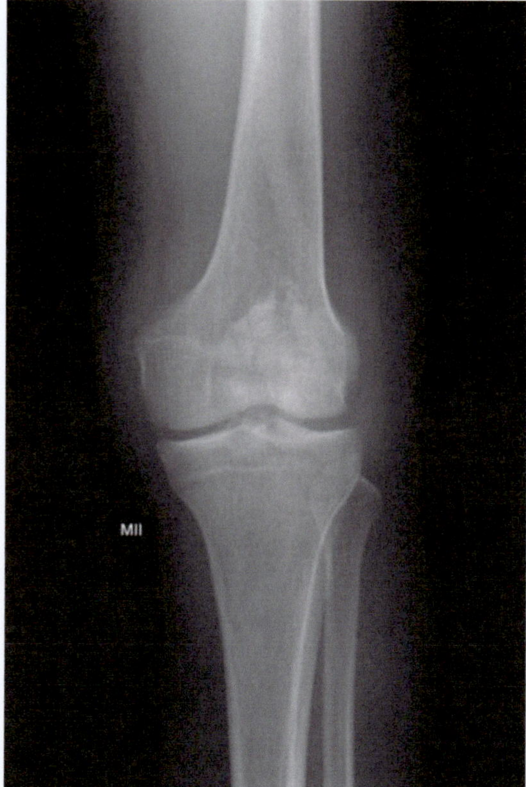

Fig. 22.1 AP X-ray

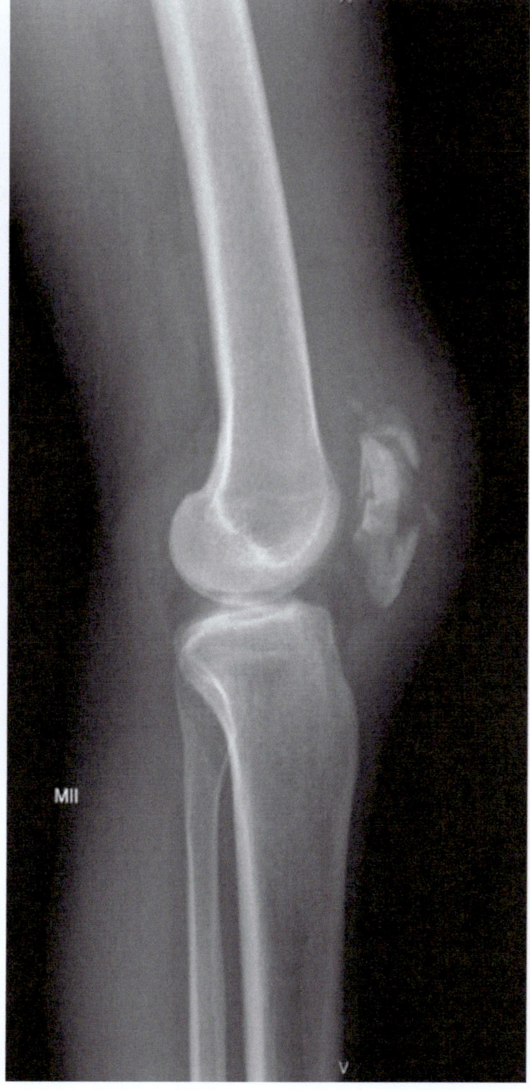

Fig. 22.2 Lateral X-ray

22.3 Proposed Treatment

22.3.1 Planning

Due to the comminuted and intraarticular nature of the fracture (AO/OTA fracture classification: 34 C3) and bone quality of the patient, open reduction and internal fixation (ORIF) of the patella were proposed. The goal was to achieve anatomical alignment and absolute stability by interfragmentary compression (e.g., lag screws, tension-band, compression plating). In the setting of absolute stability, there is minimal movement at fracture sites, and direct bone healing occurs without callus formation [1].

22.3.2 Preparation

After receiving spinal anesthesia, the patient was placed in a supine position and with a tourniquet at the level of the left thigh set to 250 mm/Hg. The limb was scrubbed with chlorhexidine, ChloraPrep→ and draped in sterile fashion. Anatomical landmarks were identified and marked appropriately, and the skin covered with 3M Ioban Incise Drape→.

22.3.3 Approach

A midline incision was performed to provide broad exposure from the superior pole of the

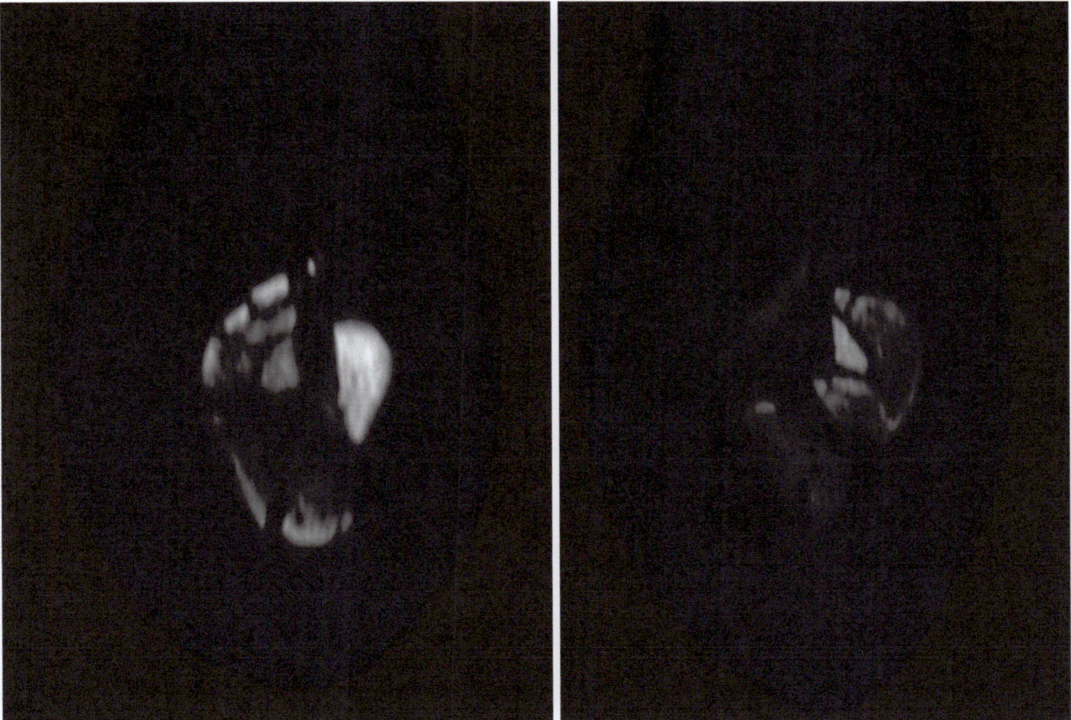

Fig. 22.3 CT Scan Coronal View

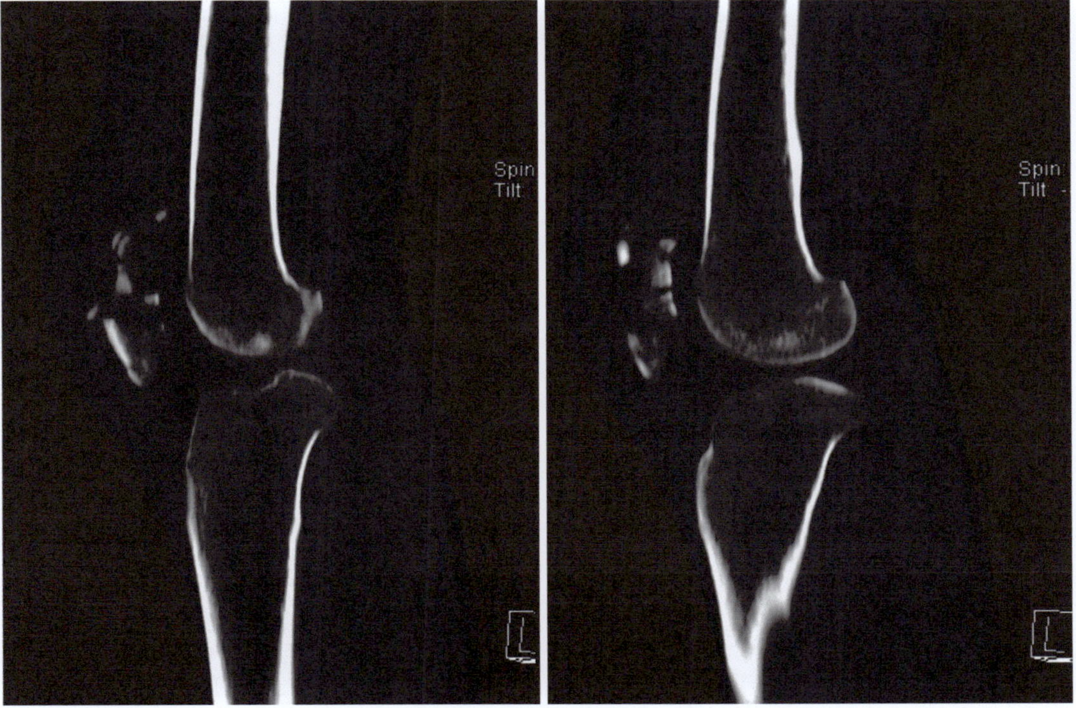

Fig. 22.4 CT Scan Sagittal View

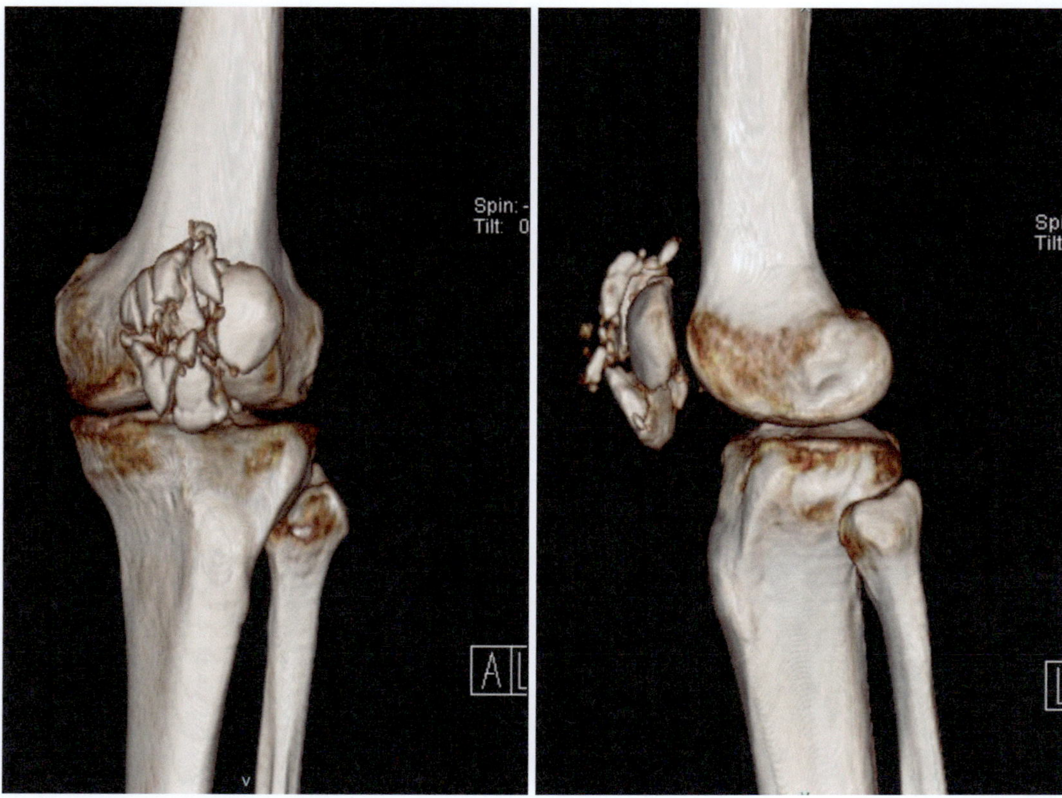

Fig. 22.5 CT Scan 3D Reconstruction

patella proximally to the tibial tubercle distally [2] (Fig. 22.6). For this case, a blunt osteotomy (Fig. 22.7) of the tibial tubercle was also utilized to aid in exposure and access to the fracture fragments. This allowed for excellent visualization of the articular surface (Fig. 22.8).

A provisional reduction was then obtained and maintained by Kirschner wires and clamps. In order to obtain absolute stability, a fixed-angle plate was used on the anterior cortex of the patella. The plate was utilized as a tension-band converting the distraction forces during knee flexion into compression forces [3] (Fig. 22.9a and b). A postoperative CT scan showed excellent anatomic reduction (Fig. 22.10a and b).

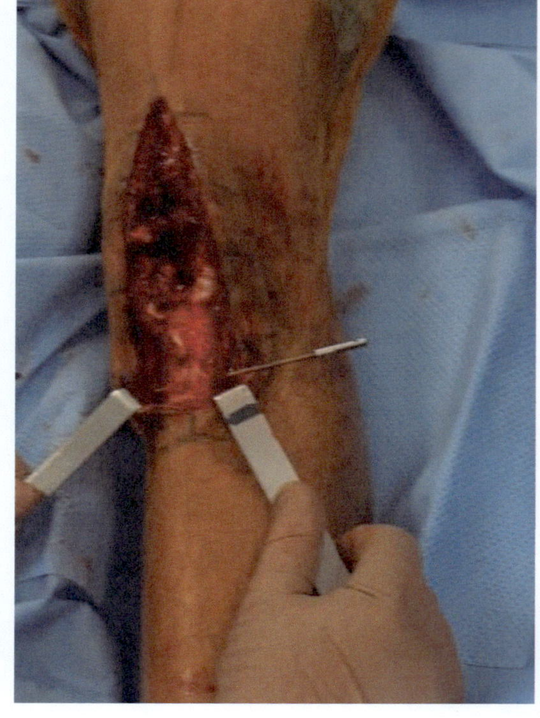

Fig. 22.6 Surgical Approach

22.3.4 Postoperative Care

Because the tension-band compresses the fracture during joint flexion, full range of motion was allowed after the ORIF. Post-operatively, the patient was placed in a hinged knee brace locked

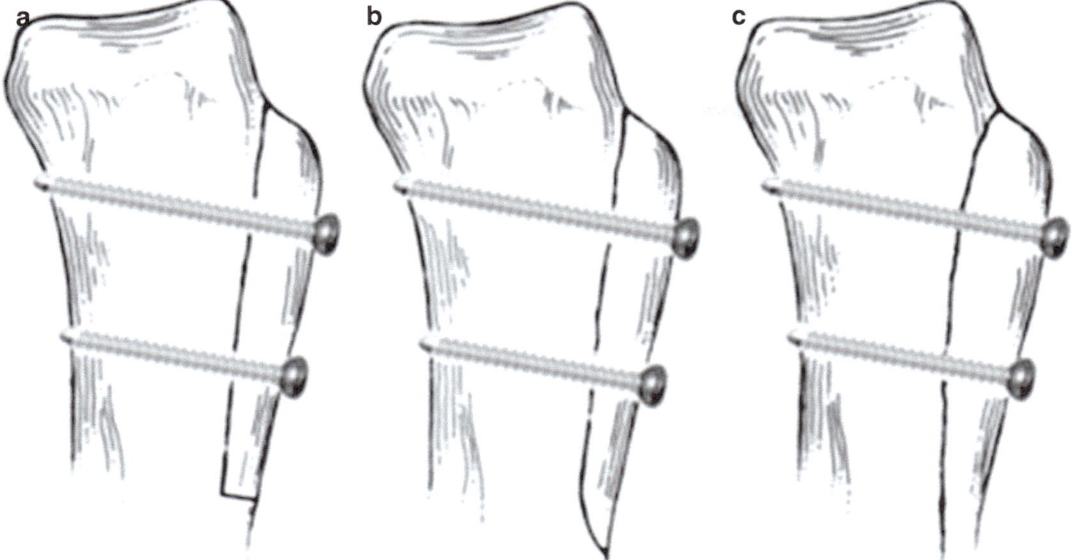

Fig. 22.7 Three types of tibial tuberosity osteotomy techniques: TTO-B (blunt) (**a**), TTO-S (sloped) (**b**), and TTO-G (greenstick) (**c**). (Reprinted with permission from Luhmann et al. 7 (Fig. 22.1) [2]. Copyright Springer, Berlin. All permission requests for this image should be made to the copyright holder)

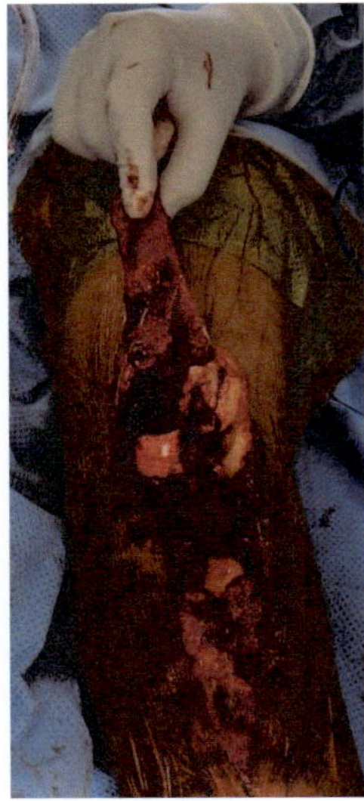

Fig. 22.8 Access to articular surface after tibial tubercle osteotomy

during partial weightbearing and unlocked at rest to allow for active range of motion at the knee for the first 4 weeks. After 4 weeks, weightbearing is advanced while in a locked brace. Physical therapy to obtain optimal range of motion at the knee is begun. Crutches and the brace are weaned between eight and 12 weeks status post-surgery.

22.4 Discussion

22.4.1 Commentary by Dr. Rajeev Garapati

This is a case of a 27-year-old male in a high energy motor vehicle accident with a comminuted patella fracture. Initial X-rays reveal a stellate patella fracture but the CT scan with 3D recons is very helpful in understanding the fracture and showing both the articular and non-articular comminution. A CT is not needed in simple patella fractures but is useful for preoperative planning in these comminuted fractures.

The goals of treatment include restoration of a functional extensor mechanism, reestablishing articular congruity and maintaining knee ROM. The patients' age and functional status play a role in deciding on the optimal treatment. If the patient was elderly or low demand, then non-operative treatment or patellectomy (complete or partial) could be considered. In evaluating the initial AP and lateral X-ray, there is not a large proximal to distal or medial to lateral separation of the patella and thus, non-operative treatment can be considered. This would leave the patient with a functioning knee but they would

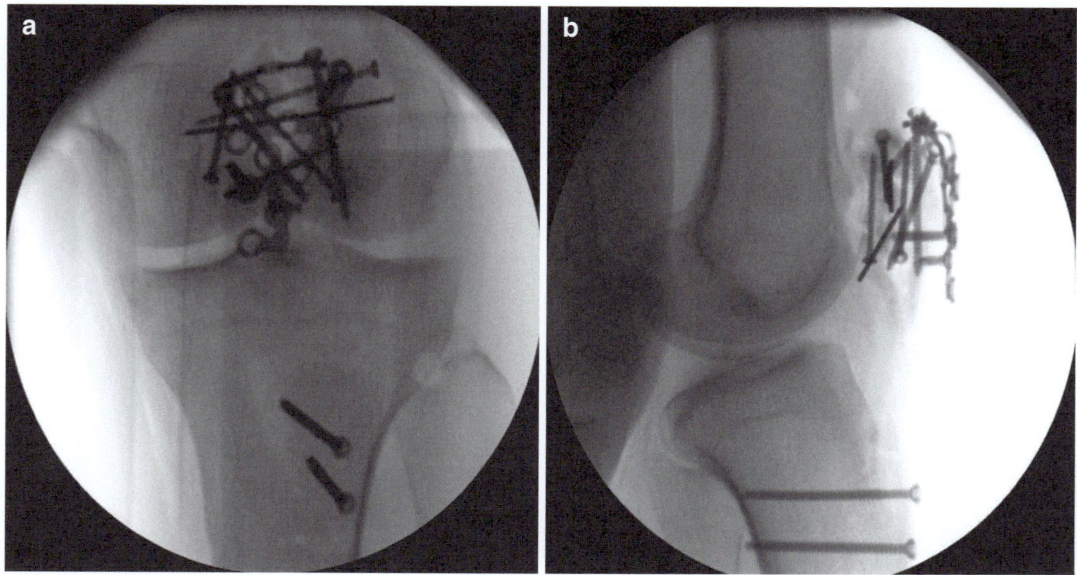

Fig. 22.9 (**a**) Intraoperative Fluoroscopy AP View. (**b**) Intraoperative Fluoroscopy Lateral View

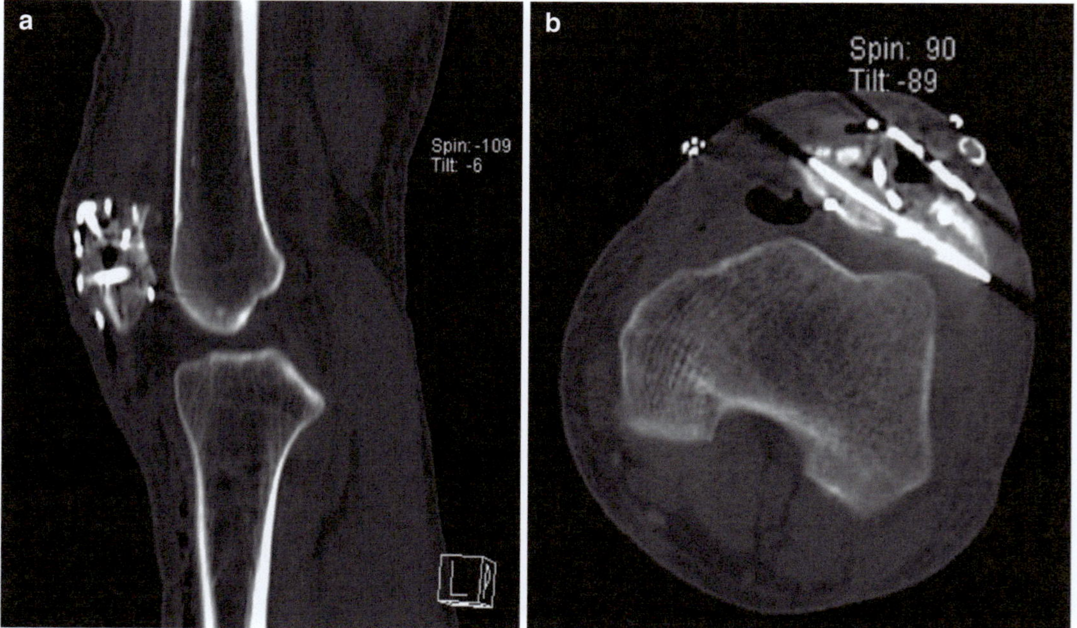

Fig. 22.10 (**a**) CT Scan Postoperative Sagittal View. (**b**) CT Scan Postoperative Axial View

have a high risk of extensor lag and patellofemoral pain and arthritis. Patellectomy is another option if the fracture is considered non-reconstructable and if the patient is low demand. This would involve excising some or all of the patella and repairing the soft tissue. Satisfactory results have been shown with maintenance of 60% of the patella but patients will have significantly increased patellofemoral contact forces and run the risk of quadriceps weakness [4].

In a young and active patient such as this, open reduction and internal fixation are the preferred treatment. A midline incision is recommended. In this case, the author has decided to perform a tibial tubercle osteotomy and elevate the extensor mechanism in order to get full visualization of the fracture. This can provide excellent exposure but is an extensile approach and involves significant soft tissue dissection. Another option for exposure is to use the midline incision and then work within and between the fracture fragments. This is soft tissue friendly but it can be difficult to completely visualize all the fragments and requires one to be adept at indirect reduction techniques. The exposure option I would have chosen would be to do a midline incision with a lateral parapatellar approach and flipping the patella medially to get direct exposure of the articular surface [5]. This preserves the major inferomedial blood supply of the patella while providing extensive visualization.

Once exposure is obtained, the author achieved anatomic reduction and absolute stability with mini-fragment screws and a mesh style plate. This is a very good option in comminuted patella fractures [6] but involves extensive exposure and the possibility of hardware irritation. Also, mesh plates and patellar plates are not always available. Simple transverse patella fractures are often fixed with longitudinal K-wires or cannulated screws with a tension-band wire. This fixation construct can be difficult to achieve in comminuted fractures and the articular side of the patella may not be able to withstand the compressive loads that are necessary in a tension-band construct. In that situation, I would reduce the major articular and non-articular fragments and fixate them with k-wires that are cut and impacted below the bony surface. Sutures can also be used to secure the soft tissue with the attached bony fragments. Once the anatomy of the patella is restored, fixation can be achieved with a cerclage wire going around the patella. I traditionally use #6 sternal wire and weave it in and out of the soft tissue circumferentially around the patella. Additionally, in order to secure the anterior fragments, one can place a figure of 8 wiring on top of the patella with the #6 sternal wire. Non-absorbable suture such as #5 fiber wire can be used instead of metal wire [7].

Post-operatively, the patient would be placed into a hinged knee brace locked in extension. The patient can be weight bearing as tolerated with the leg locked in extension. If fixation is solid, then early ROM can be started such as the author suggested. In a comminuted fracture such as this, I would keep the patient locked in extension for 4 weeks and then start unlocking the brace to start passive and active assisted ROM with physical therapy. The brace is discontinued between 8 and 12 weeks once there is bony healing and the patient has regained enough quadriceps strength to safely walk without the knee buckling. Physical therapy will continue for 2–3 months and then the patient will transition to a home exercise program. Full recovery can take 6–12 months.

22.5 Patella Fracture

Abstract

Patella fractures have a very broad spectrum of complexity, ranging from a simple transverse fracture pattern to those where joint comminution presents itself as a huge challenge for the proper treatment.

Some strategies proposed for transverse fractures such as indirect reduction by palpation with no visualization of the articular surface and fixation with modified tension-band technique may be insufficient for more complex cases.

Below we present a perspective where the direct visualization of the articular surface by means of osteotomy of the tibial tubercle and dislocation of the patella is presented as an alternative for addressing comminuted patellar fractures.

The osteotomy of the tibial tubercle allows full access to the articular surface of the patella and, if executed meticulously, can lead to a low number of complications.

22.5.1 Introduction

The patella is the largest sesamoid bone in the body and increases the moment arm of the extensor mechanism of the quadriceps by 30% [8]. Patellar fractures account for approximately 1% of all fractures [9]. The most common causes of patella fractures are traffic accidents in 78.3% of patients, followed by work-related accidents in 13.7% and domestic accidents in 11.4% [10]. Patella fractures occur as a result of either indirect or direct trauma. The indirect trauma is generated through a sudden contracture of the extensor mechanism that overcomes the resistance of the patella, usually resulting in a transverse fracture pattern. Direct trauma with the flexed knee often occurs after a fall or severe injury when hitting the bent knee against the ground or dashboard of a vehicle. The patella impacts the strong femoral trochlear or notch bone which leads to bursting of the patella. The subcutaneous position of the patella makes this bone prone to open fractures and soft tissue injuries.

Simple X-ray with anteroposterior and lateral projections can often suffice but can be complemented with an axial view. However, it should be taken into consideration that such an X-ray is difficult to obtain in a patient with an acute injury secondary to pain. In cases of comminution, a CT scan can be an excellent diagnostic method for better understanding of the fracture. Patella fractures can be classified using the Speck and Regazzoni Classification System [11].

In a patient with a history of direct trauma to the knee, a patella fracture should be ruled out. If the extensor mechanism is not intact, both a patella fracture and soft tissue disruption (i.e., patellar tendon rupture or quadriceps tendon rupture) should be worked up. The soft tissue envelope is of utmost importance due to the subcutaneous nature of the patella.

The goals of treatment for patellar fractures are a functional extensor mechanism, articular congruity, and full, painless range of motion of the knee [12]. Stellate patellar fractures are either displaced which requires open reduction and internal fixation or non-displaced which requires conservative treatment in the form of above knee cast or splint. Open reduction and internal fixation are still associated with high rates of complications [13].

Transverse fractures are best treated with a tension-band technique (classic AO Technique) which neutralizes tension forces anteriorly and converts them into stabilizing compressive forces at the posterior (articular) surface of the patella [14]. Berg described a modified tension-band technique using parallel vertical cannulated screws as an alternative for the AO foundation technique [15]. Tian compared a tension-band technique using Kirschner wires with a tension-band technique using cannulated screws; fracture reduction, time to heal, and Iowa knee scores were better for the cannulated screw group [16].

In patella fractures with comminution, small plates can be applied to the anterior cortex to provide additional stability [17]. In 1998 Berg described a tibial tubercle osteotomy and patella eversion technique for comminuted patella fractures to improve visualization and reduction of the articular surface. The osteotomy healed in all cases and did not adversely affect the clinical results [18].

22.5.2 Discussion

In the present case, a multifragmentary, displaced patella fracture in a young patient was described. Operative management of comminuted articular patella fractures (34C3 AO classification) represents a significant challenge for surgeons [19]. In this case, the presence of several comminuted fragments made reduction difficult to obtain. However, due to the patient's age and functional status, successful ORIF was necessary.

The aim of the treatment was to restore the articular congruity of the patella and to reestablishing the extensor mechanism with fixation secure enough to allow early motion. A tibial tubercle osteotomy was done to obtain direct visualization of the bone fragments to achieve an anatomical reduction of the comminuted fracture. Fixation consisted of interfragmentary lag screws and a fixed-angle plate on the anterior cortex of the patella. A satisfactory reduction was achieved. The authors believe that the tibial tubercle osteotomy aided in reduction and will not affect the outcome.

22.5.3 Alternative Treatment

As an alternative approach, a longitudinal incision in the midline on the patella, without performing an osteotomy of the tibial tubercle, can achieve an optimal exposure of the fracture site without damaging the infrapatellar branch of the saphenous nerve. The advantages and disadvantages of performing a tibial tubercle osteotomy for the treatment of comminuted patent fractures are contentious. The reduction and fixation of the fragments can also be achieved using a tension-band wiring technique, placing an eight-shaped cerclage, plus an additional cerclage in the form of a zero to increase stability [20] (Case 1).

If adequate reduction and fixation are not possible, performing a partial patellectomy is viable taking effort to minimize the amount of bone removed (Case 2). Partial patellectomy has been described as a means to preserve the moment arm of the patella resulting in less loss of strength, ligament instability, and quadriceps atrophy when compared with total patellectomy. Moreover, no significant differences in subjective and objective outcomes are described in patients who underwent partial patellectomy compared to patients who underwent open reduction and internal fixation procedure [21].

Finally, in cases with an irreparable articular cartilage defect to the patella, an osteochondral autograft can help to restore viable cartilage to the patella. A useful donor site is the proximal tibiofibular joint as it provides flat features that can be well adapted to the patella [22] (Fig. 22.11).

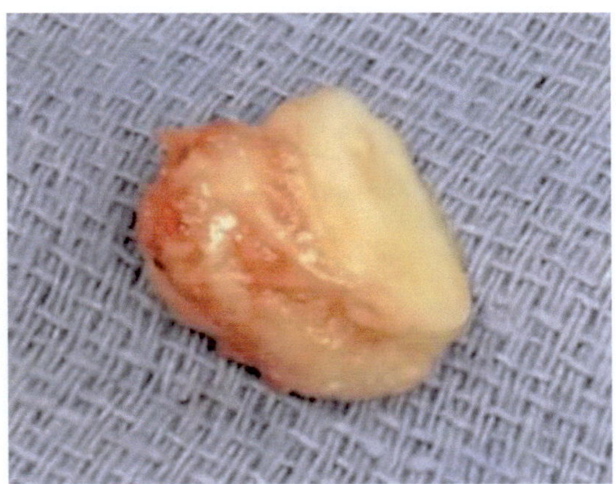

Fig. 22.11 Tibiofibular Osteochondral Autograft

22.5.4 Longer Term Consequences

Whether the fracture is displaced or non-displaced, the strong impact of the patella in the femoral trochlea or notch can often lead to significant patellar articular cartilage damage. After fracture healing, the chondromalacia and secondary osteoarthritis can emerge as a secondary problem and cause pain and disability to the patient. Patellar secondary osteoarthritis that results from cartilage injury during the initial trauma is treated according to its severity. Milder stages could be managed conservatively; however more severe case needs operative intervention. Several surgical procedures are available to treat patellar secondary osteoarthritis after significant trauma. It worth to mention two surgical techniques which are (1) patellofemoral arthroplasty and (2) arthroscopic total patellectomy.

22.5.5 Patellofemoral Arthroplasty

Patellofemoral joint arthroplasty is indicated for treatment of severe patellofemoral osteoarthritis [23–26]. However it requires expertise as revision rate is high [23] (Fig. 22.12).

22.5.6 Arthroscopic Total Patellectomy

Nowadays, total patellectomy is indicated in patients having severe patellofemoral joint osteoarthritis but in certain situations; (1) patient refuses arthroplasty, (2) arthroplasty option is not feasible due to the patient nature of work (e.g., manual laborers), (3) availability issues, (4) lack of expertise, and (5) high cost.

Patellectomy is better performed arthroscopically using an arthroscopic barrel burr with speedy recovery due to avoiding large incisions (Fig. 22.13).

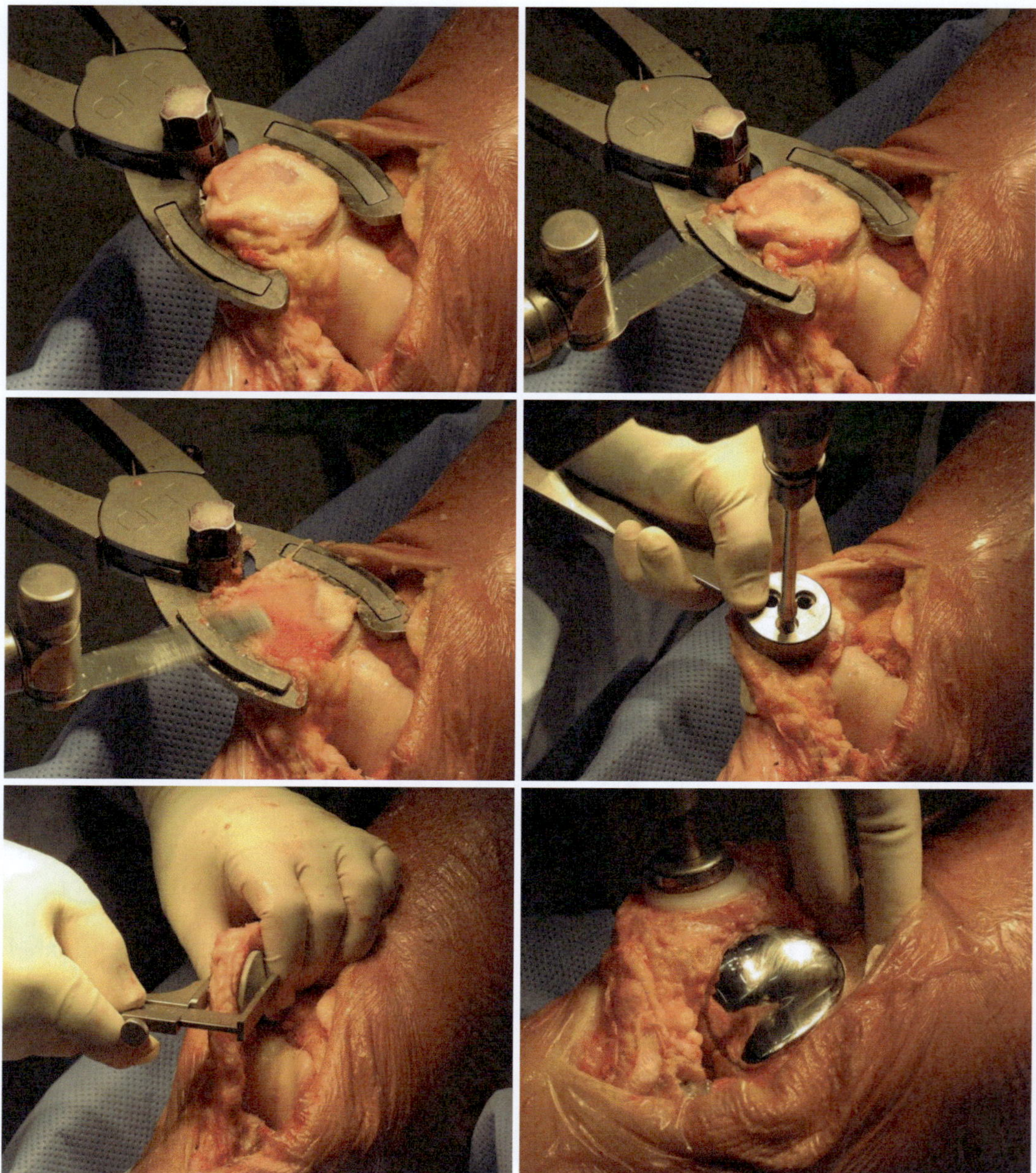

Fig. 22.12 Patellofemoral arthroplasty

22.6 Conclusions

Stellate patella fractures represent a difficult challenge to orthopedic surgeons, since they often present with comminution of the bone and underlying joint surface. An open reduction under direct visualization and reduction of the bone and articular cartilage fragments can facilitate the anatomical reconstruction of the patella. In addition, it aids in restoring the function of the extensor mechanism. In the long term, adequate ORIF of the patella preserves the function of the extensor mechanism and bone stock in the event that the patient eventually requires a knee arthroplasty. In patients with extensive comminution, a partial patellectomy remains a rea-

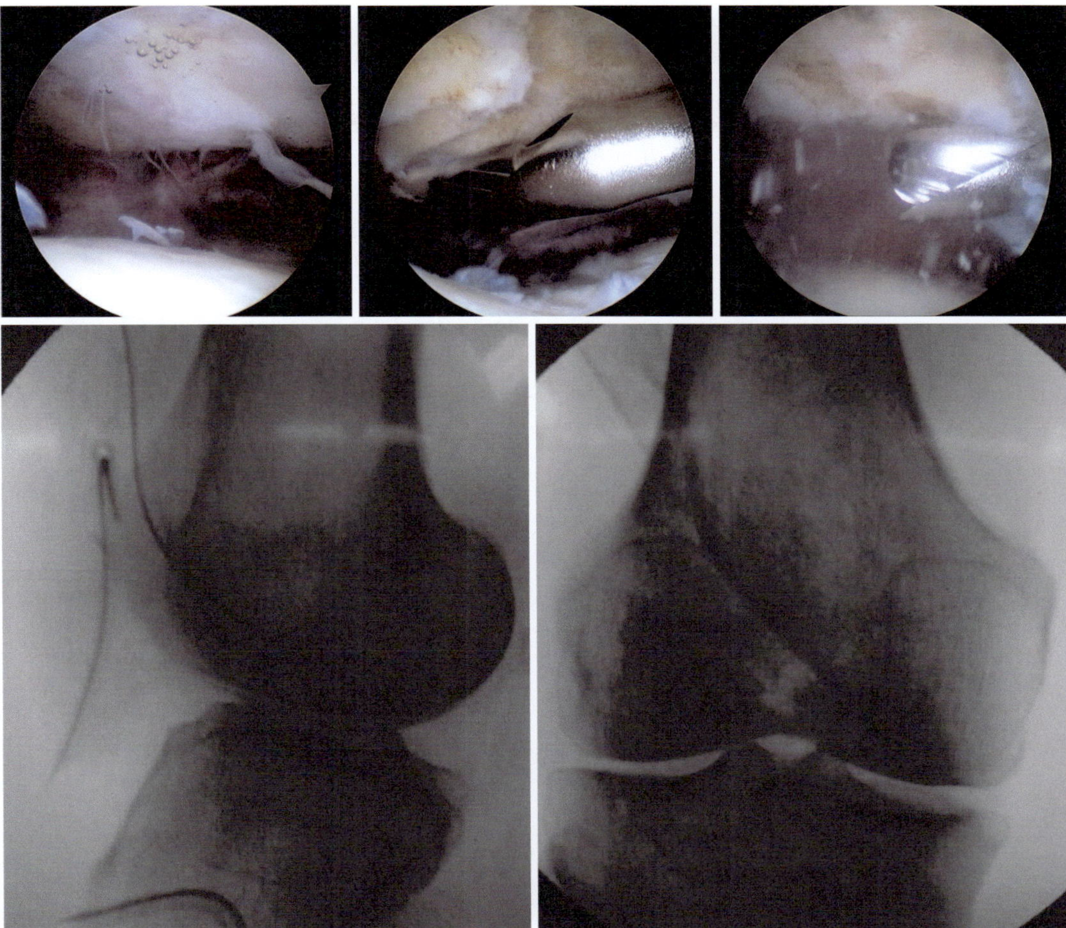

Fig. 22.13 Arthroscopic total patellectomy

sonable alternative with lower technical demand. With patients that have a stellate patella fracture and extensive articular cartilage damage, osteochondral autograft from the tibiofibular joint may provide a solution.

Take Home Message

Stellate patella fractures are a difficult challenge for surgeons who treat these injuries. The traditional approach to patella transverse fractures through a medial arthrotomy and verification of the reduction by palpation of the articular surface of the patella without visualization may be insufficient to obtain acceptable results in those patients with comminuted patella fractures. The osteotomy of the tibial tubercle allows direct visualization and better reduction of the fracture fragments allowing surgeons to achieve more anatomical reductions in these complex cases.

Fact Boxes

Stellate patella fractures often require demanding reconstruction techniques that differ from those used to treat transverse patella fractures.

In stellate patella fractures, the goals of surgical treatment are anatomical reduction with absolute stability, restoration of the extensor mechanism and articular congruity.

The osteotomy of the tibial tubercle can be technically demanding but when performed properly does not add to complications of the surgical approach.

Example Case 1 A 50-year-old woman suffered direct trauma to the anterior knee after a fall while running. The patient suffered a displaced articular patella fracture (34C1) with significant displacement of the fragments (Fig. 22.14a and b).

Surgical treatment: A midline longitudinal incision over the patella was used. Reduction and fixation with a tension-band wiring technique were performed. A cerclage wire was placed in a figure-of-eight pattern with an additional cerclage circumferentially (Fig. 22.15a and b).

Example Case 2 A 45-year-old male police officer who sustained a gunshot to the anterior knee. He was diagnosed with a stellate, displaced open patella fracture with significant comminution (Fig. 22.16a and b).

Treatment:

Initially, suture of the wound was performed acutely and antibiotic prophylaxis was provided. Definitive surgical treatment proceeded. A midline longitudinal incision over the patella was used avoiding the gunshot wound. After an unsuccessful reduction attempt of the comminuted fragments, a partial patellectomy of the inferior pole of the patella. In addition, a patellar tendon repair was performed with trans-osseous sutures and additional resorbable sutures obtaining satisfactory reduction of the superior patella fracture (Fig. 22.17a and b).

Resources Classification (Speck and Regazzoni, AO/OTA)

Speck and Regazzoni

AO/OTA Classification of patella fractures is 34

Type A (Extraarticular)

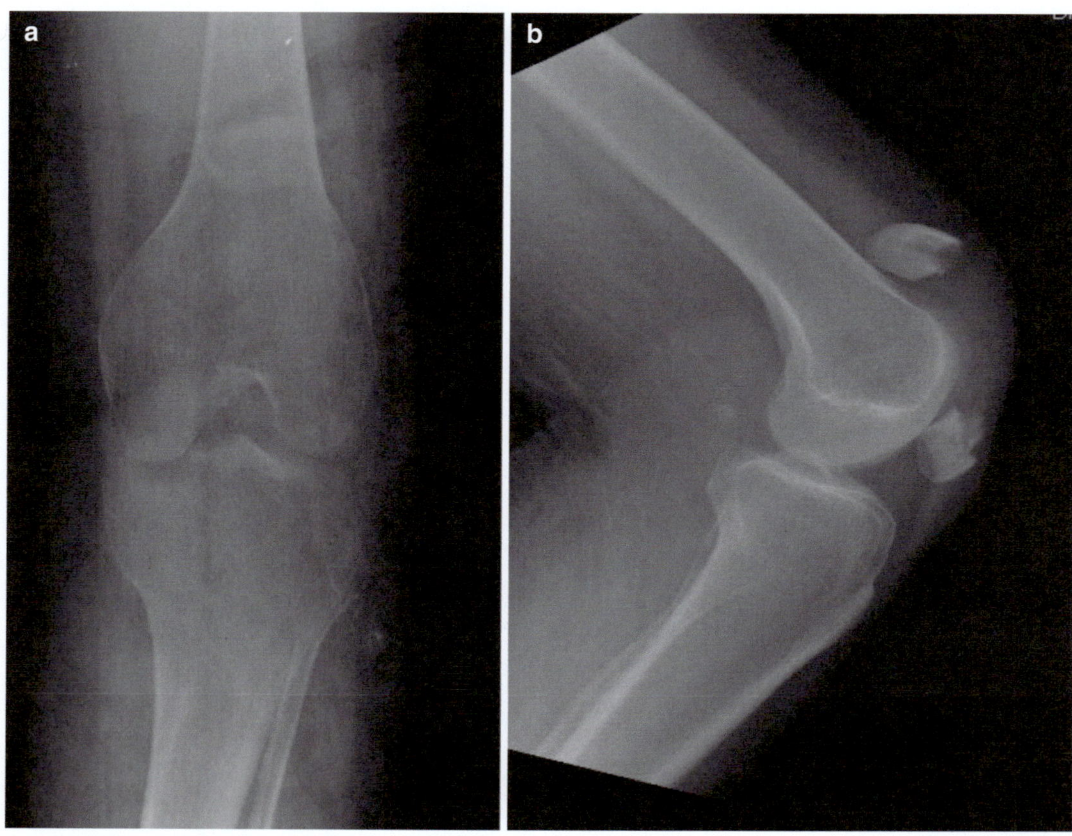

Fig. 22.14 (**a**, **b**) AP and lateral injury X-rays of knee with patella fracture

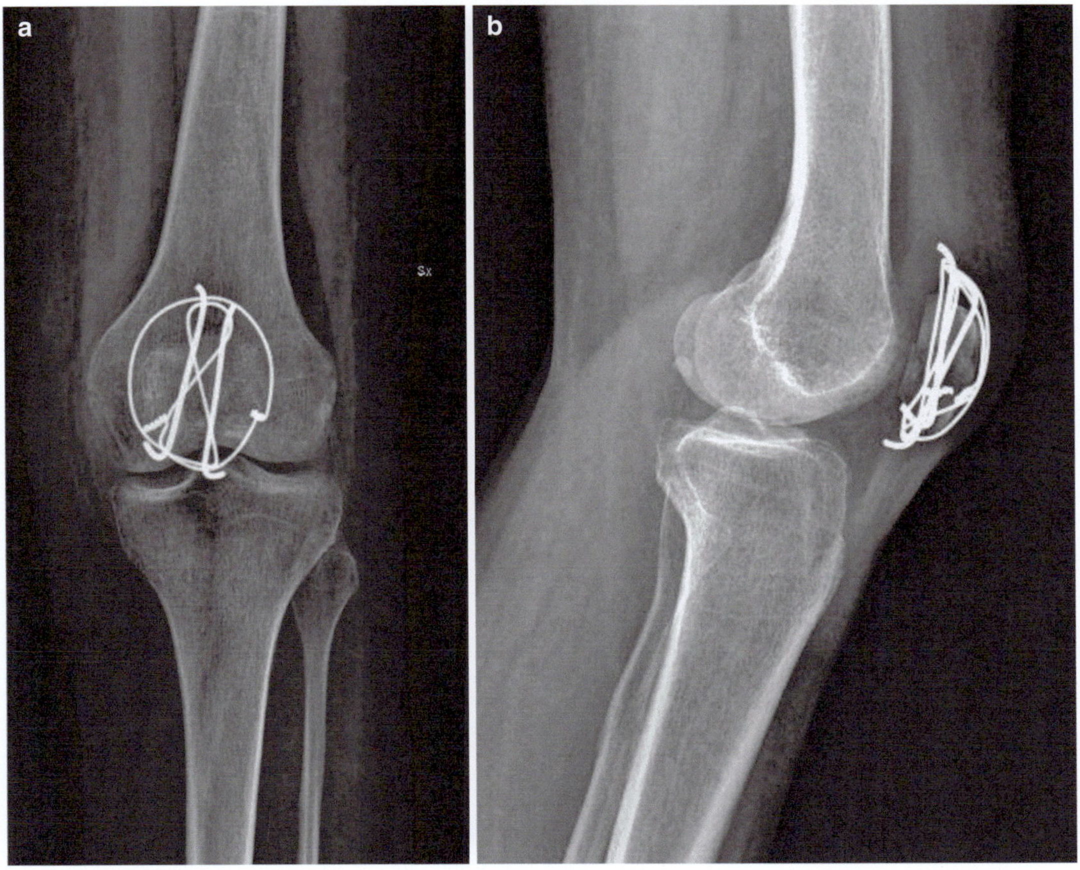

Fig. 22.15 (**a**, **b**) Postop AP and lateral X-rays demonstrating tension-band wiring and cerclage

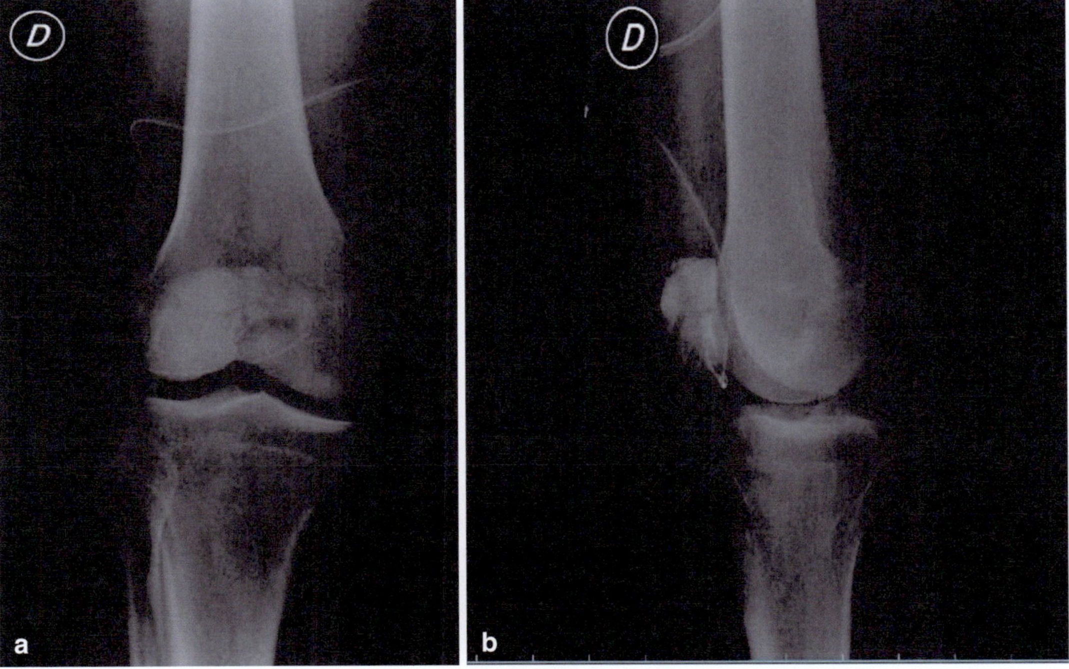

Fig. 22.17 (**a**, **b**) Postoperative X-rays following repair with transosseous sutures and partial patellectomy

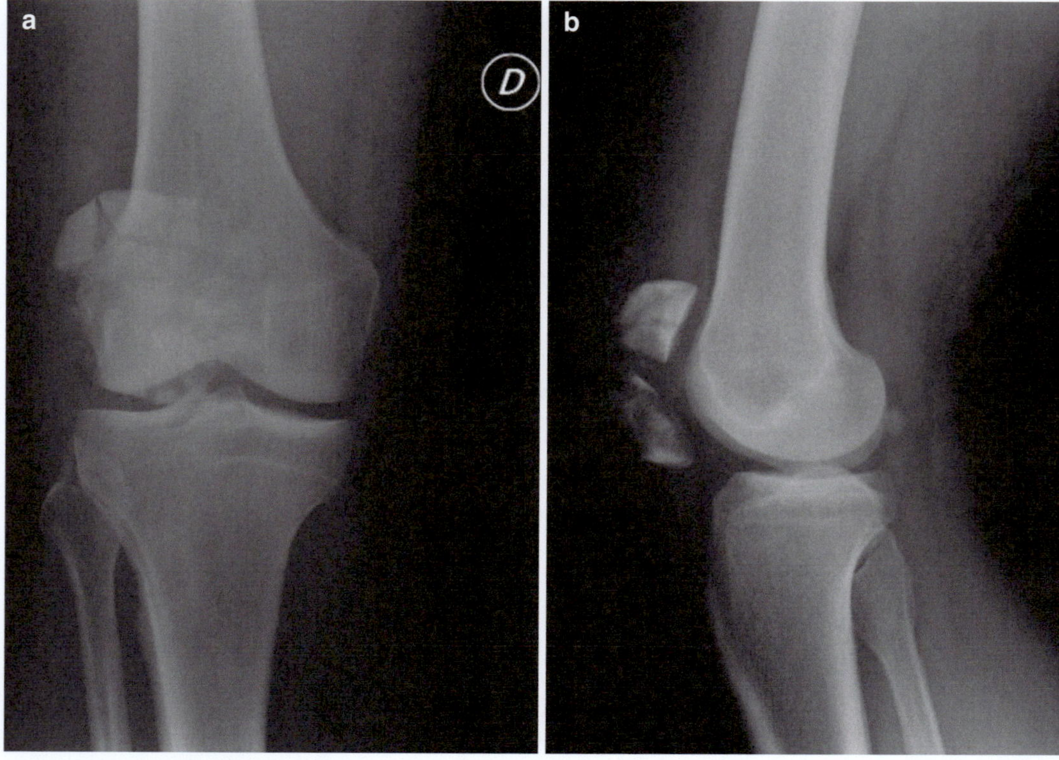

Fig. 22.16 (**a**, **b**) Comminuted stellate patella fracture

A1: Avulsion
A2: Isolated body

Type B (Partial Articular)

B1: Vertical lateral
B2: Vertical medial

Type C (Complete Articular)

C1: Transverse simple
C2: Transverse + second fragment
C3: Complex or comminuted

References

1. Perren SM. Evolution of the internal fixation of long bone fractures: the scientific basis of biological internal fixation: choosing a new balance between stability and biology. J Bone Joint Surg Br. 2002;84(8):1093–110.
2. Luhmann SJ, Fuhrhop S, O'Donnell JC, et al. Tibial fractures after tibial tubercle osteotomies for patellar instability: a comparison of three osteotomy configurations. J Child Orthop. 2011;5:19–26.
3. Hung LK, Chan KM, Chow YN, Leung PC. Fractured patella: operative treatment using the tension band principle. Injury. 1985;16(5):343–7.
4. Anand S, Hahnel JC, Giannoudis PV. Open patellar fractures: high energy injuries with a poor outcome? Injury. 2008;39:480–4.
5. Gardner MJ, Griffith MH, Lawrence BO, Lorich DG. Complete exposure of the articular surface for fixation of patellar fractures. J Orthop Trauma. 2005;19:118–23.
6. Wagner FC, Neumann MV, Wolf S, Jonaszik A, Izadpanah K, Piatek S, Südkamp NP. Biomechanical comparison of a 3.5 mm anterior locking plate to cannulated screws with anterior tension band wiring in comminuted patellar fractures. Injury. 2020;51(6):1281–7. https://doi.org/10.1016/j.injury.2020.03.030.
7. Gosal JS, Singh P, Field RE. Clinical experience of patellar fracture fixation using metal wire or non-absorbable polyester—a study of 37 cases. Injury. 2001;32:129–35.
8. Kaufer H. Mechanical function of the patella. JBJS. 1971;53(8):1551–60.
9. Boström A. Fracture of the patella: a study of 422 patellar fractures. Acta Orthop Scand. 1972;143:1–80.

10. Wild M, Windolf J, Flohé S. Fractures of the patella. Der Unfallchirurg. 2010;113(5):401–11.
11. Speck M, Regazzoni P. Classification of patellar fractures. Z Unfallchir Versicherungsmed. 1994;87(1):27–30.
12. Melvin SJ, Mehta S. Patellar fractures in adults. J Am Acad Orthop Surg. 2011;19(4):198–207.
13. Neyisci C, Erdem Y, Kilic E, Arsenishvili A, Kürklü M. A pilot study of a novel fixation technique for fixation of comminuted patellar fractures: arthroscopic-controlled reduction and circular external fixation. J Knee Surg. 2020;33(9):931–7. https://doi.org/10.1055/s-0040-1708830.
14. Müller ME, Allgöwer M, Müller ME, Schneider R, Willenegger H. Manual of internal fixation: techniques recommended by the AO-ASIF group. Springer Science & Business Media; 1991.
15. Berg EE. Open reduction internal fixation of displaced transverse patella fractures with figure-eight wiring through parallel cannulated compression screws. J Orthop Trauma. 1997;11(8):573–6.
16. Tian Y, Zhou F, Ji H, Zhang Z, Guo Y. Cannulated screw and cable are superior to modified tension band in the treatment of transverse patella fractures. Clin Orthop Relat Res. 2011;469(12):3429–35.
17. Taylor BC, Mehta S, Castaneda J, French BG, Blanchard C. Plating of patella fractures: techniques and outcomes. J Orthop Trauma. 2014;28(9):e231–5.
18. Berg EE. Extensile exposure of comminuted patella fractures using a tibial tubercle osteotomy: results of a new technique. J Orthop Trauma. 1998;12(5):351–5.
19. Böstman O, Kiviluoto O, Nirhamo J. Comminuted displaced fractures of the patella. Injury. 1981;13(3):196–202.
20. Schuett DJ, Hake ME, Mauffrey C, Hammerberg EM, Stahel PF, Hak DJ. Current treatment strategies for patella fractures. Orthopedics. 2015;38(6):377–84.
21. Bonnaig NS, Casstevens C, Archdeacon MT, et al. Fix it or discard it? A retrospective analysis of functional outcomes after surgically treated patella fractures comparing ORIF with partial patellectomy. J Orthop Trauma. 2015;29(2):80–4.
22. Espregueira-Mendes J, Andrade R, Monteiro A, Pereira H, da Silva MV, Oliveira JM, Reis RL. Mosaicplasty using grafts from the upper tibiofibular joint. Arthrosc Tech. 2017;6(5):e1979–87.
23. Lustig S. Patellofemoral arthroplasty. Orthop Traumatol Surg Res. 2014;100(1 Suppl):S35–43. https://doi.org/10.1016/j.otsr.2013.06.013. Epub 2014 Jan 9.
24. Remy F. Surgical technique in patellofemoral arthroplasty. Orthop Traumatol Surg Res. 2019;105(1S):S165–76. https://doi.org/10.1016/j.otsr.2018.05.020. Epub 2019 Jan 8.
25. Lonner JH. Patellofemoral arthroplasty. J Am Acad Orthop Surg. 2007;15(8):495–506. https://doi.org/10.5435/00124635-200708000-00006.
26. Farr J, Arendt E, Dahm D, Daynes J. Patellofemoral arthroplasty in the athlete. Clin Sports Med. 2014;33(3):547–52. https://doi.org/10.1016/j.csm.2014.03.003. Epub 2014 Apr 19.

23 Quadriceps Tendon Tear: Evaluation and Management in a 54-Year-Old Man

Jason L. Koh, Roshan Wade, and Chaitanya Waghchoure

23.1 Case

History: A 54-year-old office worker and recreational tennis player was brought to the emergency department after a slip and fall while descending stairs. He felt immediate pain and difficulty bearing weight on his right leg. He had no previous history of knee or leg issues.

On physical examination the right thigh was tender to palpation. There was swelling of the right knee. Proximal to the patella the quadriceps tendon was soft and not readily identifiable by palpation. The patient was unable to perform a straight leg raise. There was no medial or lateral tenderness at the knee and the Lachman test was negative.

J. L. Koh (✉)
Mark R. Neaman Family Chair of Orthopaedic Surgery, NorthShore University HealthSystem, Evanston, IL, USA

Orthopaedic & Spine Institute, Skokie, IL, USA

NorthShore University HealthSystem, Evanston, IL, USA

University of Chicago Pritzker School of Medicine, Chicago, IL, USA

Northwestern University McCormick School of Engineering, Evanston, IL, USA

R. Wade
Seth G.S Medical College and KEM Hospital, Acharya Donde Marg, Mumbai, Maharashtra, India

C. Waghchoure
B J Medical College, Pune, India

Mumbai, Maharashtra, India

23.2 Imaging Findings

An initial set of radiographs were obtained. There was no obvious patella baja. A possible decrease in the soft tissue density proximal to the patella could be identified (Fig. 23.1).

An MRI of the knee joint was obtained, clearly demonstrating a complete tear of the quadriceps tendon (Fig. 23.2).

23.3 Treatment

Open repair of the quadriceps tendon was performed. A tourniquet was placed on the leg but was not elevated. A longitudinal incision was made over the proximal half of the patella and extended proximally. Extensive hematoma was evacuated from the knee and acutely bleeding blood vessels were coagulated. The bed of the ruptured quadriceps tendon was prepared down to bare bone and the torn edge of the tendon was debrided until intact tendon fibers could be identified. Three 2 mm high- strength suture tapes were placed in a Krackow locking stitch of the medial, central, and lateral edges of the quadriceps tendon, extending 3–5 cm proximal to the torn edge, 2 pairs of figure-eight high strength # 2 sutures in a locking fashion on the medial and lateral retinacular tissue was also applied.

Three pilot drill holes were made in the proximal patella and tapped. The suture tapes

© ISAKOS 2022
J. L. Koh et al. (eds.), *The Patellofemoral Joint*, https://doi.org/10.1007/978-3-030-81545-5_23

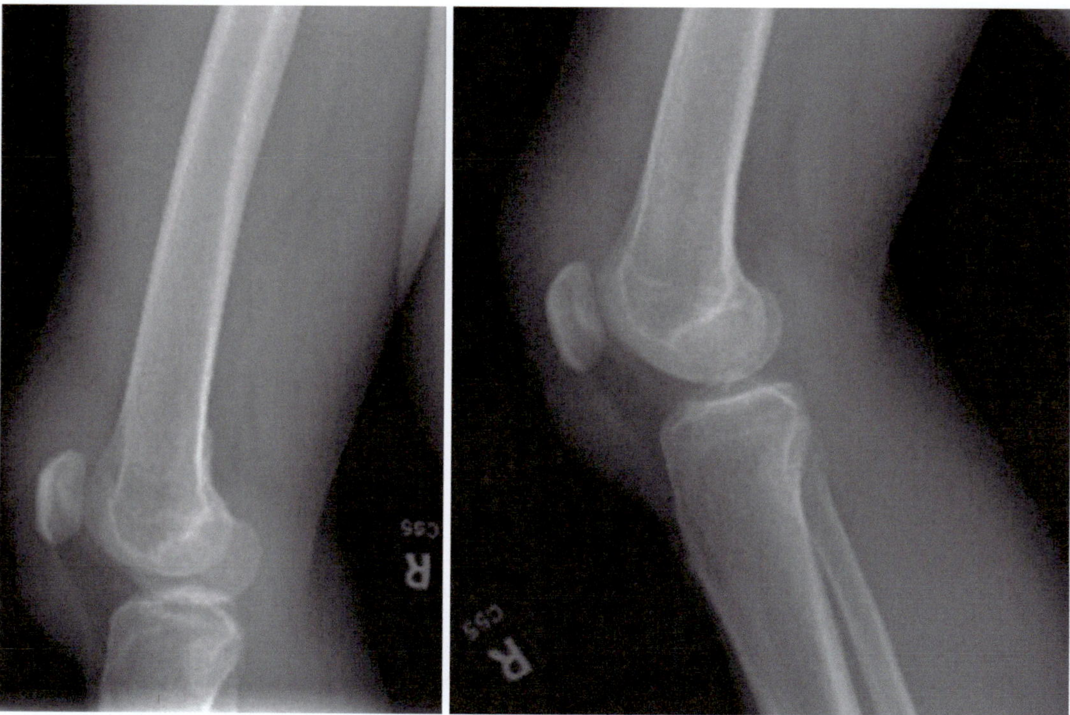

Fig. 23.1 Initial X-ray, lateral

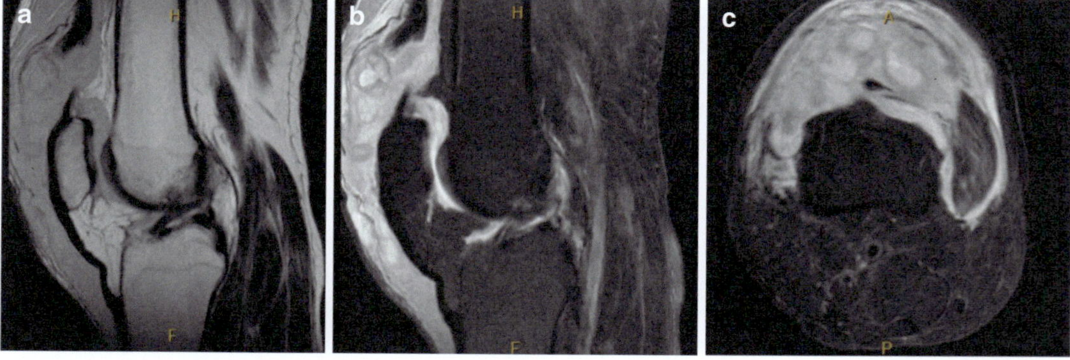

Fig. 23.2 Pre-operative MRI. (**a** and **b**) Sagittal views demonstrating quadriceps tendon tear and gap formation. (**c**) Axial view demonstrating complete tear of quadriceps tendon

were then secured to the patella using 4.75 mm PEEK interference screw-in anchors to the bony bed. The suture tapes were then tied down over the anchor followed by use of the #2 high strength eyelet sutures in a locking fashion through the tendon, reapproximating the more superficial layer of the quadriceps mechanism.

23.4 Rehabilitation

Weightbearing with a knee brace locked in extension was permitted immediately. At 2 weeks range of motion to 30° was permitted, and progressed to 60° at 4 weeks and 90° at 6 weeks. The brace was discontinued after 8 weeks. Closed chain strengthening exercises eventually permitted a return to normal function and no extensor lag with straight leg raising after approximately 12 months.

23.5 Discussion

The diagnosis of quadriceps tendon rupture can often be made clinically on examination by identification of a palpable defect proximal to the patella and inability to straight leg raise. However, particularly in the case of significant soft tissue swelling or obesity, it may be difficult to palpate the tendon, and inability to straight leg raise can be caused by multiple factors. Radiographs often do not show diagnostic findings and can be misinterpreted [1], and incorrect diagnosis of a quadriceps tendon rupture has been identified even with the use of ultrasound in up to 33% of patients [2, 3]. MRI has been shown to have excellent diagnostic sensitivity and specificity, with a positive predictive value of 1.0, and is strongly preferred when available.

The use of suture anchors to repair the quadriceps tendon rather than bone tunnels was described in 2000 in Italy [4] and in 2002 in the US [5]. The advantage in using suture anchors is decreased dissection. When compared to transosseous tunnel techniques, suture anchor repairs have been demonstrated to be essentially biomechanically equivalent in several studies [6–8], with slight differences in load to failure or gap formation. Clinical outcomes with either technique have been equivalent [9]. The use of suture tapes has been shown to be biomechanically superior to high strength suture for tendon repair [10], but both showed significant gapping (mean, 7.82 ± 3.64 mm) with an initial 150 N preload. Suture tape with knotless anchors has been shown to be biomechanically superior to either transosseous tunnels or traditional suture anchors in a cadaver model [11]; however, no clinical outcome data is available for suture tape compared to high strength suture repair.

23.6 Perspective: Roshan Wade, Chaitanya Waghchoure

Quadriceps tendon rupture more commonly occurs in middle age population and is seen at the osteotendinous junction. In young population, it occurs at the midtendon or the musculotendinous junction. Spontaneous tears in middle and old age correlate with the hypovascular zone which has been consistently identified between 1 and 2 cm from the superior pole of patella [12]. The quadriceps tendinopathy which is mainly seen in old age presents with persistent anterior knee pain, pain with activities or even deficit of extensor mechanism and could lead to partial or complete tears [13]. Clinically, the palpable gap of the quadriceps rupture may be obscured in the presence of significant swelling and the diagnosis can be missed. Usually radiographs are relatively normal. Occasionally, osteophytes at the superior pole of suggestive of associated tendinopathy could be seen on lateral radiographs. However, MRI scan is always confirmatory and defines the severity of the tear. Often partial tears can be treated conservatively with immobilization in full extension for 6 weeks followed by protected weight bearing, range of movement, and gradual quadriceps strengthening. Early surgical repair is indicated in complete tears.

Surgical repair techniques include end-to-end repair, transosseous suture repair, suture anchor fixation, and augmentation with autografts and allografts. Acute tears need to be repaired within 72 h to prevent tendon retraction. In case of acute midtendon tears, we prefer end-to-end suture repair using locked Krackow stitch in both proximal and distal ends. For the more common osteotendinous tears, we use both traditional transosseous suture repair or suture anchor fixation. In transosseous technique, 3 parallel vertical tunnels of 4 mm each are made in the patella to pull through the suture tapes weaved across the quadriceps tendon, which are finally secured at the inferior pole of the patella (Fig. 23.3). Advantage of transosseous suture repair is that its inexpensive as compared to the use of suture anchors and also it is easier to remove the sutures

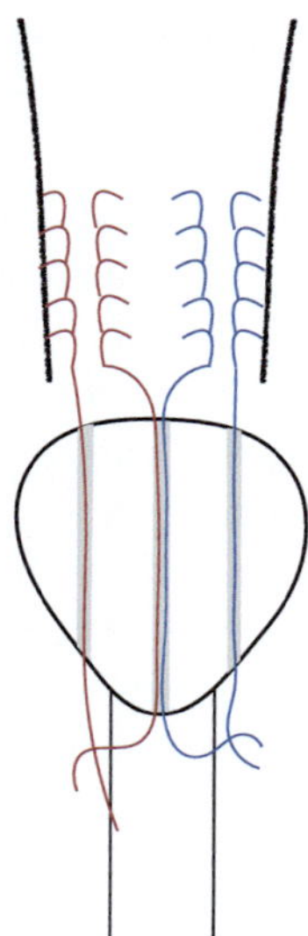

Fig. 23.3 Transosseous repair (Courtesy by Dr. Roshan Wade)

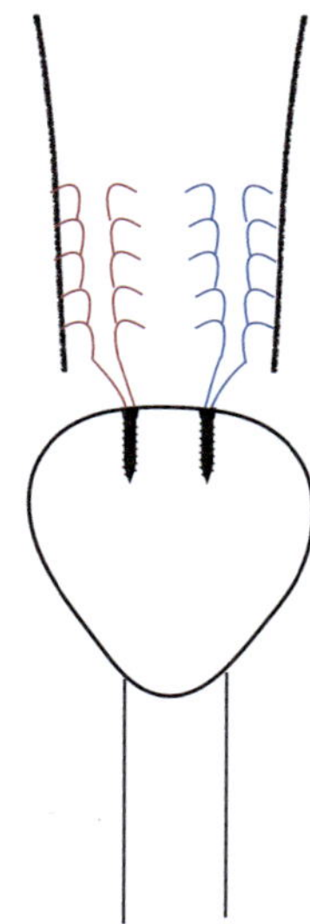

Fig. 23.4 Suture anchor repair (Courtesy by Dr. Roshan Wade)

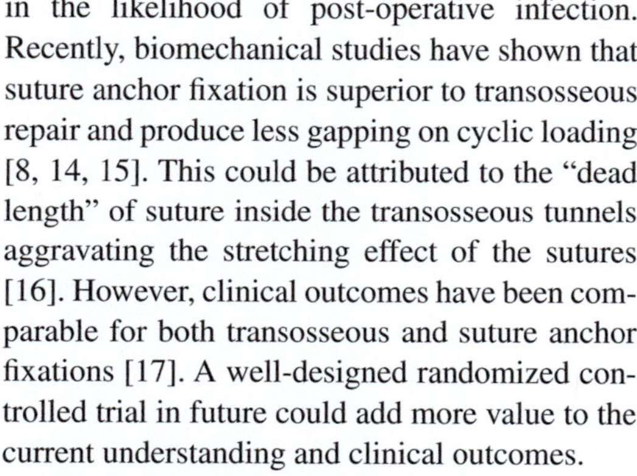

in the likelihood of post-operative infection. Recently, biomechanical studies have shown that suture anchor fixation is superior to transosseous repair and produce less gapping on cyclic loading [8, 14, 15]. This could be attributed to the "dead length" of suture inside the transosseous tunnels aggravating the stretching effect of the sutures [16]. However, clinical outcomes have been comparable for both transosseous and suture anchor fixations [17]. A well-designed randomized controlled trial in future could add more value to the current understanding and clinical outcomes.

Similar to the author, for both transosseous and suture anchor fixation techniques, we prefer UltraTape over high strength suture in order to securely grasp the quadriceps tendon over a length of 3–5 cm using a Krackow stitch. Use of Modified Mason-Allen pattern of stitch is an effective alternative [18]. Apart from the simplicity of the procedure, use of suture anchor offers the advantage of limited surgical exposure, avoiding the need to violate patellar tendon unlike transosseous suture fixation at lower pole of patella and less gapping on cyclic loading as mentioned earlier. The only drawback is difficulty in removal of the anchors if complicated by a post-operative infection.

In the current case scenario provided, the patient is in his mid-50s with a complete tear of quadriceps tendon at the osteotendinous junction. We would manage this patient with early surgical repair using suture anchor technique (Fig. 23.4) in similarity with the author's technique. Intra-operatively, we like to assess the "safe zone" of range which permits early movements without putting resistance at the repair site. Usually it is less than 90°. Post-operatively, toe-touch walking to full weight bearing is allowed using the long knee brace as per the pain tolerance, the range of motion is encouraged in the "safe zone" for the first 6 weeks, progressively including closed chain exercises, straight leg raising, core, and quadriceps strengthening.

Additionally, it is important to rule out pre-existing tendinopathy which could alter the management. The presence of poor quality tendon warrants debridement as well as augmentation using a semitendinosus autograft as described by Chahla et al. [19] (Fig. 23.5). In this technique, a transverse tunnel in patella is created to pass the graft and secured additionally with suture anchors to the superior pole of patella after soft tissue tunneling of the graft. The free ends of the graft are tunneled through the medial and lateral aspects of quadriceps tendon in criss-cross fashion and finally secured to the tendon with non-absorbable sutures.

In conclusion, early diagnosis and surgical repair of quadriceps tendon rupture are crucial.

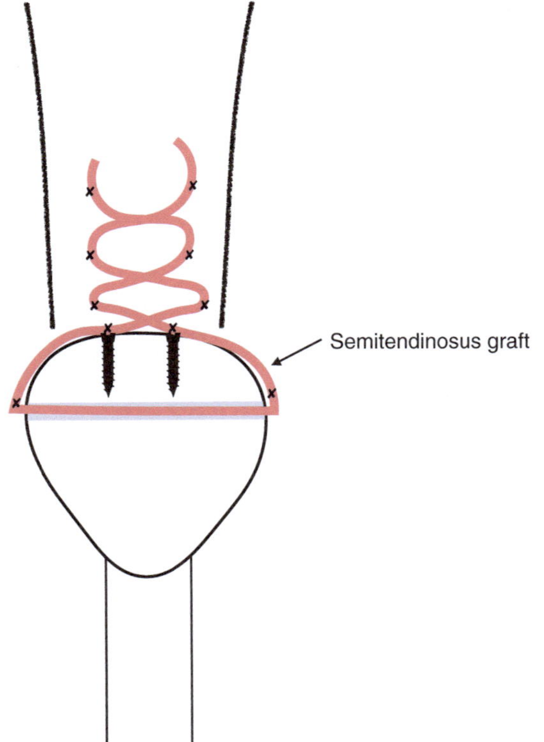

Fig. 23.5 Semitendinosus augmentation of suture anchor repair (Courtesy by Dr. Roshan Wade)

Technique of repair depends mainly on the location of tear and the tendon quality either using transosseous or suture anchor fixation with or without graft augmentation.

Key Points

Diagnosis of quadriceps tendon injuries is improved by the use of MRI.

Use of suture anchors for quadriceps tendon repair demonstrates biomechanically and clinically equivalent results to transosseous tunnel repair.

Suture tape is biomechanically superior to high strength suture for tendon repair, but no comparable clinical data is currently available.

References

1. Kaneko K, DeMouy EH, Brunet ME, Benzian J. Radiographic diagnosis of quadriceps tendon rupture: analysis of diagnostic failure. J Emerg Med. 1994;12(2):225–9.
2. Swamy GN, Nanjayan SK, Yallappa S, Bishnoi A, Pickering SA. Is ultrasound diagnosis reliable in acute extensor tendon injuries of the knee? Acta Orthop Belg. 2012;78(6):764–70.
3. Perfitt JS, Petrie MJ, Blundell CM, Davies MB. Acute quadriceps tendon rupture: a pragmatic approach to diagnostic imaging. Eur J Orthop Surg Traumatol. 2014;24(7):1237–41.
4. Maniscalco P, Bertone C, Rivera F, Bocchi L. A new method of repair for quadriceps tendon ruptures. A case report. Panminerva Med. 2000;42(3):223–5.
5. Richards DP, Barber FA. Repair of quadriceps tendon ruptures using suture anchors. Arthroscopy. 2002;18(5):556–9.
6. Lighthart WA, Cohen DA, Levine RG, Parks BG, Boucher HR. Suture anchor versus suture through tunnel fixation for quadriceps tendon rupture: a biomechanical study. Orthopedics. 2008;31(5):441.
7. Hart ND, Wallace MK, Scovell JF, Krupp RJ, Cook C, Wyland DJ. Quadriceps tendon rupture: a biomechanical comparison of transosseous equivalent double-row suture anchor versus transosseous tunnel repair. J Knee Surg. 2012;25(4):335–9.
8. Sherman SL, Copeland ME, Milles JL, Flood DA, Pfeiffer FM. Biomechanical evaluation of suture anchor versus transosseous tunnel quadriceps tendon repair techniques. Arthroscopy. 2016;32(6):1117–24.
9. Plesser S, Keilani M, Vekszler G, Hasenoehrl T, Palma S, Reschl M, et al. Clinical outcomes after treatment of quadriceps tendon ruptures show equal results independent of suture anchor or transosseus repair technique used—a pilot study. PLoS One. 2018;13(3):e0194376.
10. Roessler PP, Burkhart TA, Getgood A, Degen RM. Suture tape reduces quadriceps tendon repair gap formation compared with high-strength suture: a cadaveric biomechanical analysis. Arthroscopy. 2020;36(8):2260–7.
11. Kindya MC, Konicek J, Rizzi A, Komatsu DE, Paci JM. Knotless suture anchor with suture tape quadriceps tendon repair is biomechanically superior to transosseous and traditional suture anchor-based repairs in a cadaveric model. Arthroscopy. 2017;33(1):190–8.
12. Yepes H, Tang M, Morris SF, Stanish WD. Relationship between hypovascular zones and patterns of ruptures of the quadriceps tendon. J Bone Joint Surg Am. 2008;90(10):2135–41.
13. Maffulli N, Papalia R, Torre G, Denaro V. Surgical treatment for failure of repair of patellar and quadriceps tendon rupture with ipsilateral hamstring tendon graft. Sports Med Arthrosc. 2017;25:51–5.
14. Petri M, Dratzidis A, Brand S, et al. Suture anchor repair yields better biomechanical properties than transosseous sutures in ruptured quadriceps tendons. Knee Surg Sports Traumatol Arthrosc. 2015;23:1039–45.
15. Ettinger M, Dratzidis A, Hurschler C, et al. Biomechanical properties of suture anchor repair compared with transosseous sutures in patellar tendon ruptures: a cadaveric study. Am J Sports Med. 2013;41:2540–4.
16. Bushnell BD, Byram IR, Weinhold PS, et al. The use of suture anchors in repair of the ruptured patel-

lar tendon: a biomechanical study. Am J Sports Med. 2006;34:1492–9.
17. Ciriello V, Gudipati S, Tosounidis T, Soucacos PN, Giannoudis PV. Clinical outcomes after repair of quadriceps tendon rupture: a systematic review. Injury. 2012;43(11):1931–8.
18. Bushnell BD, Whitener GB, Rubright JH, Creighton RA, Logel KJ, Wood ML. The use of suture anchors to repair the ruptured quadriceps tendon. J Orthop Trauma. 2007;21(6):407–13.
19. Chahla J, DePhillipo NN, Cinque ME, et al. Open repair of quadriceps tendon with suture anchors and semitendinosus tendon allograft augmentation. Arthrosc Tech. 2017;6(6):e2071–7.

24 Chronic Patella Tendon Tear in a 24-Year-Old Man: Revision Procedure

Jason L. Koh, Sabrina M. Strickland, and Petri Sillanpää

24.1 Case Presentation (Jason L. Koh)

A 24-year-old with a 3 year history of a failed patella tendon repair of the left knee presents with persistent weakness and occasional giving way during walking and other activities. A previous repair 3 years ago failed after a fall at 6 weeks postop, and the revision was complicated by infection. He undergone extensive physical therapy and is able to do body weight squats, but is unable to do single leg squats, run, or jump, or have a reciprocal gait on stairs.

J. L. Koh
Mark R. Neaman Family Chair of Orthopedic Surgery, Evanston, IL, USA

NorthShore University HealthSystem, Evanston, IL, USA

Orthopedic & Spine Institute, Skokie, IL, USA

University of Chicago, Pritzker School of Medicine, Chicago, IL, USA

Northwestern University McCormick School of Engineering, Evanston, IL, USA

S. M. Strickland (✉)
Hospital for Special Surgery, New York, NY, USA
e-mail: StricklandS@HSS.EDU

P. Sillanpää
Department of Orthopedic Surgery, Pihlajalinna Koskisairaala Hospital, Tampere, Finland
e-mail: petri.sillanpaa@tuni.fi

24.2 Physical Examination

The patient has a significantly abnormal gait. There are no rashes or lesions about the knee. There is a well healed anterior incision. Passive knee range of motion is 0–130°, which is equal to the contralateral side. However, active extension is limited by a 20° extensor lag with straight leg raise. Knee extension strength is 4/5. There is mild crepitus and there is obvious patella alta.

24.3 Radiographic Examination

X-rays of the left knee show no significant degenerative changes. There is marked patella alta with a Caton-Deschamps ratio of approximately 3 (Fig. 24.1a). There are calcifications in the patellar tendon. X-rays of the contralateral knee demonstrate Caton-Deschamps ratio of 1.5 (Fig. 24.1b). AP radiographs demonstrate increased patella alta on the figure.

An MRI demonstrated extreme patella alta and elongated patella tendon tissue with no obvious tendon defect (Fig. 24.2).

24.4 Proposed Treatment

The decision was made to proceed with patella tendon shortening and reinforcement with autograft semitendinosus tendon and high strength suture tape.

© ISAKOS 2022
J. L. Koh et al. (eds.), *The Patellofemoral Joint*, https://doi.org/10.1007/978-3-030-81545-5_24

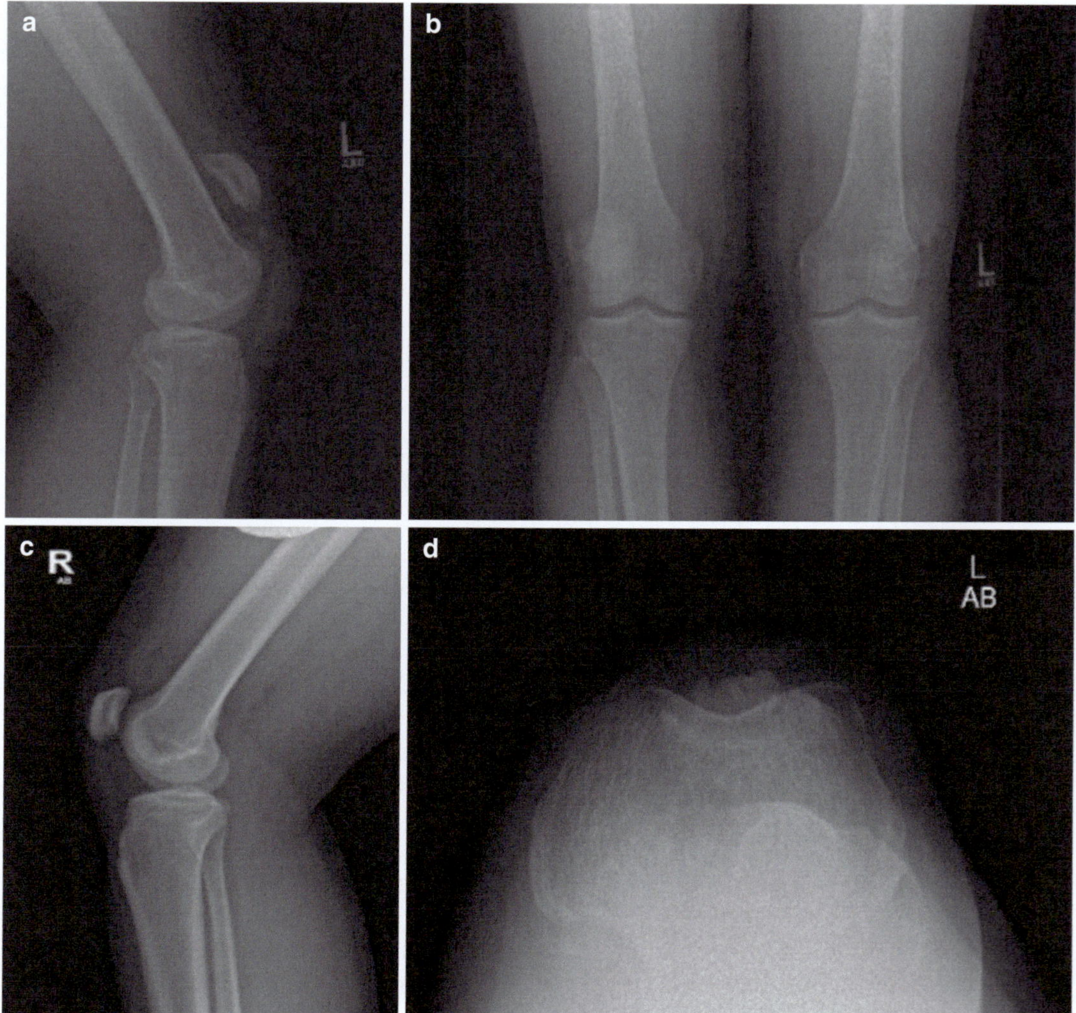

Fig. 24.1 (**a**) Left knee radiograph showing marked patella alta and calcifications in the patella tendon. (**b**) Right knee radiograph showing patella alta in non-injured right knee. (**c**) AP radiograph showing relative increased patella alta on left knee. (**d**) Sunrise view showing that patella remains proximal to trochlea in flexion

The extensor mechanism was exposed and extensive releases were performed to mobilize the quadriceps tendon and patella to move distally. The inferior pole of the patella was noted to be somewhat fragmented and of poor quality bone. The semitendinosus tendon was harvested and sized at 30 cm long by 4.5 mm diameter. A 4.5 mm transverse tunnel was made across the tibial tubercle. High strength suture tapes were placed through the tunnel along with the semitendinosus graft. The suture tapes were then passed around the proximal patella and quadriceps to reduce the patella to the height of approximately 68 mm of tendon length equal to that as measured preoperatively on the opposite side. Passing the tapes proximal to rather than through the patella was chosen due to concern about further damaging to the distal patella bone. The medial and lateral borders of the patella were then exposed adjacent to the patellar tendon and the hamstring graft was secured to the prepared bed of the patella medially and laterally using all suture anchors with Krackow locking stitches into the tendon. The redundant patella tendon was partially split, imbricated and sutured to itself at the appropriate length using high strength suture as described by

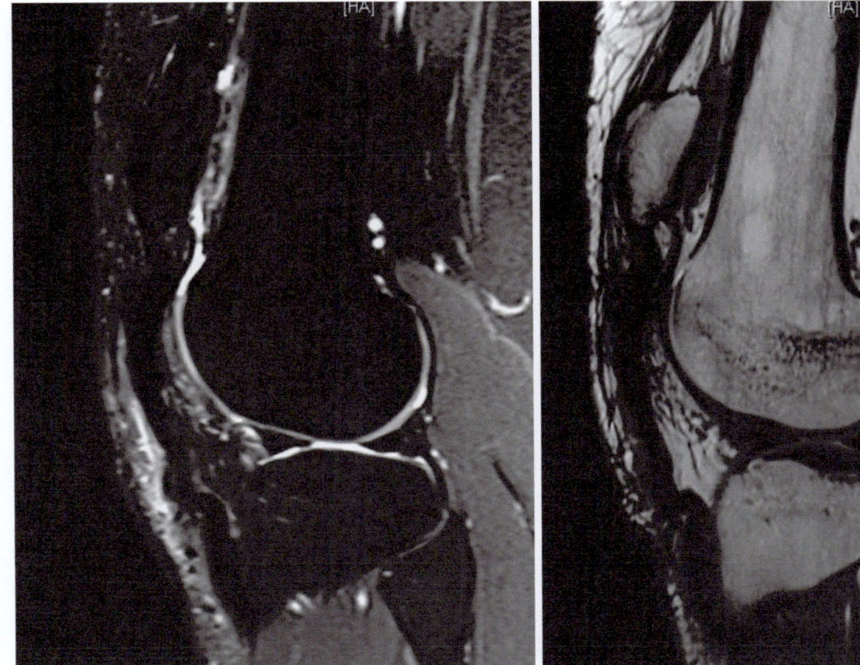

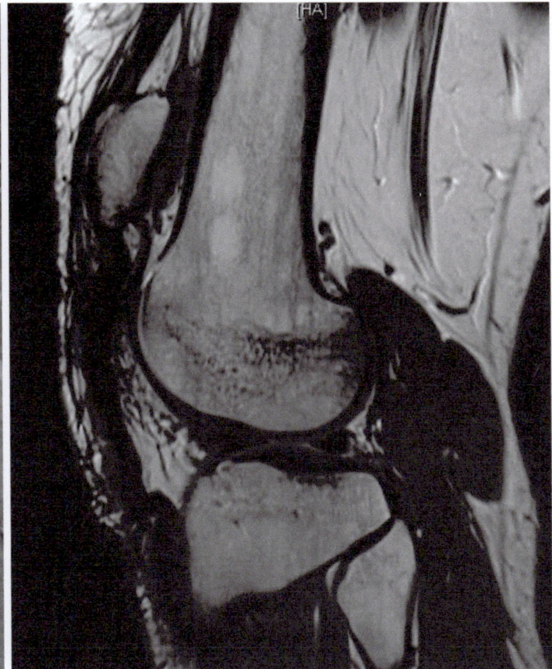

Fig. 24.2 Sagittal MRI demonstrating elongated patella tendon tissue

Andrish [1]. The remaining hamstring tendon was sutured to the imbricated tissue.

24.5 Rehabilitation and Outcome

Given the previous failure of repair, the knee was locked in extension for 4 weeks followed by 30° progression every 2 weeks to 90°. By 6 months, the patient had no extensor lag and was able to reciprocate stairs and ladders. Quadriceps strength had returned to 90% of the contralateral side, and he was planning to attend the fire academy to become a firefighter. Postop films demonstrated relative maintenance of the patient's normal patella height (Fig. 24.3).

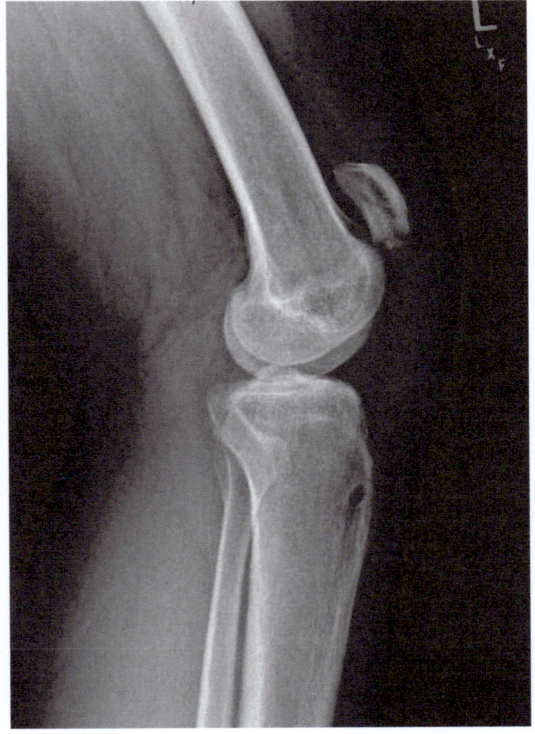

Fig. 24.3 Maintenance of comparable patella height. Note transverse tunnel in tubercle for graft and cerclage augmentation

24.6 Discussion

Patella tendon repairs have a significant risk of failure due to re-rupture or repair elongation as a result of the very high loads on this relatively narrow tissue. Grasping suturing techniques have limited strength. Augmentation of the repair with wire, Dall-Miles cables, and high strength suture passed through the tibial tubercle and patella have all been evaluated, with high strength suture having equivalent gap formation to 18 gauge wire when loaded [2]. Stronger high strength suture tape should have equivalent or superior results, and was used in this procedure.

Often patella tendon tissue may be deficient due to multiple surgeries, tissue degeneration, or infection. Use of semitendinosus tendon which when quadrupled has load to failure similar to 20 mm patella tendon should provide sufficient graft strength. Challenges remain with regard to fixation of the graft to the patella and tibia.

Imbrication of redundant patella tendon tissue has been described by Andrish [1] in the treatment of pediatric patella alta, where a tubercle distalization may not be appropriate due to open physes. This technique has been effective in many cases, and in this situation, allowed for the preservation of intact tissue while shortening the tendon.

Take Home Message

Use of high strength suture or suture tape through a tunnel in the tubercle and either through or proximal to the patella can act as an internal splint to limit gap formation and elongation of the patella tendon repair.

Key Points

- Cerclage augmentation of the patella to the tibial tubercle can minimize gap formation or elongation of patella tendon repair.
- Semitendinosus tendon can be used to augment deficient patella tendon tissue.
- Pants-over-vest imbrication of redundant tissue can effectively and safely shorten the patella tendon.

24.7 Commentary by Sabrina Strickland

24.7.1 Proposed Treatment

An extensile incision should be made to free up the quad tendon from adhesions to the anterior femur. The patella should be mobilized to allow as little tension as possible to decrease patella alta. In this case there appears to be good tissue at the proximal aspect of the patellar tendon and poor quality tissue distally and, therefore, I would excise the distal tissue and reattach the patellar tendon to the tubercle using suture anchors. The patellar height can be assessed fluoroscopically and compared to pre-operative images of the contralateral knee to set the patellar tendon length. The semitendinosus tendon should be harvested while keeping the attachment on the medial tibia. I typically find that anchors to the patella distally do not achieve adequate purchase in these chronic cases and therefore would Krakow the tendon to the medial edge of the patellar tendon once the length has been set as well as attempt a suture anchor that relies on cortical fixation such as an all suture anchor in the inferomedial aspect of the patella. The hamstring would then be run over the proximal aspect of the patellar tendon and sutured laterally to the patellar tendon with suture anchor augmentation if possible (Fig. 24.4).

The hamstring tendon would then be secured to the tibia just lateral to the tibial tubercle with interference screw fixation sized to the diameter of the tendon.

I avoid suture augmentation as the suture can act like a cheese wire and cut through the tibial tubercle, in some cases I do pass sutures through the native patellar tendon proximally and fix to suture anchors distally.

This image depicts the hamstring tendon running up and over the patellar tendon.

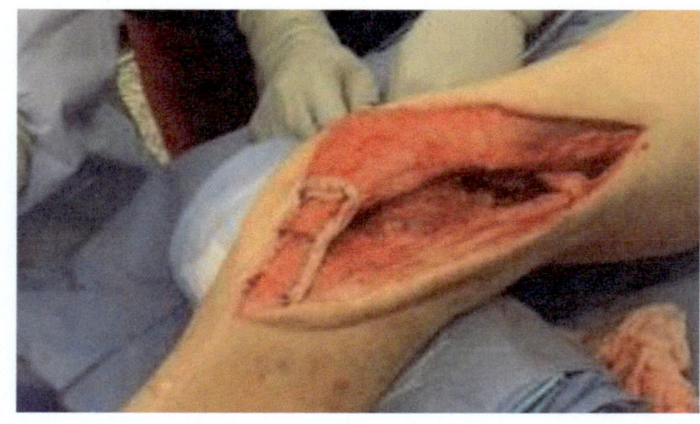

Fig. 24.4 Augmented patella tendon repair

24.7.2 Rehabilitation and Outcome

Due to the revision status of this surgery the knee should be locked in extension for 6 weeks with crutches for ambulation. Gentle active flexion with passive extension was allowed at 1 week limiting the range of motion to 45° and advancing gradually to 90 at 6 weeks. At 6 weeks patient could transition out of the crutches, unlock the brace when sitting yet continue to protect the extensor mechanism while ambulating especially outdoors with a brace for an additional 4 weeks.

24.7.3 Discussion

Unfortunately, patellar tendon repairs are fraught with complications. Repairs may result in scarring and patella baja or lengthening and patella alta. There may be loss of continuity of the tendon as well as re-rupture. In rare cases there may be wound issues requiring skin graft.

Numerous case reports in the literature have suggested augmentation of the repair with suture loops, wire and even external fixators to limit strain on the tendon during healing.

In severe cases I have resorted to allograft replacement of the patellar tendon with a bone block distally and a patellar bone block proximally while suturing the quad tendon graft to the native quad tendon proximally to minimize strain on the reparative tissue.

Take Home Message

- Use fluoroscopy intraoperatively to assess patellar height.
- Protect repair for a prolonged period of time post operatively.

Key Points

Suture anchor fixation in the patella may be poor.

Semitendinosus tendon can be used to augment repair with fixation via suture anchor on the lateral tibia.

Prolonged protection of weighted flexion is suggested while graft heals and remodels.

24.8 Perspective: Petri Sillanpää

Patella tendon rupture in a young patient is a rare, but devastating injury. This case was a complicated one with postoperative loss of repair strength resulting in an elongated patellar tendon with extensive scar tissue formation seen in MRI. In radiographs, extremely high riding patella was noted. Biomechanically, the patella alta deformity explains the extension loss. Immediate concern is how long this patella position has been present and how much quadriceps shortening has occurred.

Alternatives to shorten and fixate patellar tendon are not simple. Andrish's imbrication method is useful as it does not require full-thickness tendon detachment. I personally use it for pediatric patients to correct patella alta. Obviously, the method cannot carry the loads to the extensor mechanism in this kind of a revision case. Additionally, a cerclage might be used, from tibial tubercle to distal patella, to unload the tendon repair or reconstruction. A semitendinosus tendon was chosen to act as a load carrier instead of metal wire. That is my preferred method in young and middle-aged patients too. In elderly population, by adding a metal wire, from tibial tubercle to distal patella, immediate ROM might be allowed to enhance their limited rehabilitation capacity. In that case, a 4 mm cannulated screw can be placed horizontally, and the cerclage wire passes the cannulated screw, preserving wire cut-out from the bone when the extensor mechanism is loaded immediately after surgery. I might have considered this also in this kind of a revision case. The advantage is the early knee ROM which is possible while the reconstruction itself is unloaded. For the patellar tendon reconstruction, a hamstring autograft is necessary. Alternatively, allograft can be used and if the distal patellar pole would have been fractured, bone-patellar tendon-bone allograft is useful for reconstruction and fixation can be performed with staples and sutures, for example.

In a case that metal cerclage wire through cannulated screws would have been used with immediate free knee ROM, the cerclage wire typically needs to removed, after 6–8 weeks, when knee flexion gains 90°. The cerclage and the screws can be removed from small incisions, as the wire passes easily out from the cannulated screws. The downside of this additional metal hardware is the necessity for implant removal, though. In some cases, such as poly-trauma and older people, early mobilization might be an advantage due to limited compliance for immobilization and hardware removal might not be necessary. The treatment approach utilized by Dr. Koh resulted in a good outcome, and biological methods with no metal hardware would have been my option too, in a young patient as described here. Most commonly, patellar tendon ruptures occur in older people and therefore an approach that allows full weight-bearing as soon as possible is a viable option. Sometimes chronic patellar tendon rupture with elongation results in quadriceps shortening, and extensive tension can be felt in quadriceps, even limiting the ability to pull patella distally for the aimed amount. Therefore, lengthening of the quadriceps tendon by V-Y plasty, for example, might be necessary to relocate the patella in anatomic position in relation to trochlea.

References

1. Patel RM, Gombosh M, Polster J, Andrish J. Patellar tendon imbrication is a safe and efficacious technique to shorten the patellar tendon in patients with patella Alta. Orthop J Sports Med. 2020;8(10):2325967120959318.
2. Flanigan DC, Bloomfield M, Koh J. A biomechanical comparison of patellar tendon repair materials in a bovine model. Orthopedics. 2011;34(8):e344–8.

Part V

Tendinopathies of the Patellofemoral Joint: Case-Based Evaluation and Treatment

25 Tendinopathies of the Patellofemoral Joint: A Case-Based Approach: Osgood–Schlatter's Disease in a 12-Year-Old

Juan Pablo Martinez-Cano, Sheanna Maine, and Marc Tompkins

25.1 Case 1

A 12-year-old boy presented with 6-months history of progressive anterior knee pain in his right knee. Pain increases with physical activity and sometimes makes him stop from doing exercise. He plays soccer 4–5 times a week and kicks with his right foot.

In the physical examination, the knees are well aligned, with no effusion, range of movement is from 0 to 140°. There is tenderness and swelling on his right knee tibial tuberosity, which seems oversized (Fig. 25.1). There is severe tightness of hamstrings, gastrocsoleus, iliotibial band, and rectus femoris.

X-rays show irregularity of the tibial tuberosity apophysis, with separation and fragmentation, especially in his right knee (Fig. 25.2).

Labs tests showed normal complete blood test count and insufficiency of total 25-hydroxyvitamin D (24 ng/mL). Conservative treatment for this patient included: no physical exercise or efforts during 2 months, 1400 UI daily supplementation of vitamin D and 10 sessions of physical therapy (modalities and stretches). Two months after, pain had improved and patient was allowed to return to soccer practice and physical activity.

J. P. Martinez-Cano (✉)
Fundación Valle del Lili, Department of Orthopaedic Surgery, Universidad Icesi, Cali, Colombia
e-mail: juan.martinez.ca@fvl.org.co

S. Maine
Department of Orthopaedic Surgery, Redcliffe Hospital, Queensland Children's Hospital, Brisbane Private Hospital, North West Private Hospital, Brisbane, Australia

M. Tompkins
Department of Orthopaedic Surgery, University of Minnesota, Minneapolis, MN, USA

25.1.1 Definition

Osgood Schlatter's disease (OSD) is a traction apophysitis of the tibial tuberosity caused by the repetitive pull of the patellar tendon on the tibia. This condition was described separately and simultaneously by Robert Osgood and Carl Schlatter in 1903, as a lesion where the tibial tubercle separates due to repetitive strain by the patellar tendon and becomes tender [1, 2]. Kicking, jumping, sprinting sports, history of previous Sever's disease and lower limbs muscle tightness are among the risk factors that have been associated with OSD [3, 4].

It is one of the leading causes of anterior knee pain in children and adolescents. Usually, patient history and physical examination are enough for diagnosis. It can be easily recognized when you find local pain, swelling, and tenderness on the tuberosity, in one or both knees.

© ISAKOS 2022
J. L. Koh et al. (eds.), *The Patellofemoral Joint*, https://doi.org/10.1007/978-3-030-81545-5_25

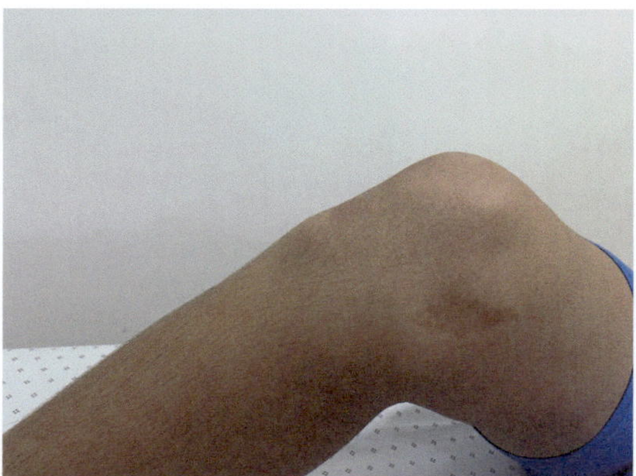

Fig. 25.1 Right knee of the 12-year-old patient with OSD, lateral view

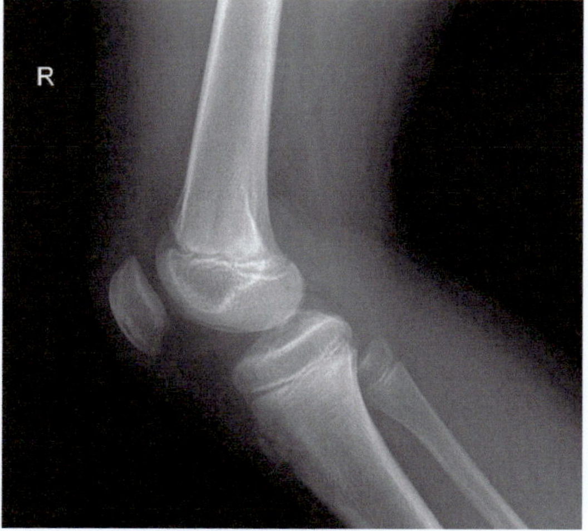

Fig. 25.2 Lateral X-ray of this patient, showing the fragmentation and separation of the tibial tuberosity apophysis

25.1.2 Pathophysiology

The maturation of the tibial tuberosity has four stages described by Ehrenborg [5]. These are shown in Fig. 25.3. First, the cartilaginous stage (0–11 years), followed by the apophyseal stage (11–14 years), the epiphyseal stage (14–18 years), and finally, the bony stage (>18 years). Most OSD cases are seen during the apophyseal stage [7].

The natural history of OSD begins with repetitive overuse traction on the tibial tuberosity apophyseal cartilage causing cartilaginous damage of the tubercle during the apophyseal maturation stage. As it continues to pull, there may be tears of the secondary ossification center, an opened shell-like separation and fragmentation [8]. This may lead to nonunion of this secondary ossification center, followed by bone enlargement in the tuberosity [9] (Fig. 25.4).

Patients initially report pain that appears only after physical activities, but if it does not resolve early, it progresses to permanent pain on the tibial tuberosity that may interfere with daily activities [10]. It may be bilateral in up to 30% of cases [11]. The outcome is usually a tuberosity that heals with a prominence at the tubercle, after fusion of the apophysis. It might also progress into an ununited free ossicle that continues to hurt with palpation and kneeling [6].

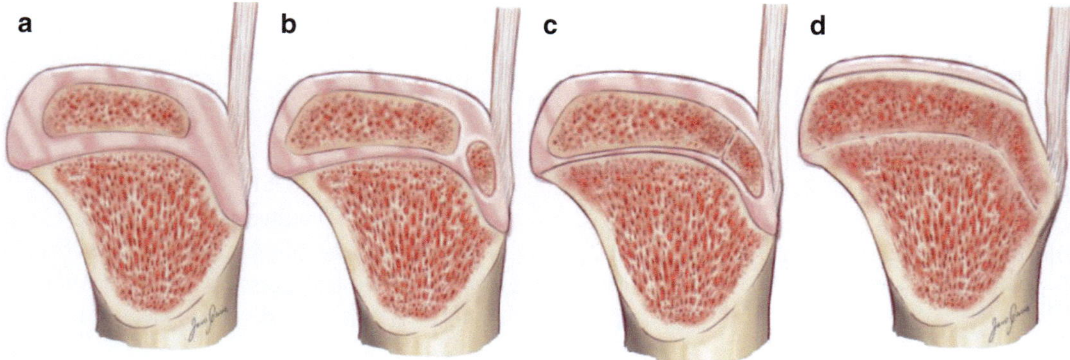

Fig. 25.3 Radiological stages in the maturation of the tibial tuberosity. (**a**) Cartilaginous stage (0–11 years), (**b**) Apophyseal stage (11–14 years), (**c**) Epiphyseal stage (14–18 years), (**d**) Bony stage (>18 years). (Adapted and modified from [5, 6])

25.1.3 Conservative Treatment

More than 90% of patients improve after conservative management [9]. Rest, NSAIDs, ice, activity modification,and stretching are the basis of non-operative treatment [4]. Exercise should be discontinued until symptoms improve. This can take a few weeks or several months, depending on the stage and severity of the disease. Most have some pain relief after 3–4 weeks of rest, but it usually takes more time to have complete pain relief [12]. Rathleff et al. showed, in a prospective cohort study, that only 16% of patients were able to return to play after a 12-week program including activity modification and knee strengthening; return improved to 69% after 12 months [13].

If there is pain, regardless of conservative treatment, other options might be considered. For instance, Vitamin D should be measured and supplemented if insufficiency or deficiency is found. Smida et al. found decreased levels of Vitamin D in a whole cohort of 80 children with OSD [14]. These patients received 6000 IU of Vitamin D per day and after 3 months of treatment, 84.2% of them had no pain. The limitation is that they had no control group for this study. Vitamin D may play a role and future studies should help to clarify the routine use of supplementation for OSD. Though, Vitamin D has shown to help fractures heal faster in osteoporotic women compared to placebo that could explain why it can be useful for OSD symptoms, probably favoring the healing of those fragmented tuberosities in OSD [15].

Another treatment that can be considered in patients with open physes, with no response to initial treatment, is using a brace or cast for a short period of time. Duperron et al. studied a case series of 35 patients with OSD treated, immobilized with cruro-maleolar resin cast for 4 weeks. The median time to restart sport was 11 weeks, with 66% of patients returning before or at 12 weeks of treatment. Delay of return to sport was statistically associated with the presence of a radiological ossicle after the immobilization

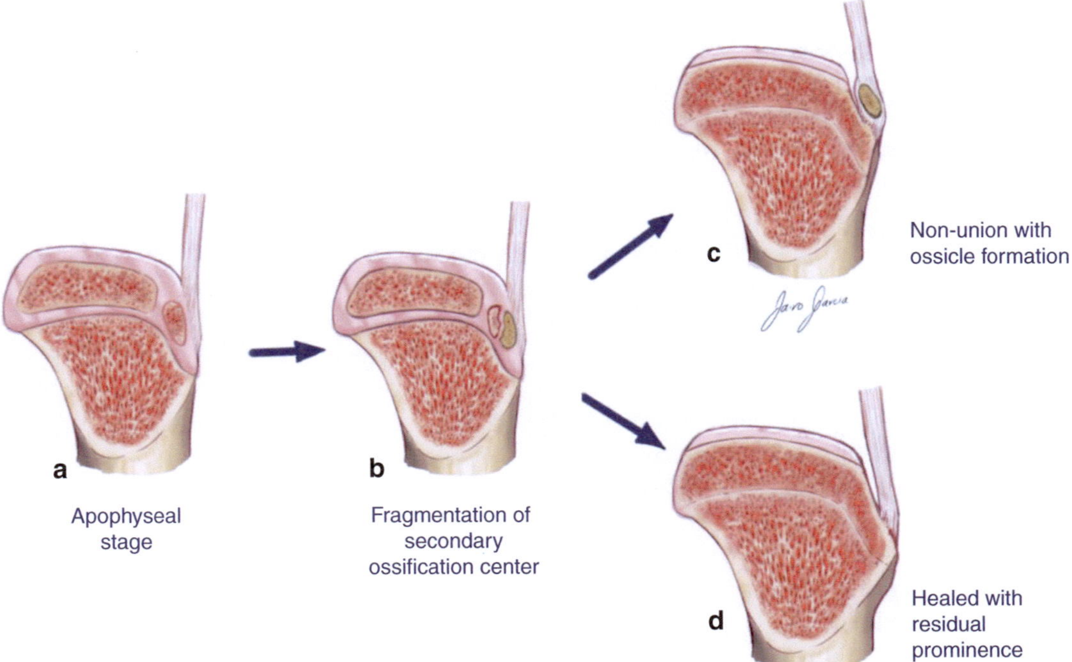

Fig. 25.4 Natural history of Osgood–Schlatter disease. (**a**) The secondary ossification center of tibial tuberosity appears. (**b**) Fragmentation of the bony ossification center. (**c**) Ununited free ossicle. (**d**) Complete healing and fusion of the apophysis with prominent tuberosity. (Adapted and modified from [5, 6])

(12.3 weeks without ossicle versus 23.8 weeks with ossicle, $P = 0.03$). Fifty percent of ossicles were not present after cast [16]. Unfortunately, there was no control group to compare the use of cast with standard conservative treatment. Use of a cast, therefore, remains controversial for OSD and some authors suggest instead a removable brace to maintain range of movement [17].

Finally, injection of hyperosmolar dextrose with lidocaine, over the apophysis and patellar tendon origin, may be useful for symptom reduction and asymptomatic return to sport participation [18].

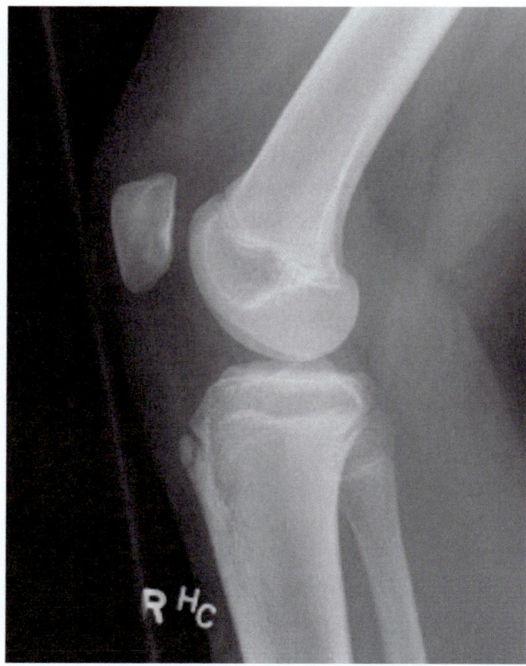

Fig. 25.5 Lateral X-ray from the knee of a patient with a symptomatic ossicle after OSD that required open excision

25.1.4 Operative Treatment

Surgical treatment should be considered if there are persistent and clinically significant symptoms after physeal closure. This is frequently associated with persistent ossicles that can be surgically removed.

Young adults or adolescents with closed physes that continue to have symptoms are surgical candidates. There have been described techniques to treat them surgically with open removal of the prominent tibial tuberosity using osteotome and rongeurs. The tuberosity is resected down to the insertion of the tendon and the intratendinous ossicles are also excised [19]. Figure 25.5 shows a symptomatic ossicle.

This treatment has shown good and excellent results. In young military recruits, 85% had a final follow-up Kujala score >85 points, 35% had no pain, and 50% had less than 30 mm of pain on the Visual Analog Score. For 87% of them, there were no restrictions in everyday activities and 75% resumed the same level of previous sports activity [20]. In a another study, Flowers et al. showed 88% complete pain relief, after tibial tuberosity excision. This included the removal of any osseocartilaginous material remaining in the substance of the tendon [7].

Arthroscopic treatment has also been described [21]. In this technique the tibial tuberosity prominence and ossicles may be addressed, as well as additional intra-articular pathology. It has the advantage of leaving no scar in front of the tuberosity that could be associated with kneeling pain [22].

Again, these surgical treatment studies are case-series, with no control group, which is a methodological disadvantage that decreases validity of their findings. There is room, in OSD research, for new studies with a design that can help answer questions about the efficacy of conservative and surgical treatment. Fortunately, OSD is a self-limited disease that improves in 90% of cases.

25.2 Case 2

A 12-year-old boy with anterior knee pain in both knees during the last 9 months. He plays soccer 4–5 days a week; he sometimes experiences symptoms while playing. He has done 50 sessions of physical therapy with no improvement.

Figure 25.6 shows the enlargement of both tibial tuberosities, especially the right one. He had pain on palpation of the tuberosities. There was moderate tightness of hamstrings, gastrocsoleus, iliotibial band, and rectus femoris.

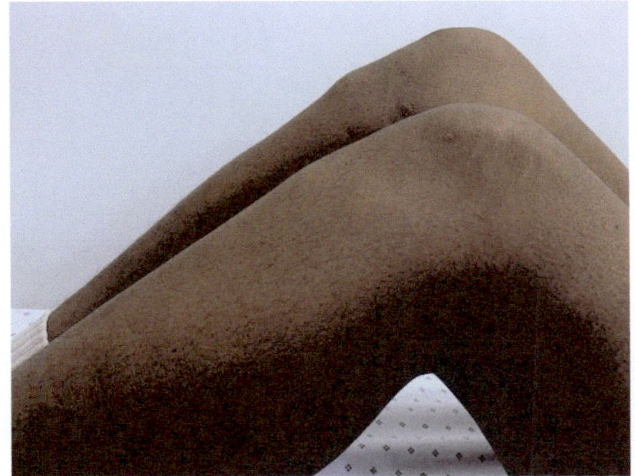

Fig. 25.6 Lateral view of both knees from case 2, a 12-year-old patient with OSD, showing the typical prominence of tibial tuberosities for this pathology

Conservative management included: 6 weeks of rest, stretching exercises, and 1000 UI of Vitamin D daily. He returned to clinic feeling better in his left knee and some residual pain in his right knee. Figure 25.7 shows his X-rays. Total Vitamin D, at this follow-up visit, was still insufficient: 27.9 ng/mL, despite supplement given.

At this stage, the patient and his family desired something more, but wished to avoid surgery. Vitamin D supplementation was increased to 2000 UI day and an inguinocrural cast was placed in his right knee for 3 weeks. After this period of time, the cast was taken off and the patient had no more pain in both his knees. He returned to playing soccer 4 weeks later, with no further symptoms. Vitamin D supplementation was continued for 3 more months.

Take Home Message

- Osgood–Schlatter is a self-limited condition related to growth and overuse.
- In over 90% of the cases, OSD improves with conservative treatment, which includes rest, ice, and activity modification.

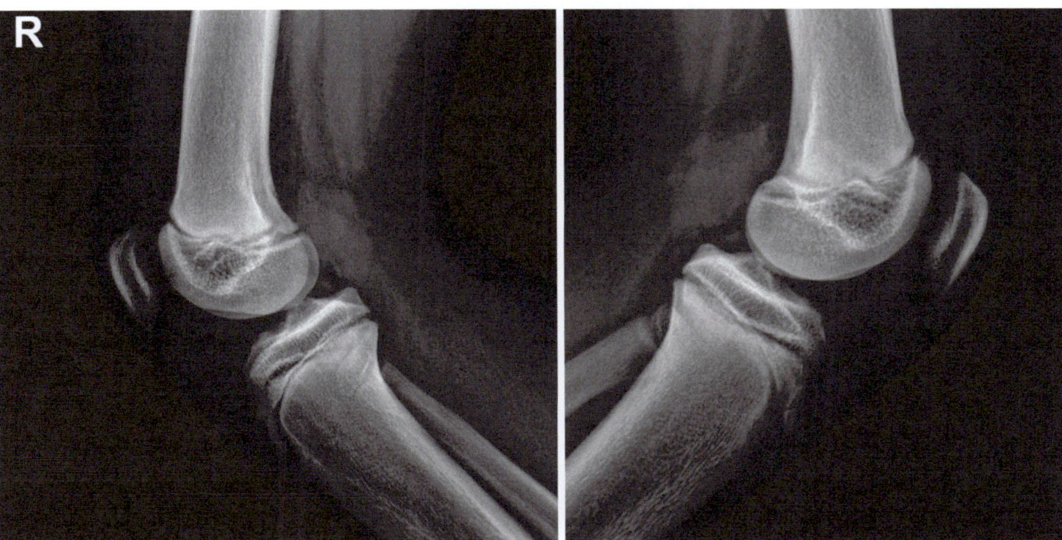

Fig. 25.7 Lateral X-rays of right and left knees for case 2, with opened shell-like separation of the tuberosities in this OSD patient

25.3 2–3 Fact Boxes

Fact Box 1

Osgood–Schlatter usually presents during the puberty growth spurt:

Boys: 12–15 years old.

Girls: 8–12 years old.

Fact Box 2

Surgery is indicated after physeal closure when symptoms persist, and are significant.

It may be arthroscopic or open surgery, and any free ossicle should be removed.

25.4 Algorithm for the Treatment of Osood–Schlatter's Disease

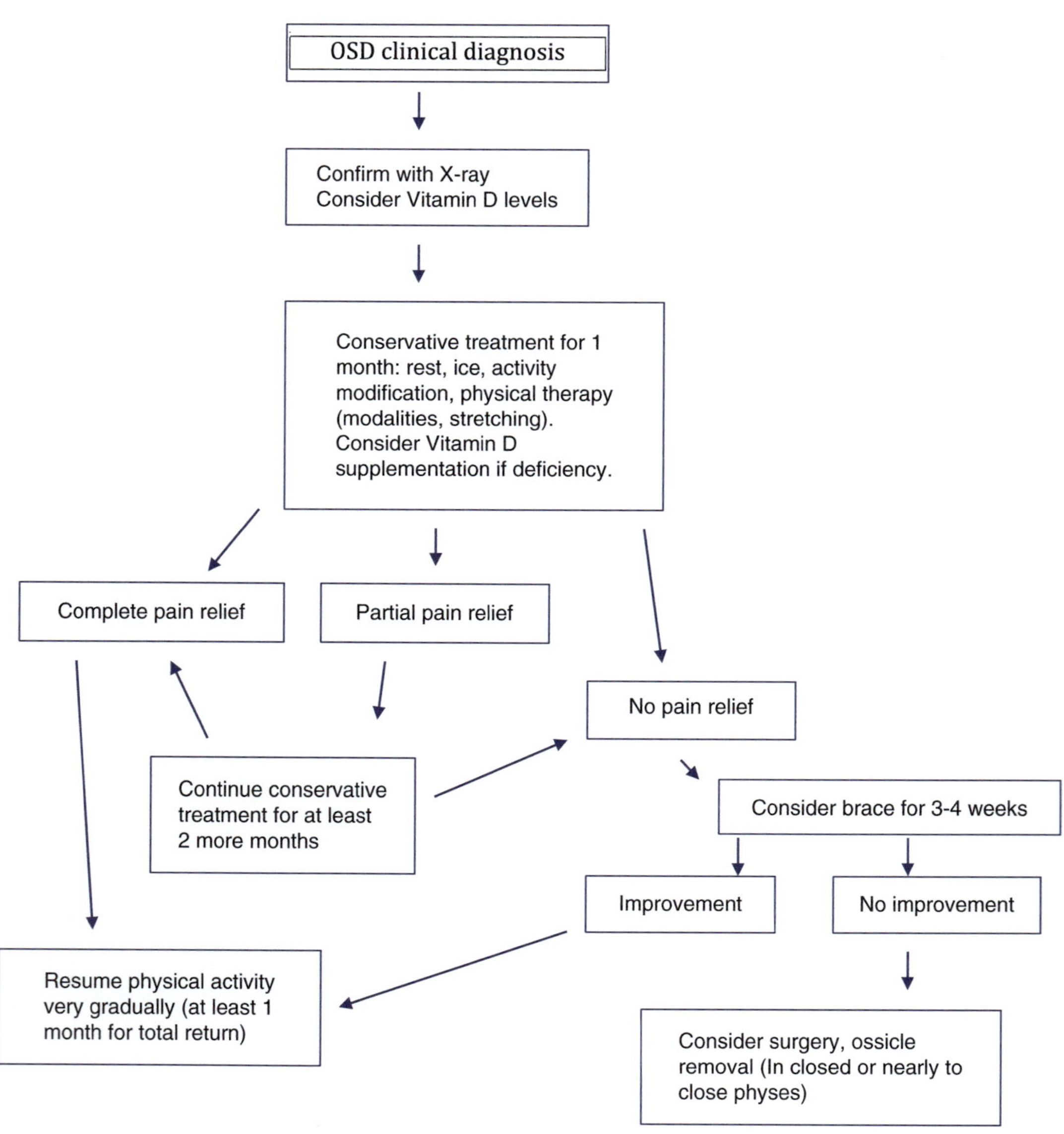

References

1. Osgood RB. Lesions of the tibial tubercle occurring during adolescence. 1903. Clin Orthop Relat Res. 1993;(286):4–9.
2. Schlatter C. Verletzungen des schnabelformigen Fortsatzes der oberen Tibiaepiphyse. Beitr Klin Chir. 1903;38:874–87.
3. Watanabe H, Fujii M, Yoshimoto M, Abe H, Toda N, Higashiyama R, et al. Pathogenic factors associated with Osgood-Schlatter disease in adolescent male soccer players: a prospective cohort study. Orthop J Sports Med. 2018;6(8):2325967118792192.
4. Smith JM, Varacallo M. Osgood Schlatter's disease (tibial tubercle apophysitis). Treasure Island: StatPearls Publishing; 2019.
5. Ehrenborg G, Lagergren C. Roentgenologic changes in the Osgood–Schlatter lesion. Acta Chir Scand. 1961;121:315–27.
6. Gholve PA, Scher DM, Khakharia S, Widmann RF, Green DW. Osgood Schlatter syndrome. Curr Opin Pediatr. 2007;19(1):44–50.
7. Flowers MJ, Bhadreshwar DR. Tibial tuberosity excision for symptomatic Osgood-Schlatter disease. J Pediatr Orthop. 1995;15(3):292–7.
8. Hirano A, Fukubayashi T, Ishii T, Ochiai N. Magnetic resonance imaging of Osgood-Schlatter disease: the course of the disease. Skeletal Radiol. 2002;31(6):334–42.
9. Ladenhauf HN, Seitlinger G, Green DW. Osgood-Schlatter disease: a 2020 update of a common knee condition in children. Curr Opin Pediatr. 2020;32(1):107–12.
10. Vaishya R, Azizi AT, Agarwal AK, Vijay V. Apophysitis of the Tibial tuberosity (Osgood-Schlatter disease): a review. Cureus. 2016;8(9):e780.
11. Maher PJ, Ilgen JS. Osgood-Schlatter disease. BMJ Case Rep. 2013;2013
12. Indiran V, Jagannathan D. Osgood-Schlatter disease. N Engl J Med. 2018;378(11):e15.
13. Rathleff MS, Winiarski L, Krommes K, Graven-Nielsen T, Holmich P, Olesen JL, et al. Activity modification and knee strengthening for osgood-schlatter disease: a prospective cohort study. Orthop J Sports Med. 2020;8(4):2325967120911106.
14. Smida M, Kandara H, Jlalia Z, Saied W. Pathophysiology of Osgood-Schlatter disease: does vitamin D have a role? Vitam Miner. 2018;7(2):1–6.
15. Doetsch AM, Faber J, Lynnerup N, Watjen I, Bliddal H, Danneskiold-Samsoe B. The effect of calcium and vitamin D3 supplementation on the healing of the proximal humerus fracture: a randomized placebo-controlled study. Calcif Tissue Int. 2004;75(3):183–8.
16. Duperron L, Haquin A, Berthiller J, Chotel F, Pialat JB, Luciani JF. Étude d'une cohorte de 30 patients immobilisés avec une résine cruro-malléolaire pour une maladie d'Osgood-Schlatter. Sci Sports. 2016;31(6):323–35.
17. Frank JB, Jarit GJ, Bravman JT, Rosen JE. Lower extremity injuries in the skeletally immature athlete. J Am Acad Orthop Surg. 2007;15(6):356–66.
18. Topol GA, Podesta LA, Reeves KD, Raya MF, Fullerton BD, Yeh HW. Hyperosmolar dextrose injection for recalcitrant Osgood-Schlatter disease. Pediatrics. 2011;128(5):e1121–8.
19. Pihlajamaki HK, Visuri TI. Long-term outcome after surgical treatment of unresolved osgood-schlatter disease in young men: surgical technique. J Bone Joint Surg Am. 2010;92(Suppl 1 Pt 2):258–64.
20. Pihlajamaki HK, Mattila VM, Parviainen M, Kiuru MJ, Visuri TI. Long-term outcome after surgical treatment of unresolved Osgood-Schlatter disease in young men. J Bone Joint Surg Am. 2009;91(10):2350–8.
21. Lui TH. Endoscopic management of Osgood-Schlatter Disease. Arthrosc Tech. 2016;5(1):e121–5.
22. Circi E, Atalay Y, Beyzadeoglu T. Treatment of Osgood-Schlatter disease: review of the literature. Musculoskelet Surg. 2017;101(3):195–200.

26 Patellar Tendinopathy in a 21-Year-Old Long Jumper

Timothy L. Miller, Michael Baria, Nicola Maffulli, Seth L. Sherman, and Adam Money

26.1 Case Description:

The patient is a 22 year-old female Division I collegiate track and field athlete who specializes in long jump. She presented following referral from her primary care team physician during the mid-indoor track and field season. She described recurrent left anterior knee pain worsening with sports participation over the previous 2–3 years. She reported that prior to the last 3 months her pain occurred only with running, jumping, or other athletic activity, but now the pain was affecting most activities of daily including climbing stairs and kneeling. She described the pain as aching to sharp at times and was occasionally associated with a general sense of instability. She denied any falls or direct trauma to her knee other than repetitive landings in a sand pit for long jump training and competitions. She denied any incidents of patellar subluxation or dislocation. She described tenderness and enlargement of the anterior knee just inferior to the patella but no effusions or mechanical symptoms including locking or catching. She denied any limping with daily activities but reported limping regularly after intense practice sessions. She had been previously treated with oral and topical anti-inflammatory medications, ice, physical therapy exercises, stretching, topical ultrasound, iontophoresis, kinesiotaping, and an infrapatellar strap with minimal improvement. She wished to continue to her collegiate athletic career and complete her senior year of eligibility the following year.

The patient's physical examination demonstrated neutral overall standing alignment, no internal tibial torsion, minimal lateral offset of the tibial tubercle, and slight patella alta. Passive range of motion measured from 2° of hyperextension to 141 degrees of flexion. The patient's Q-angle was

T. L. Miller (✉)
The Ohio State University Wexner Medical Center Jameson Crane Sports Medicine Institute, Columbus, OH, USA

Ohio State University, Columbus, OH, USA
e-mail: timothy.miller@osumc.edu

M. Baria
The Ohio State University Wexner Medical Center Jameson Crane Sports Medicine Institute, Columbus, OH, USA

Department of Physical Medicine and Rehabilitation, Columbus, OH, USA

N. Maffulli
University of Salerno School of Medicine, Surgery and Dentistry, Salerno, Italy

Centre for Sports and Exercise Medicine, Barts and The London School of Medicine and Dentistry, London, UK

Institute of Science and Technology in Medicine, Keele University School of Medicine, Stoke on Trent, UK

S. L. Sherman · A. Money
Department of Orthopedic Surgery, Stanford University, Stanford, CA, USA

© ISAKOS 2022
J. L. Koh et al. (eds.), *The Patellofemoral Joint*, https://doi.org/10.1007/978-3-030-81545-5_26

measured at 17°. She reported pain anteriorly at the patellar tendon with deep squat testing. Her left knee extension strength was 5 out of 5 and equal to the opposite side though slight atrophy was noted at the vastus medialis obliquus muscle compared with the unaffected right lower extremity. No extensor lag was present with active knee extension though pain was reproduced with resisted knee extension between 0 and 60 degrees of flexion. There was no crepitation noted with active or passive range of motion, and no joint effusion was present. Mild soft tissue swelling was noted over the proximal patellar tendon with exquisite tenderness present at the proximal patellar tendon when the knee was in extension and slightly decreased with knee flexion—a positive Bassett's sign. Observed range of motion did not demonstrate a J-sign, and passive lateral patellar tilt did not demonstrate increased tightness of the lateral patellar retinaculum. The patient's lateral patellar glide demonstrated 2 quadrants with no apprehension on lateral patellar translation. The remainder of the patient's knee examination was within normal limits though further evaluation demonstrated signs of generalized ligamentous laxity including slight knee hyperextension bilaterally and bilateral wrist hyperflexion (Beighton score = 4/9).

The patient's imaging evaluation included plain radiographs in the 45-degree flexion posterior-anterior (Rosenberg), lateral, and 30-degree flexed patellofemoral (Merchant) views. Radiographs demonstrated normal overall alignment with a congruent patellofemoral joint, mild patella alta (Caton-Deschamps Index = 1.21). Normal morphology of the patellar inferior pole was present with indistinctness of the proximal patellar tendon noted (Fig. 26.1). No diagnostic ultrasound was performed as the patient presented with an MRI of her knee having already been completed. The MRI demonstrated severe patellar tendinosis with a high-grade partial avulsion tear at the inferior pole of the patella that associated with edema within the infrapatellar fat pad. The partial avulsion tear encompassed approximately the lateral ½ to 2/3 of the patellar tendon origin with the anterior-posterior (A-P) thickness of the tendon expanded to 14 mm. Early enthesopathy of the inferior patellar pole was also evident (Fig. 26.2a–c).

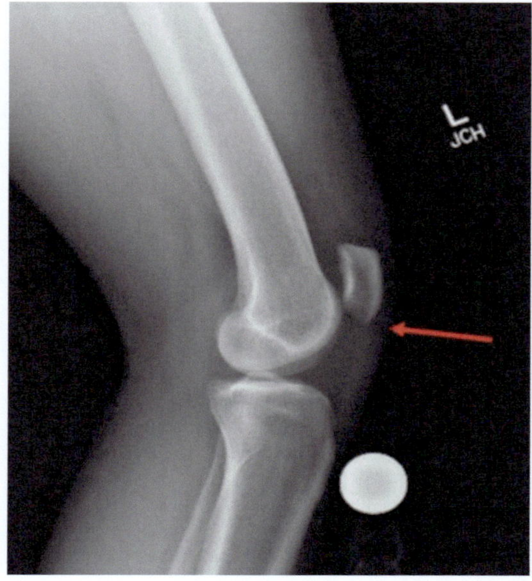

Fig. 26.1 Lateral radiograph of the left knee demonstrating patella alta and indistinctness of the proximal patellar tendon origin

26.2 Author 1 Comments and Proposed Treatment Plan

A variety of treatment options may be considered for this patient. These include continued conservative management with physical therapy, non-steroidal anti-inflammatory drugs (NSAIDs), low-intensity pulsed ultrasound, extracorporeal shockwave therapy, and transdermal glyceryl nitrate application. Minimally invasive treatments including injection of sclerosing agents, orthobiologics treatments such as platelet-rich plasma, application of stem (stromal) cell therapies, hyaluronic acid injections, ultrasound-guided percutaneous tendon debridement (Fig. 26.3) or scraping [1, 2] could also be considered. Finally, arthroscopic debridement (Fig. 26.4) and open surgical debridement (Fig. 26.5) with or without tendon to bone repair could be considered in a patient with chronic recalcitrant symptoms and partial patellar tendon tearing [3].

Given the increasing severity of the patient's symptoms, her lack of improvement with conservative treatment, and the extent of injury evident

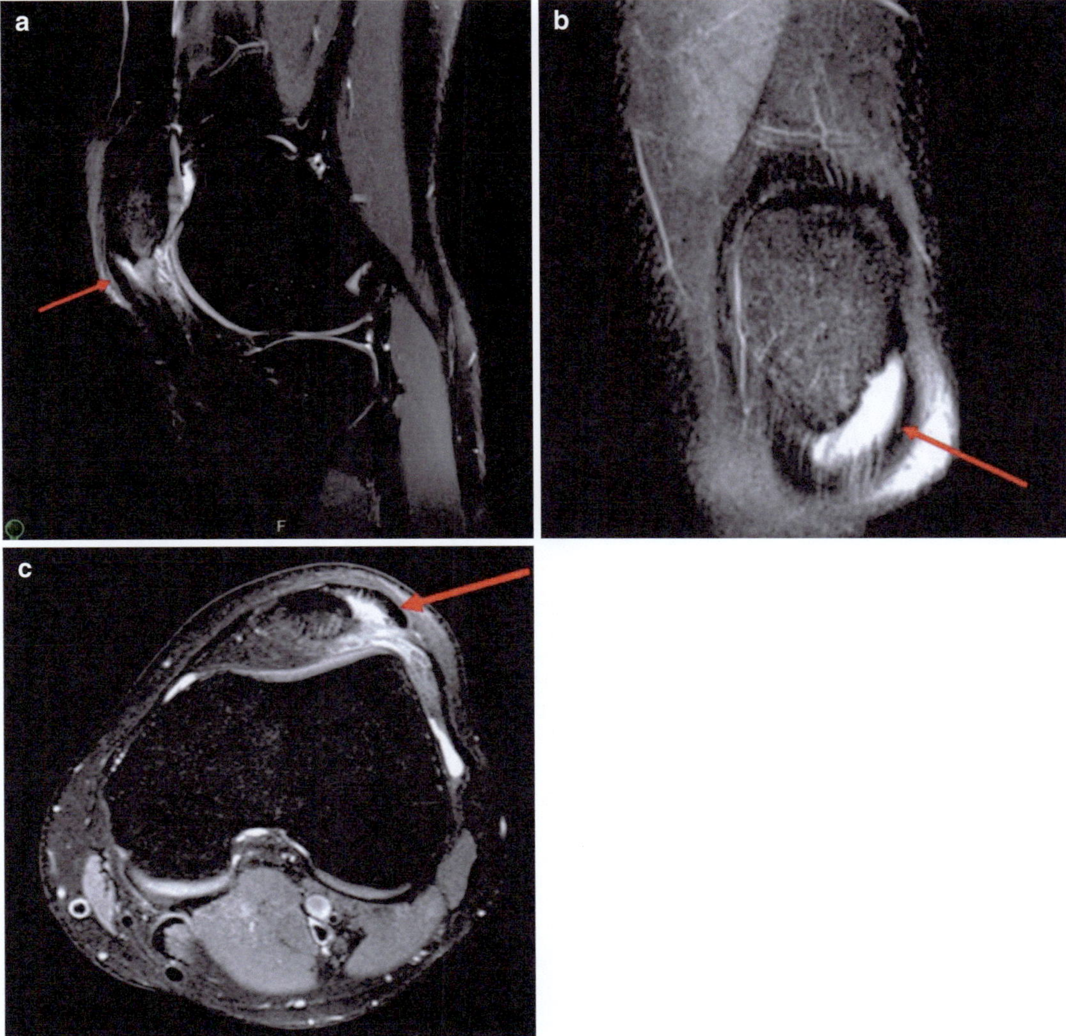

Fig. 26.2 T2 sagittal (**a**), coronal (**b**), and axial (**c**) MRI slides demonstrating proximal expansion and high-grade partial tearing of the patellar tendon origin with associated infrapatellar fat pad edema and early enthesopathy of the inferior patellar pole

on imaging, the proposed treatment plan for this patient would include:

1. Cessation of athletic activity including jumping.
2. Continued physical therapy exercises including eccentric strengthening of the quadriceps musculature.
3. Open surgical debridement of the proximal patellar tendon origin, the inferior patellar pole, and the infrapatellar fat pad.
4. Surgical repair of the central and lateral patellar tendon origin with suture anchors inserted into the inferior patellar pole.

This proposed treatment plan is based on the chronicity of the athlete's symptoms, her goal of returning to elite level jumping activities, failure of conservative treatment, and the extent of the tendon tear present on MRI.

MRI grading has been used by other authors to determine the effectiveness of treatment and the

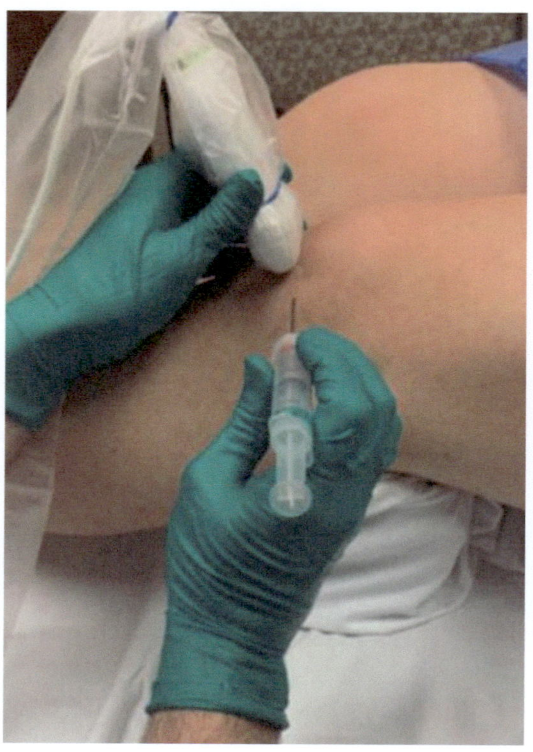

Fig. 26.3 Setup for ultrasound-guided percutaneous patellar tendon scraping. (Figure borrowed with permission from Baria et al. *Techniques in Orthopaedics*, 2020 [1])

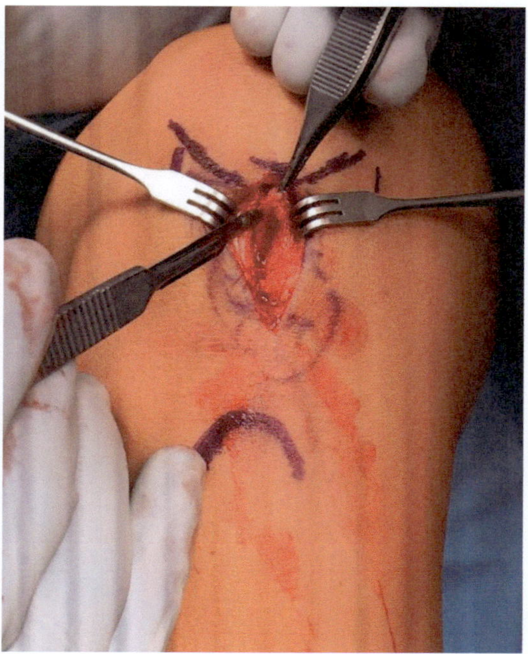

Fig. 26.5 Intraoperative photograph of open patellar tendon debridement. A section of the damaged central proximal patellar tendon origin is excised in an ellipse pattern

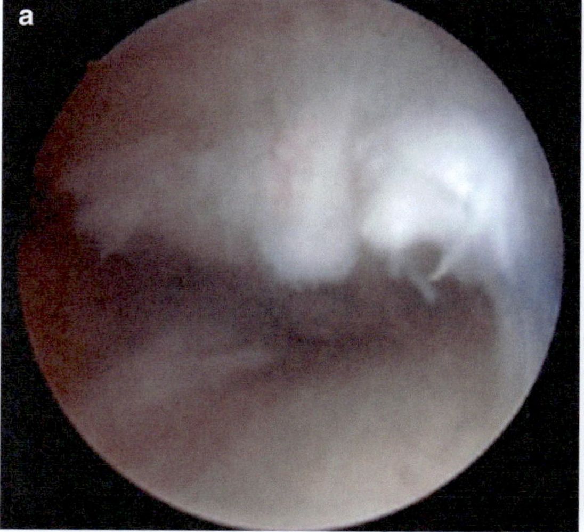

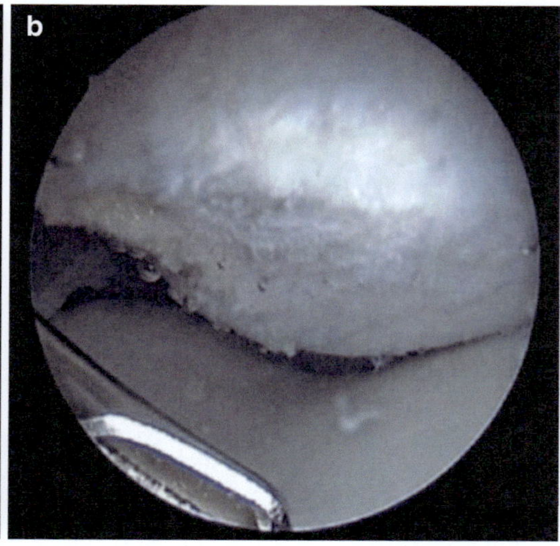

Fig. 26.4 Arthroscopic view from an inferolateral portal demonstrating tendinosis and partial posterior patellar tendon tear off of the inferior patellar pole (**a**). Arthroscopic view of the inferior patellar pole and proximal patellar tendon following arthroscopic debridement (**b**)

Table 26.1 The Popkin-Golman MRI classification system for patellar tendinopathy and partial patellar tendon tears [7]

Grade	Extent of tendon tear
1	No tear. Tendon A-P thickness < 8 mm
2	Tear thickness < 25%
3	Tear thickness > 25% but <50%
4	Tear thickness > 50%

Table 26.2 The Blazina classification for functional limitations associated with patellar tendonitis [8]

Stage	Functional limitations
1	Pain after athletic activities
2	Pain at the beginning of sports activity, disappearing after warm-up and sometimes returning with fatigue
3	Pain at rest and during athletic activity; inability to participate in sports
4	Rupture of the patellar tendon

potential need for surgery [4, 5]. Ogon et al. prospectively evaluated 30 patients who underwent arthroscopic patellar tendon release to treat refractory patellar tendinopathy [6]. These authors preoperatively assessed the patients' MRIs for evidence of tendon thickening, edema of the infrapatellar fat pad, bone marrow edema of the inferior pole of the patella, and infrapatellar bursitis. The results of the study indicated that athletes with preoperative infrapatellar fat pad edema required significantly longer time to return to sports following surgery than those without fat pad edema [7].

Recently, Golman and associates correlated their MRI classification system of patellar tendinopathy—the Popkin-Golman MRI Classification System (Table 26.1) with the Blazina clinical classification system for patellar tendonitis (Table 26.2). By combining these two systems the authors were able to develop treatment protocols for individuals with patellar tendinopathy and partial patellar tendon tears [7, 8]. Their MRI grading system assessed the amount of cross-sectional structural damage of the patellar tendon that was present on axial cut T2 MRI. This study found that an A-P tendon thickness of greater than 8.8 mm correlated with a partial patellar tendon tear, and athletes with tendon thickness greater than 11.5 mm and tearing of >50% of the tendon thickness were unlikely to improve with non-surgical treatment [7]. This female collegiate long jumper's MRI demonstrated an A-P thickness of the proximal patellar tendon of 14 mm (Fig. 26.2a) and approximately 80% tendon detachment from the inferior patellar pole (Fig. 26.2c).

Both open and arthroscopic debridement have demonstrated high rates of symptom resolution for patellar tendinopathy when there is minimal tendon detachment [9, 10]. However, open surgical debridement with tendon repair is recommended for this athlete given the extent of the athlete's partial tear and concern for further tendon detachment with debridement alone [7]. Use of suture anchor fixation for the tendon repair is suggested because of the smaller incision required, the less invasive nature of the fixation compared with transosseous fixation, and the ability to avoid the remaining components of the extensor mechanism (specifically the quadriceps tendon). Additionally, there is increasing evidence that suture anchor fixation provides equal if not superior pullout strength and resistance to repair-site gapping when compared with transosseous techniques [11–13].

26.3 Author 2 Comments and Proposed Treatment:

This patient's situation requires 2 distinct treatment decisions: (1) how to effectively manage her while in-season and (2) how to provide more definitive treatment for long-term relief. The in-season treatment should be minimally invasive and avoid direct manipulation/violation of the tendon and paratenon. This reduces post-procedure pain, allows for expedient return to training, and minimizes the risk of an adverse event such as tendon rupture. For this reason, a peritendinous procedure that eliminates the neovascularity and neoneural proliferation is well-suited for in-season treatment. Ingrowth of neovascularity and painful nerve fibers is the first histopathologic change in tendinopathy, and their elimination has been demonstrated to reduce tendon pain [14–16]. Neovessel/neoneural ablation can be achieved with either high volume image guided injections (HVIGI) or tendon scraping [14, 17]. The HVIGI requires volumes of fluid, approximately 50 mL. In our experience this can lead to increased patient discomfort. For this reason, we favor using the tendon scraping approach. After local anesthesia is obtained with 5–10 cc of 1% lidocaine with epinephrine, a 14-gauge needle is used to separate the infrapatellar fat pad from the posterior surface of the tendon, thus disrupting the neovessel/neoneural ingrowth. In our experience, patients have 24–48 h of post-procedure soreness but are able to begin training safely within 1 week and experience symptom relief within 2 weeks. Our athletes have returned to full competition without adverse events 2–4 weeks post-tendon scraping. Neovessel ablation can also be achieved by intra-tendon sclerosant injections, but direct injection into the tendon is not ideal for in-season athletes [18]. Athletes should avoid undergoing in-season platelet-rich plasma (PRP) injections because of direct tendon violation, significant post-injection pain, and an expected 1–3 months to see clinical improvement. Another reasonable option for in-season treatment is extracorporeal shockwave therapy (ESWT). This reduces tendon pain without violating tendon structure [19]. Both tendon scraping and ESWT are used in our practice, but we believe scraping offers two distinct advantages. It is complete after a single procedure as compared to ESWT which often requires multiple sessions. Additionally, scraping is covered by insurance, whereas ESWT is most often an out-of-pocket expense.

At the conclusion of the season, we would present this athlete with 2 more definitive treatment options. If they desired to avoid surgery, we would recommend 2 intra-tendinous leukocyte-rich PRP injections, 2–4 weeks apart, followed by 1 month of guided rehabilitation and return to sport progression [20]. If the athlete failed to improve within 6–12 weeks of the PRP injections, we would recommend surgical consultation. Alternatively, after struggling with chronic patellar tendinosis, many patients are hesitant to try more non-operative treatments that require several months to determine their efficacy. If the patient expressed this hesitation, we would move immediately to surgical consultation in the off-season.

26.4 Author 3 Comments and Proposed Treatment:

This patient seems to present with a typical patellar tendinopathy (PT)—"Jumper's knee" [8]. Typically more than one-third of athletes presenting for treatment for PT are unable to return to sport within 6 months [21]. A major issue is whether these patients should continue to practice sport. The present evidence suggests that unless she abstains from jumping and other painful activities, no amount and no type of conservative management means will help. If this patient wishes to continue to train and compete in the present season, she should undergo high load eccentric and concentric exercises, with high load isometric exercises just before each competition jump. The latter may produce a marked short lived analgesic effect which may allow the athlete to continue with competition.

If the athlete wishes to attempt conservative management, then a relative rest from running and jumping should be implemented for three consecutive months. She should have heavy load eccentric exercises, and consider ESWT. In such elite athletes, no validated ESWT protocols exist. I suggest that they have one or two sessions per week for 3–4 weeks, for up to 10 sessions of ESWT. I normally suggest radial shock wave therapy at the maximum intensity tolerated by the patient for 2500 hits. The patient has to undertake heavy load eccentric exercises in any case for 3 months. Regenerative medicine is appealing and popular, but the results are often over-emphasized by the lay press. We undertake high volume image guided injections at the interface between the Hoffa's body and the patellar tendon, in the area of maximum neovascularity, followed by eccentric exercise.

Surgery may be considered after failure of appropriate well performed conservative management for 3–6 months. It is not, and should not, be considered a short cut. Although we have described endoscopic and minimally invasive techniques, we normally undertake an open procedure, with excision of the peritendon, removal or drilling of the inferior patellar pole when this is involved in the pathological process, multiple longitudinal tenotomies, and excision of the tendinopathic area. We stress to the patients, their families, coaches, and managers that the recovery period is long. We find that the most common cause of failure of well performed surgery, be it endoscopic or open, is failure to appreciate that you cannot accelerate biology, and that it takes an average of 9 months to reap the benefits of surgery. Failure to realize this begs for a bad result and recurrence of the condition. The addition of orthobiologics to the procedure does not result in any benefits, and only gives a sense of false security.

26.5 Author 4 Comments and Proposed Treatment

Treatment must begin with conservative measures. An exhaustive course of non-operative management should be completed prior to consideration for surgery. Approximately 90% of athletes will respond to this approach and may avoid operative intervention [22]. We would recommend a "core-to-floor" rehabilitation protocol with a focus on flexibility, eccentric and isometric strengthening, identifying potential muscular imbalances (gluteus medius), and/or rectifying altered biomechanics. In conjunction with physical therapy, anti-inflammatory medications can be used to reduce soft tissue swelling or intra-articular effusion. Ibuprofen, Naprosyn, Celecoxib, etc. are all reasonable options. We prefer to use Celecoxib for the convenience of once daily dosing and for its favorable gastrointestinal profile. An alternative anti-inflammatory option is diclofenac gel or similar topical options applied locally to the patellar tendon. A neoprene sleeve may be utilized for proprioception and gentle compression. A Cho-Pat strap may be substituted once localized tenderness is substantially reduced. If the athlete responds to treatment, they should undergo rigorous return-to-play testing similar to an ACL reconstruction return to play protocols with consider of isometric testing to ensure corrected biomechanics prior to return to sport.

If initial attempts at rehabilitation and oral/topical anti-inflammatory medication did not provide relief, we would then consider injections. We prefer to use PRP as studies have demonstrated it as an effective treatment of patellar tendinopathy [23]. Anecdotally, the senior author has seen success with this approach in our own clinic. Based on best available evidence, we choose to use leukocyte-rich PRP as it potentially accelerates the recovery in patellar tendinopathy [24]. This is administered under ultrasound guidance. Multiple injections may also be beneficial. We recommend a two-week holiday off any anti-inflammatory prior to administration of PRP and immediately following the injection. We then allow graduated return to activity 10–14 days after injection. Return to sport is based on symptom improvement and functional progression. Athletes must meet criteria for advancement to linear, lateral, and then complex movement patterns prior to return to sport.

If the athlete fails to respond to non-surgical measures over a prolonged timeframe, we consider surgical intervention. We prefer an arthroscopic technique with high anterolateral and anteromedial portals. We will excise any hypertrophic fat pad and inflamed plica sheet or shelf. We will then arthroscopically debride the proximal posterior patella tendon to healthy tissue. A 70° arthroscopy may be useful for visualization in this situation. If the extra-articular distal patella pole impinges the posterior patellar tendon, an arthroscopic power rasp utilized under fluoroscopic guidance can resect the prominence. This creates a healthy bed of bleeding bone adjacent to the debrided tendon. Use of the power rasp is an efficient method of bony resection and limits the potential for iatrogenic damage. A spinal needle may be utilized under arthroscopic visualization to trephinate the tendon along the point of maximal tenderness. This may also be an opportunity to localize and inject leukocyte-rich PRP during dry arthroscopy at the end of the case.

Post-operatively, the patient is weight bearing as tolerated and we allow for range of motion immediately. The patient is then placed into a hinged knee brace for several weeks to protect the patella tendon until the athlete regains quadriceps control. Our protocol includes blood-flow restriction therapy, which potentially reduces muscular atrophy and promotes hypertrophy during the period of healing and protection [25]. Patients progress through the rehabilitation protocol with robust return to activity testing at each stage of recovery (i.e., run, sprint, lateral movements). We typically limit impact activities for approximately 3 months to avoid risk of catastrophic tendon rupture. Return to play can occur any time after 4 months post-debridement. However, this minimum time criteria must take into account functional progression prior to full clearance.

26.6 Key Points on Athletes with Patellar Tendinopathy

Patellar tendinopathy is an overuse injury that affects the proximal patellar tendon, the inferior pole of the patella, and the infrapatellar fat pad [26, 27]. The condition occurs along a continuum of severity ranging from mild inflammation and tendonitis to severe tendon degeneration with partial to complete patellar tendon rupture. It is a common and sometimes debilitating condition in running and jumping athletes. Devising an effective treatment plan for this condition should be based on three components of the patient's presentation. The severity and chronicity of the patient's symptoms, the level of effect the condition has on athletic activities, and the degree of structural damage evident on imaging are the key factors for determining an appropriate treatment strategy for athletes with patellar tendinopathy [28, 29].

The initial treatment for patellar tendinopathy for nearly all levels of severity should include rest from the causative activity, icing of the affected area, NSAIDs, and a physical therapy program that promotes flexibility, increased bloodflow, and eccentric strengthening of the extensor mechanism. When standard conservative treatments fail to improve the athlete's symptoms within 3 months, imaging with radiographs, ultrasound, and/or MRI should be obtained to assess the severity of the damage at the patellar tendon, the inferior patellar pole, and the infrapatellar fat pad. If a moderate level of severity is present with minimal tendon detachment, topical and percutaneous techniques including injections of orthobiologic or cell-based therapies may be considered. If more extensive tendon disruption is present or if bony and/or fat pad edema is identified, surgical treatment with arthroscopic or open debridement and repair should be considered [30].

Fact Box 1

- Three anatomic sites contribute to the presence of patellar tendinopathy and must be assessed and treated in order to fully address the source of the patient's symptoms. [26, 27]
  1. Proximal patellar tendon
  2. Inferior patellar pole
  3. Infrapatellar fat pad

Fact Box 2

- The anterior to posterior (A-P) thickness of the patellar tendon is considered abnormal if >7 mm [6]
- A-P tendon thickness >8.8 mm correlates with a partial patellar tendon tear [7]
- Athletes with tendon thickness >11.5 mm are unlikely to improve with non-surgical treatment [7]

Fact Box 3

- Devising an effective treatment strategy for this condition should be based on three components of the patient's presentation [28]
  1. Symptom severity
  2. Level of effect on athletic activities
  3. Degree of structural damage on MRI
- Surgery is rarely required and should be reserved for patients who have failed conservative treatment, if >50% of the tendon thickness is affected, or if a high-grade tear in present [30]

Resources and Relevant Websites

1. Walrod BJ. Evaluation and treatment of soft tissue overuse injuries. Endurance sports medicine: a clinical guide. 2016. pp. 93–111.
2. https://www.webmd.com/fitness-exercise/jumpers_knee#1.

References

1. Baria MR, Plunkett E, Miller MM, Borchers J, Miller TL, Magnussen RA. Ultrasound-guided percutaneous tendon scraping: a novel technique for treating patellar tendinopathy. Techn Orthop. 9000; Epub ahead of print.
2. Baria MR, Vasileff WK, Miller M, Borchers J, Miller TL, Magnussen RA, et al. Percutaneous ultrasonic tenotomy effectively debrides tendons of the extensor mechanism of the knee: a technical note. Knee. 2020;27(3):649–55.
3. Figueroa D, Figueroa F, Calvo R. Patellar tendinopathy: diagnosis and treatment. J Am Acad Orthop Surg. 2016;24(12):e184–92.
4. Coupal TM, Munk PL, Ouellette HA, Al-Shikarchy H, Mallinson PI, Choudur H. Popping the cap: the constellation of MRI findings in patellofemoral syndrome. Br J Radiol. 2018;91(1089):20170770.
5. Hutchison MK, Houck J, Cuddeford T, Dorociak R, Brumitt J. Prevalence of patellar tendinopathy and patellar tendon abnormality in male collegiate basketball players: a cross-sectional study. J Athl Train. 2019;54(9):953–8.
6. Ogon P, Izadpanah K, Eberbach H, Lang G, Sudkamp NP, Maier D. Prognostic value of MRI in arthroscopic treatment of chronic patellar tendinopathy: a prospective cohort study. BMC Musculoskelet Disord. 2017;18(1):146.
7. Golman M, Wright ML, Wong TT, Lynch TS, Ahmad CS, Thomopoulos S, et al. Rethinking patellar tendinopathy and partial patellar tendon tears: a novel classification system. Am J Sports Med. 2020;48(2):359–69.
8. Blazina ME, Kerlan RK, Jobe FW, Carter VS, Carlson GJ. Jumper's knee. Orthop Clin North Am. 1973;4(3):665–78.
9. Kaeding CC, Pedroza AD, Powers BC. Surgical treatment of chronic patellar tendinosis: a systematic review. Clin Orthop Relat Res. 2007;455:102–6.
10. Coleman BD, Khan KM, Kiss ZS, Bartlett J, Young DA, Wark JD. Open and arthroscopic patellar tenotomy for chronic patellar tendinopathy. A retrospective outcome study. Victorian Institute of Sport Tendon Study Group. Am J Sports Med. 2000;28(2):183–90.
11. Lanzi JT Jr, Felix J, Tucker CJ, Cameron KL, Rogers J, Owens BD, et al. Comparison of the suture anchor and transosseous techniques for patellar tendon repair: a biomechanical study. Am J Sports Med. 2016;44(8):2076–80.
12. Capiola D, Re L. Repair of patellar tendon rupture with suture anchors. Arthroscopy. 2007;23(8):906.e1–4.
13. O'Dowd JA, Lehoang DM, Butler RR, Dewitt DO, Mirzayan R. Operative treatment of acute patellar tendon ruptures. Am J Sports Med. 2020;48(11):2686–91.
14. Alfredson H. Ultrasound and Doppler-guided mini-surgery to treat midportion Achilles tendinosis: results of a large material and a randomised study comparing two scraping techniques. Br J Sports Med. 2011;45(5):407–10.
15. Kaeding C, Best TM. Tendinosis: pathophysiology and nonoperative treatment. Sports Health. 2009;1(4):284–92.
16. Hall MM, Rajasekaran S. Ultrasound-guided scraping for chronic patellar tendinopathy: a case presentation. PM R. 2016;8(6):593–6.
17. Morton S, Chan O, King J, Perry D, Crisp T, Maffulli N, et al. High volume image-guided injections for patellar tendinopathy: a combined retrospective and prospective case series. Muscles Ligaments Tendons J. 2014;4(2):214–9.
18. Sunding K, Willberg L, Werner S, Alfredson H, Forssblad M, Fahlström M. Sclerosing injections and ultrasound-guided arthroscopic shaving for patellar tendinopathy: good clinical results and decreased tendon thickness after surgery-a medium-term follow-up study. Knee Surg Sports Traumatol Arthrosc. 2015;23(8):2259–68.
19. Vetrano M, Castorina A, Vulpiani MC, Baldini R, Pavan A, Ferretti A. Platelet-rich plasma versus focused shock waves in the treatment of Jumper's knee in athletes. Am J Sports Med. 2013;41(4):795–803.
20. Andriolo L, Altamura SA, Reale D, Candrian C, Zaffagnini S, Filardo G. Nonsurgical treatments

of patellar tendinopathy: multiple injections of platelet-rich plasma are a suitable option: a systematic review and meta-analysis. Am J Sports Med. 2019;47(4):1001–18.

21. Cook JL, Khan KM, Harcourt PR, Grant M, Young DA, Bonar SF. A cross sectional study of 100 athletes with Jumper's knee managed conservatively and surgically. The Victorian Institute of Sport Tendon Study Group. Br J Sports Med. 1997;31(4):332–6.
22. Ogon P, Maier D, Jaeger A, Suedkamp NP. Arthroscopic patellar release for the treatment of chronic patellar tendinopathy. Arthroscopy. 2006;22(4):462.e1–5.
23. Jeong DU, Lee CR, Lee JH, Pak J, Kang LW, Jeong BC, et al. Clinical applications of platelet-rich plasma in patellar tendinopathy. Biomed Res Int. 2014;2014:249498.
24. Dragoo JL, Wasterlain AS, Braun HJ, Nead KT. Platelet-rich plasma as a treatment for patellar tendinopathy: a double-blind, randomized controlled trial. Am J Sports Med. 2014;42(3):610–8.
25. DePhillipo NN, Kennedy MI, Aman ZS, Bernhardson AS, O'Brien LT, LaPrade RF. The role of blood flow restriction therapy following knee surgery: expert opinion. Arthroscopy. 2018;34(8):2506–10.
26. Lorbach O, Diamantopoulos A, Kammerer KP, Paessler HH. The influence of the lower patellar pole in the pathogenesis of chronic patellar tendinopathy. Knee Surg Sports Traumatol Arthrosc. 2008;16(4):348–52.
27. Rees J, Houghton J, Srikanthan A, West A. The location of pathology in patellar tendinopathy. Br J Sports Med. 2013;47(9):e2–e.
28. Khan KM, Maffulli N, Coleman BD, Cook JL, Taunton JE. Patellar tendinopathy: some aspects of basic science and clinical management. Br J Sports Med. 1998;32(4):346–55.
29. Maffulli N, Cook JL, Khan KM. Re: Recalcitrant patellar tendinosis in elite athletes: surgical treatment in conjunction with aggressive postoperative rehabilitation. Am J Sports Med. 2006;34(8):1364; author reply –5.
30. Gemignani M, Busoni F, Tonerini M, Scaglione M. The patellar tendinopathy in athletes: a sonographic grading correlated to prognosis and therapy. Emerg Radiol. 2008;15(6):399–404.

Patellar Tendinitis at Osgood–Schlatter's Lesion of a 32 Year Old

27

Orlando D. Sabbag, Miho J. Tanaka, Adam Money, and Seth L. Sherman

27.1 Case Vignette

The patient is a 32-year-old male nephrologist with an active lifestyle. He had a history of Osgood–Schlatter's disease as an adolescent. Symptoms gradually improved into his late teenage years but have worsened over the past decade. He reports persistent pain and prominence over the tibial tubercle ossicle. The patient has difficulty with kneeling, running, and some daily life activities. Pain ranges from 0 to 6 out of 10. He rates his knee 60% of normal. He has tried prolonged activity modification, exercise programs, medical management, and supportive wear about the knee.

On physical exam, the patient has bilateral varus alignment and a non-antalgic gait. He is able to double and single limb squat with good symmetry but with pain over the distal patella tendon in deep flexion. Seated evaluation demonstrates focal tenderness and prominence over the tibial tubercle ossicle. There was no tenderness over the proximal patella tendon or fat pad. There was a normal Q-angle and no J-sign. Supine evaluation demonstrated no effusion, full range of motion, no apprehension or guarding with patella deviation, normal patella tilt, and stable ligaments. Pre-operative radiographs (Fig. 27.1) demonstrate large, fragmented, and prominent tibial tubercle ossicle.

Given the failure of extensive conservative management, surgery was indicated. The patient was taken to the operating room under light general anesthesia (LMA). He was positioned supine. Tourniquet was placed but not inflated. Longitudinal incision was made in the midline centered over the tibial tubercle prominence. The paratenon and distal patella tendon were split to expose the fragmented tibial tubercle ossicle. The ossicle was carefully resected along with adjacent abnormal and inflammatory tissue. The tibia posterior proximal to the tendon insertion was smoothed with a manual rasp. The region was copiously irrigated and hemostasis was obtained. Fluoroscopy demonstrated complete resection of the tibial tubercle ossicle (Fig. 27.2). The patella tendon was closed with interrupted vicryl suture with the knee in flexion. The paratenon was closed separately with absorbable suture.

The patient was allowed to weight bear as tolerated with a hinged knee brace locked in extension. Immediate range of motion was initiated. The brace was unlocked and discontinued with

O. D. Sabbag
Bone and Joint Surgery Associates, San Antonio, TX, USA
e-mail: Info@TexasOrthoCenter.com

M. J. Tanaka (✉)
Department of Orthopaedic Surgery, Massachusetts General Hospital, Harvard Medical School, Boston, MA, USA
e-mail: mtanaka5@mgh.harvard.edu

A. Money · S. L. Sherman
Department of Orthopedic Surgery, Stanford University, Stanford, CA, USA
e-mail: shermans@stanford.edu

© ISAKOS 2022

J. L. Koh et al. (eds.), *The Patellofemoral Joint*, https://doi.org/10.1007/978-3-030-81545-5_27

quadriceps control. Progressive functional rehabilitation protocol was utilized avoiding high impact for 3 months to protect the distal tendon insertion. After several months, the patient normalized daily life activities and was able to perform recreational sport without pain. VAS 0–2/10 and he rated his knee 95% of normal at final follow-up. He was taking no medications for pain or swelling and utilized a knee sleeve as needed for activity.

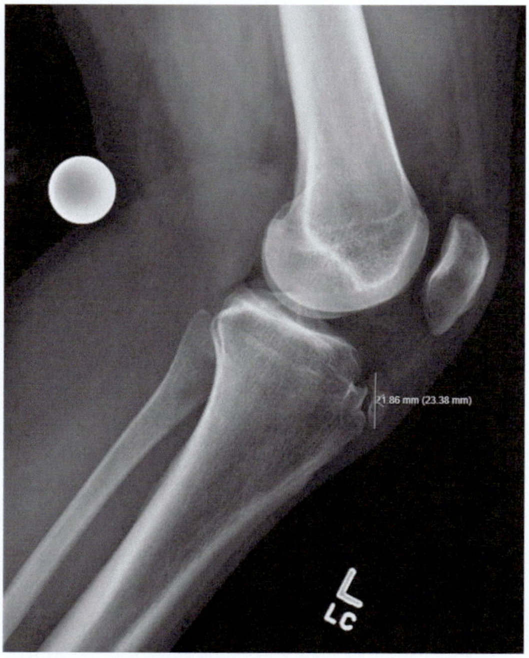

Fig. 27.1 Large tibial tubercle ossicle from refractory Osgood–Schlatter's disease in a skeletally mature knee

27.2 Background

Osgood–Schlatter disease (OSD), or tibial tuberosity traction apophysitis, is one of the most common causes of anterior knee pain in skeletally immature adolescent athletes, with an estimated prevalence of approximately 10% (11% in boys and 8% in girls) [1]. It usually affects girls between the ages of 8 and 13, and boys between the ages of 10 and 15, corresponding to their rapid phase of skeletal growth [2]. Rapid growth can result in excessive pull of the patellar tendon on the tibial tuberosity apophysis, leading to overload at the tenoperiosteal junction [3]. Shortening of the rectus femoris muscle has been associated as a risk factor [1].

27.3 Evaluation

Patients with OSD commonly complain of insidious onset of a dull ache and swelling over the patellar tendon attachment to the tibial tuberosity. Symptoms often occur in association with activi-

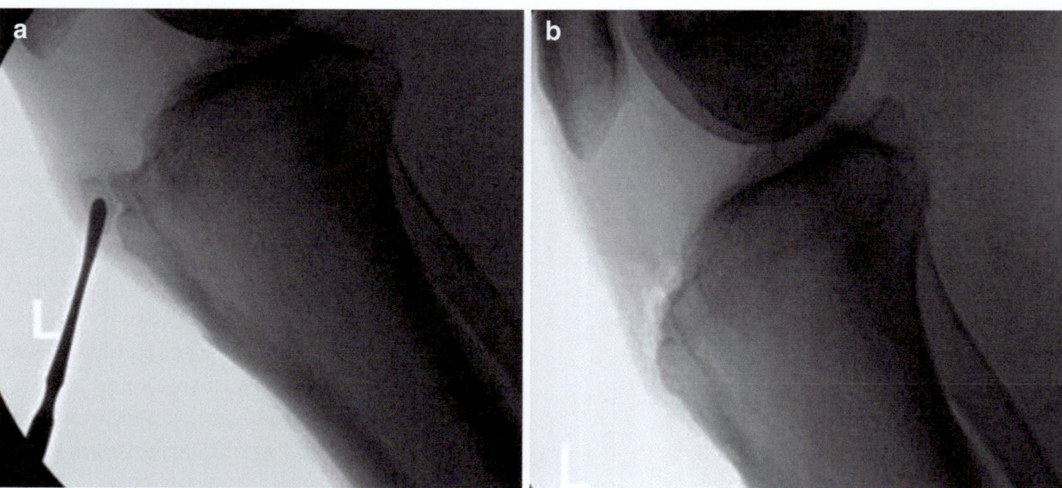

Fig. 27.2 Tibial tubercle ossicle (**a**) before and (**b**) after open surgical excision performed through a distal patella tendon split

ties such as running, jumping, and climbing stairs, and can be exacerbated with knee extension against resistance. Bilateral symptoms can be present in 20–30% of patients although symptom intensity is not often symmetric [4]. On examination, patients exhibit tenderness over the tibial tuberosity, which can have increased prominence.

Radiographs of the affected knee may demonstrate distal tendon thickening, a superficial ossicle at the patellar tendon insertion, and soft tissue swelling anterior to the tibial tuberosity. In chronic cases, tibial tuberosity overgrowth or fragmentation of the tibial tuberosity apophysis can be observed. While advanced imaging is not necessary for diagnosis, ultrasound and MRI can assist with staging and prognosis of clinical course. MRI may also assist in excluding additional diagnoses in the differential such as patellar tendinitis, tumor, and infection.

27.4 Nonsurgical Treatment

OSD typically resolves with skeletal maturity as the tibial tuberosity apophysis closes. Symptomatic management includes the use of non-steroidal anti-inflammatory medication, cryotherapy, infrapatellar padding, and a brief period of immobilization in a brace. Activity modification with rest or participation in non-impact sports, such as cycling and swimming, is recommended. A period of physical therapy with emphasis in hamstring and quadriceps stretching and strengthening can help accelerate return to activities. Some authors advocate the use of analgesic injections, such as dextrose and lidocaine, in those with recalcitrant symptoms [5]. The use of steroid injections has mostly been abandoned given concerns for ensuing patellar tendon atrophy and rupture [6].

27.5 Surgical Treatment

Although the natural history of OSD is that of gradual symptom resolution, up to 10% of patients may continue to experience symptoms into adulthood [7–10]. A comparative study assessing disability levels in college-aged athletes with a history of OSD compared to a healthy cohort demonstrated greater disability levels with sports and activities of daily life in the affected group [11]. Additionally, some studies have reported that in patients reporting no activity limitations, 40% can still experience some discomfort with kneeling [12]. Surgical treatment is indicated in patients with debilitating symptoms not responding to conservative management and in those with persistent symptoms after achieving skeletal maturity. Surgical options include open, arthroscopic, and bursoscopic excision of loose bony fragments and debridement of the tibial tuberosity. Excision of the loose bony fragments and resection of the tibial tuberosity prominence was first described through an open approach [10, 13–16]. The procedure is performed through a longitudinal infrapatellar midline incision to expose the tibial tuberosity. The patellar tendon is identified, and the medial and lateral expansions are released. The tendon is then elevated with care to preserve its distal insertion. Any osseocartilaginous material in the undersurface of the tendon is thoroughly debrided. Occasionally, a longitudinal split in the distal portion of the tendon may be necessary in order to excise intratendinous ossicles. Resection of the tibial tuberosity prominence is accomplished with osteotomes or a saw.

In recent years, minimally invasive arthroscopic debridement techniques have been described [17, 18]. This approach avoids an incision directly over the patellar tendon which can become symptomatic and also allows the surgeon to address possible concomitant intra-articular pathology. Debridement is performed through standard inferomedial and inferolateral parapatellar portals. These portals can be raised slightly if needed for improved visualization of the anterior interval. A motorized shaver and radiofrequency (RF) ablation device are used to perform and anterior interval release with care to preserve the anterior meniscal horns and intermeniscal ligament. With the knee in extension to facilitate visualization, the surgeon works down the anterior tibial slope to debride all bony lesions.

Retrievers are used to remove loose bony fragments, and an arthroscopic burr can be utilized to shell out larger bony fragments.

As an alternative to arthroscopic debridement, excision of bony lesions can be accomplished through a soft tissue tunnel into the bursal space [19]. This can be accomplished with low anterolateral (L-AL) and anteromedial (L-AM) portals created separately after performing standard diagnostic arthroscopy. This technique has the proposed advantage of preserving the infrapatellar fat pad which can be violated during standard arthroscopic debridement. To create the L-AL portal, the soft spot between the lateral border of the patellar tendon and Gerdy's tuberosity is incised and a straight mosquito clamp is used to create a soft tissue tunnel into the bursal space. An arthroscopic sheath with a blunt trocar is introduced and then the trocar exchanged for the arthroscope. The L-AM portal is created either with trans-illumination technique or under direct visualization using a spinal needle. With the knee in extension, a working space is created with resection of the bursal tissue using a motorized shaver. Using a combination of the arthroscopic shaver, RF ablator, burr, curettes, and retrievers, excision of bony fragments and resection of bony lesions are accomplished.

27.6 Post-surgical Rehabilitation

Following open or arthroscopic/bursoscopic surgery, patients are able to resume full weight bearing with assistance of mobility aids as needed. The patient may resume immediate unrestricted passive motion, except in cases in which there is concern for weakening of the patellar tendon insertion on the tibial tuberosity. Post-operative immobilization is not necessary, but a hinged knee brace locked in full extension may be considered for ambulation in cases of quadriceps weakness or to allow the patellar tendon to heal. Quadriceps strengthening exercises can be started immediately and advanced as tolerated. A formal course of physical therapy is not required and is recommended on a case to case basis in those not able to progress with the program. Return to unrestricted activity is expected at 6 weeks for most patients and 12 weeks in those requiring tendon to bone healing.

27.7 Outcomes

Open surgical management of OSD has yielded successful results, with >80% of patients reporting good to excellent results in various short to mid-term studies [10, 13–16, 20, 21]. A long-term study by Pihlamajaki et al. in 178 consecutive military recruits found that 87% of patients reported no restrictions with activities of daily life and 75% were able to return to pre-operative athletic activities at a median duration of follow-up of 10 years [22]. In this study, patients with persistent symptoms longer than 4 years affecting participation in military training or service, and who had radiographic evidence of OSD were treated with open surgical debridement. Despite achieving good functional results after surgery, 62% of patients in the cohort reported at least some discomfort with kneeling, 32% reported at least some discomfort with squatting, and 16% reported moderate to severe pain with activities. The rates of post-operative tenderness with kneeling were superior in this cohort to those reported with conservative treatment of OSD in prior studies [12]. Additionally, one third of patients were found to have persistent or new radiographic ossicles at final follow-up. Yet, the presence of these ossicles was not correlated with post-operative symptoms. Re-operation rate was low, with only 1% of patient requiring additional procedures for persistent symptoms.

Persistent pain with kneeling has been associated with the surgical scar resulting from an open approach. Alternative incisions, including a vertical anterolateral and a transverse incision 1 cm proximal to the tibial tuberosity, have been proposed in order to avoid this issue [20, 22]. Yet, although these alternative open approaches have been reported to be safe and successful, it remains unclear whether they are effective in avoiding peri-incisional scarring and tenderness.

Arthroscopic debridement techniques have been proposed to address some of the described limitations of the open approach and also to offer a better cosmetic appearance [17, 18, 23]. A study by Circi and Beyzadeoglu reported satisfactory functional outcomes and pain relief in 11 skeletally mature athletes at a mean of 66 months post-operatively after arthroscopic debridement for unresolved OSD. All patients were able to return to the same pre-operative level of athletic competition. The study reported no post-operative complications or re-operations. The arthroscopic technique has the additional advantage of allowing the surgeon to address concomitant intra-articular pathology, such as an intra-articular ossicle associated with OSD [24]. Despite these promising results, larger long-term studies and comparative outcome studies are needed in order to assess the proposed advantages of the arthroscopic technique over open surgical debridement.

More recently, the direct bursoscopic technique has been described as an alternative to standard arthroscopy in order to address the technical challenges of performing a thorough debridement while avoiding injury to critical intra-articular structures. During the anterior interval release necessary with standard arthroscopic portals, the anterior horn of the menisci and the intermeniscal ligament can be damaged. Additionally, resection of the infrapatellar fat pad could lead to scarring and post-operative discomfort. A study by Eun et al. reported satisfactory outcomes in a cohort of 18 military recruits treated with bursoscopic ossicle excision at a mean follow-up of 45 months [19]. However, 21% of the patients were unable to return to their duty and 21% had persistent difficulties with kneeling after surgery. One third of the patients reported persistent tibial tuberosity prominence. There were no re-operations reported. A proposed limitation of the bursoscopic approach is its limited working space and visualization which could result in incomplete debulking of the tibial tuberosity prominence as evidenced in this study. Although this technique was found to be safe and effective, there is paucity of data on its long-term outcomes and proposed superiority over other techniques.

27.8 Complications

The overall complication rate after surgical debridement has been reported to be as high as 5% [22]. Reported complications include infection, hematoma, and deep venous thrombosis (DVT). Some authors have also reported persistent post-operative scar tenderness after 3 months in 10% of patients treated with open debridement using a midline incision [16]. In the skeletally immature, care should be taken to avoid injury to the tibial tuberosity apophysis which could result in early closure. Tibial tuberosity internal fixation, bone grafting, or drilling of the apophysis is not recommended in skeletally immature patients given concerns for premature fusion of the anterior proximal tibial physis, which could result in genu recurvatum [25, 26]. Moreover, comparative studies have found these techniques to produce inferior results compared to standard open or arthroscopic debridement [13, 16].

27.9 Conclusion

OSD is a common cause of anterior knee pain in the growing adolescent athlete. Most patients responding well to conservative management and have resolution of symptoms after physeal closure. In patients in whom conservative management fails or symptoms persist or recur in adulthood, surgical resection can be considered. Arthroscopic and bursoscopic debridement techniques have been proposed as safe and effective alternatives to open debridement in order to minimize the risk of post-operative scarring and peri-incisional tenderness with kneeling. Future studies are needed to further refine the indications and techniques for surgical management and to identify factors prognostic of successful outcomes.

References

1. Lucena GLD, Gomes CDS, Guerra RO. Prevalence and associated factors of Osgood-Schlatter syndrome in a population-based sample of Brazilian adolescents. Am J Sports Med. 2010;39(2):415–20. https://doi.org/10.1177/0363546510383835.
2. Gholve PA, Scher DM, Khakharia S, Widmann RF, Green DW. Osgood Schlatter syndrome. Curr Opin Pediatr. 2007;19(1):44–50. https://doi.org/10.1097/mop.0b013e328013dbea.
3. Demirag B, Ozturk C, Yazici Z, Sarisozen B. The pathophysiology of Osgood-Schlatter disease: a magnetic resonance investigation. J Pediatr Orthop B. 2004;13:379–82.
4. Indiran V, Jagannathan D. Osgood–Schlatter disease. N Engl J Med. 2018;378(11) https://doi.org/10.1056/nejmicm1711831.
5. Topol GA, Podesta LA, Reeves KD, Raya MF, Fullerton BD, Yeh HW. Hyperosmolar dextrose injection for recalcitrant Osgood–Schlatter disease. Pediatrics. 2011;128(5):e1121–8.
6. Rostron PK, Calver RF. Subcutaneous atrophy following methylprednisolone injection in Osgood–Schlatter epiphysitis. J Bone Joint Surg Am. 1979;61(4):627–8.
7. Woolfrey BF, Chandler EF. Manifestations of Osgood- Schlatter's disease in late teen age and early adulthood. J Bone Joint Surg Am. 1960;42-A:327–32.
8. Høgh J, L.B.T.s.o.O.-S.s.d.i.a.I.O.
9. Kujala UM, Kvist M, Heinonen O. Osgood–Schlatter's disease in adolescent athletes. Retrospective study of incidence and duration. Am J Sports Med. 1985;13(4):236–41.
10. Mital MA, Matza RA, Cohen J. The so-called unresolved Osgood–Schlatter lesion: a concept based on fifteen surgically treated lesions. J Bone Joint Surg Am. 1980;62(5):732–9.
11. Ross MD, V.D.D.l.o.c.-a.m.w.a.h.o.O.-S.d.J.S.C.R.-.
12. Krause BL, Williams JP, Catterall A. Natural history of Osgood-Schlatter disease. J Pediatr Orthop. 1990;10:65–8.
13. Binazzi R, Felli L, Vaccari V, Borelli P. Surgical treatment of unresolved Osgood–Schlatter lesion. Clin Orthop Relat Res. 1993;289:202–4.
14. Cser I, Lenart G. Surgical management of complaints due to independent bone fragments in Osgood–Schlatter disease (apophysitis of the tuberosity of the tibia). Acta Chir Hung. 1986;27(3):169–75.
15. Weiss JM, Jordan SS, Andersen JS, Lee BM, Kocher M. Surgical treatment of unresolved Osgood–Schlatter disease: ossicle resection with tibial tubercleplasty. J Pediatr Orthop. 2007;27(7):844–7.
16. Flowers MJ, Bhadreshwar DR. Tibial tuberosity excision for symptomatic Osgood–Schlatter disease. J Pediatr Orthop. 1995;15(3):292–7.
17. Beyzadeoglu T, Inan M, Bekler H, Altintas F. Arthroscopic excision of an ununited ossicle due to Osgood–Schlatter disease. Arthrosc J Arthrosc Relat Surg. 2008;24(9):1081–3.
18. DeBerardino TM, Branstetter J, Owens BD. Arthroscopic treatment of unresolved Osgood–Schlatter lesions. Arthrosc J Arthrosc Relat Surg. 2007;23(10):1127.e1121–3.
19. Eun SS, Lee SA, Kumar R, Sul EJ, Lee SH, Ahn JH, Chang MJ. Direct bursoscopic ossicle resection in young and active patients with unresolved Osgood-Schlatter disease. Arthrosc J Arthrosc Relat Surg. 2015;31(3):416–21.
20. El-Husseini TF, Abdelgawad AA. Results of surgical treatment of unresolved Osgood–Schlatter disease in adults. J Knee Surg. 2010;23(2):103–7.
21. Orava S, Malinen L, Karpakka J, Kvist M, Leppilahti J, Rantanen J, Kujala UM. Results of surgical treatment of unresolved Osgood-Schlatter lesion. Ann Chir Gynaecol. 2000;89:298–302.
22. Pihlajamaki HK, Visuri TI. Long-term outcome after surgical treatment of unresolved Osgood–Schlatter disease in young men. J Bone Joint Surg Am. 2009;91(10):2350–8.
23. Circi E, Beyzadeoglu T. Results of arthroscopic treatment in unresolved Osgood-Schlatter disease in athletes. Int Orthop. 2016;41(2):351–6. https://doi.org/10.1007/s00264-016-3374-1.
24. Choi W, Jung K. Intra-articular large ossicle associated to Osgood-Schlatter disease. Cureus. 2018;10(7):e3008. https://doi.org/10.7759/cureus.3008.
25. Jeffreys TE. Genu recurvatum after Osgood–Schlatter's disease; report of a case. J Bone Joint Surg Br. 1965;47:298–9.
26. Lynch MC, Walsh HP. Tibia recurvatum as a complication of Osgood–Schlatter's disease: a report of two cases. J Pediatr Orthop. 1991;11(4):543–4.

Part VI

Advances and the Future Treatment of Patellofemoral Disorders

Advances in Patellofemoral Disorders

28

Justin T. Smith, Betina B. Hinckel, Miho J. Tanaka, Elizabeth A. Arendt, Renato Andrade, and João Espregueira-Mendes

Numerous advances continue to be made and have contributed to the improvement of the diagnosis and treatment of patellofemoral (PF) disorders. These include technological advances in imaging and computational modeling, devices for implantation, and developments in the field of orthobiologics. Developments in dynamic/three-dimensional computed tomography (CT) imaging as well as utilization of the Porto Patellofemoral Testing Device (PPTD) have the potential to serve as enhanced tools to improve diagnosis of PF disease. Similarly, investigation into finite element analysis of the PF joint has encouraged a greater appreciation of the implications that anatomical variants can have on PF mechanics. Finally, advances in internal bracing have emerged as a potentially useful augment in the treatment of patellar instability.

28.1 Instrumented Laxity Evaluation

Physical examination plays a critical role in the accurate diagnosis of PF disorders, however it is limited by its qualitative nature and variability among examiners [1–3]. Standard imaging modalities currently fail to incorporate a dynamic assessment of the injury [4]. Although there have been previous attempts at instrumented quantita-

J. T. Smith
Steadman Hawkins Clinic of the Carolinas, Greenville, SC, USA

B. B. Hinckel (✉)
Department of Orthopaedic Surgery, William Beaumont Hospital, Royal Oak, MI, USA

M. J. Tanaka
Department of Orthopaedic Surgery, Massachusetts General Hospital, Harvard Medical School, Boston, MA, USA

E. A. Arendt
Department of Orthopaedic Surgery, University of Minnesota, Minneapolis, MN, USA

R. Andrade
Clínica do Dragão, Espregueira-Mendes Sports Centre—FIFA Medical Centre of Excellence, Porto, Portugal

Dom Henrique Research Centre, Porto, Portugal

Porto Biomechanics Laboratory (LABIOMEP), Faculty of Sports, University of Porto, Porto, Portugal

J. Espregueira-Mendes
Clínica do Dragão, Espregueira-Mendes Sports Centre—FIFA Medical Centre of Excellence, Porto, Portugal

Dom Henrique Research Centre, Porto, Portugal

School of Medicine, Minho University, Braga, Portugal

ICVS/3B's–PT, Government Associate Laboratory, Braga, Portugal

3B's Research Group—Biomaterials, Biodegradables and Biomimetics, University of Minho, Headquarters of the European Institute of Excellence on Tissue Engineering and Regenerative Medicine, Barco, Portugal

© ISAKOS 2022
J. L. Koh et al. (eds.), *The Patellofemoral Joint*, https://doi.org/10.1007/978-3-030-81545-5_28

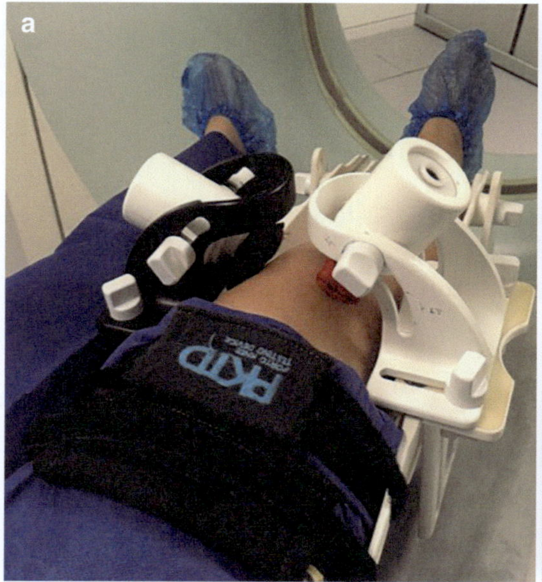

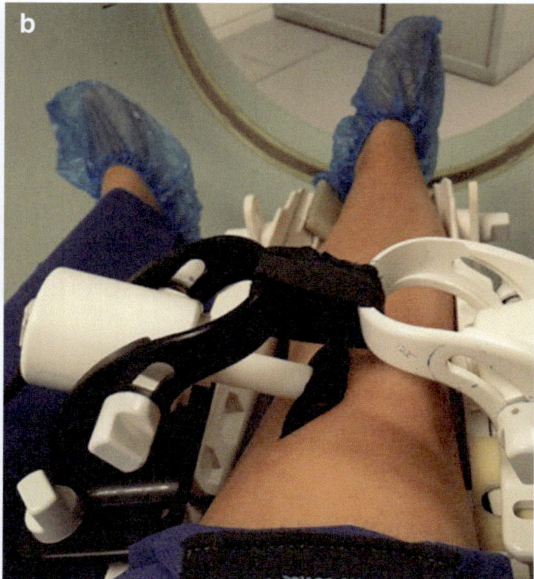

Fig. 28.1 Porto Patella Testing Device (PPTD) setup for stress-testing within imaging equipment (**a**) initial setup without any stress to obtain the position of the patella at rest, and (**b**) 30° of lateral translation stress with medial actuator

tive assessments [5–12], their outcomes and measurement methods demonstrated a significant amount of heterogeneity [13].

To address this need, the Porto Patellofemoral Testing Device (PPTD) has emerged to provide a standardized tool to quantify patellar position and displacement [14] (Figs. 28.1 and 28.2). This device has the advantage to combine stress-testing system simultaneously during an MRI or CT scan. Leal et al. have demonstrated that the PPTD offers excellent reliability, accuracy, precision, and low variability as compared to manual physical exam [14, 15]. In addition, it provides a better understanding of the pathophysiology of the various PF disorders.

Patients with idiopathic unilateral acute knee pain (AKP) with morphologically equivalent knees demonstrate increased patellar lateral displacement after stressed lateral force with the PPTD in their painful knee [15]. Patients with objective patellar instability (OPI; patients that had a patellar dislocation event with or without the presence of anatomical risk factors) and potential patellar instability (PPI; patients with risk factors but that did not have a patellar dislocation event) display the same curve pattern (steep increase close to the final displacement), but with patients with PPI showing higher stiffness than patients with OPI, as would be expected if their medial soft tissue stabilizers function better than the OPI group where the medial restraints have presumably been injured. For maximum lateral displacement, values for patients with PPI are closer to the values for patients with patellofemoral pain (PFP; patients with PF pain without anatomical risk factors) because both presumably have intact medial soft tissue stabilizers and therefore can tolerate greater force application than the patients with OPI. These results suggest that the force–displacement curve pattern is directed by the anatomy and the presence of risk factors while the amount of displacement is related to the integrity of the medial patellar restraints [16].

More research utilizing this device is needed, but data incorporating objective PF laxity and stiffness may be used to better define surgical indications in the setting of instability and to evaluate the surgical outcomes of patellar stabilization techniques. This device also offers new insights into the origin of unilateral anterior knee pain, admittedly still in an investigative mode.

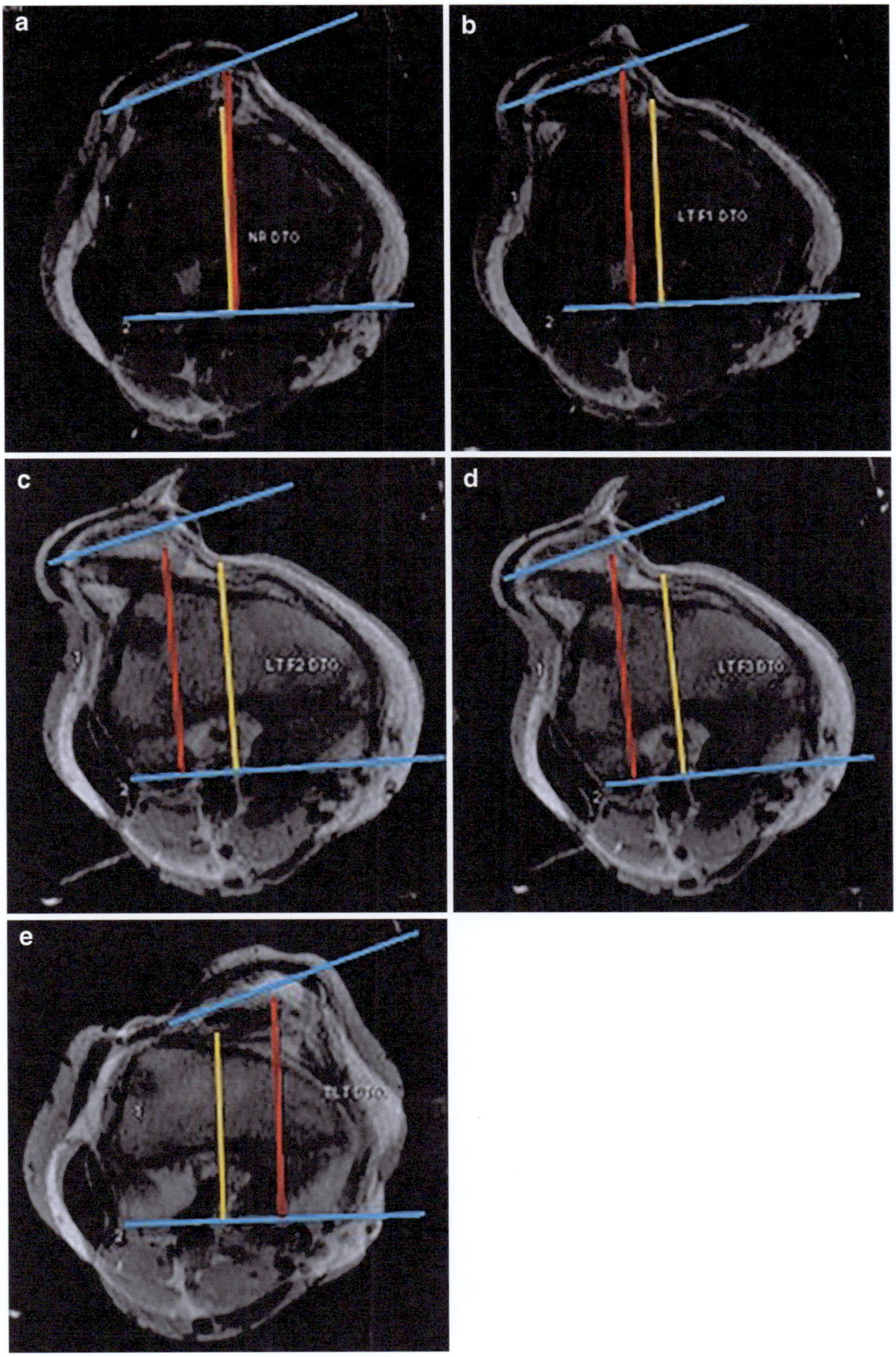

Fig. 28.2 Porto Patellofemoral Testing Device (PPTD) sequential stress testing of the right knee with a medial patellofemoral ligament tear. (**a**) Rest position (−1 mm, 18°); (**b**) Lateral transition of 0.2 bar, the patella moved 10 mm and −1° (9 mm, 17°); (**c**) Lateral transition of 0.4 bar, the patella moved 15 mm and −2° (14 mm, 16°); (**d**) Lateral transition of 0.6 bar, the patella moved 15 mm and −1° (14 mm, 17°); (**e**) Lateral tilt up to pain threshold, the patella moved 18 mm medially and increased 1° of tilt (−19 mm, 19°). From A (rest) to B (0.2 bar) to C (0.4 bar) there is low stiffness and high stiffness from C to D (0.6 bar)

28.2 Dynamic Computed Tomography (CT)

The use of dynamic CT has begun to improve our ability to quantify the contribution of each pathoanatomic variant on patellar tracking throughout knee range of motion (ROM) and provides a better understanding of the biomechanical effects that corrective surgical techniques have on patellar tracking. The images can be directly evaluated or a 3D computational model can be reconstructed based on the images acquired (Fig. 28.3).

Tanaka et al. found that knee flexion angle during imaging is a critical factor when measuring tibial tuberosity-trochlear groove distance (TTTG) to evaluate patellofemoral instability. The mean TTTG distance, which is often utilized to indicate osteotomy, varied by a mean 5.7 mm between 5 and 30° of flexion in each knee with symptomatic instability although this relationship was not completely linear. Measurements of patellar lateralization and tilt mirrored this pattern, suggesting that TTTG distance influences patellar tracking throughout knee range of motion [17].

In regard to abnormal tracking with patellar lateralization, higher grades of J-sign (>2 quadrants, or when the entire patella is lateral to the trochlear groove) have been found to be predictive of symptomatic patellar instability while

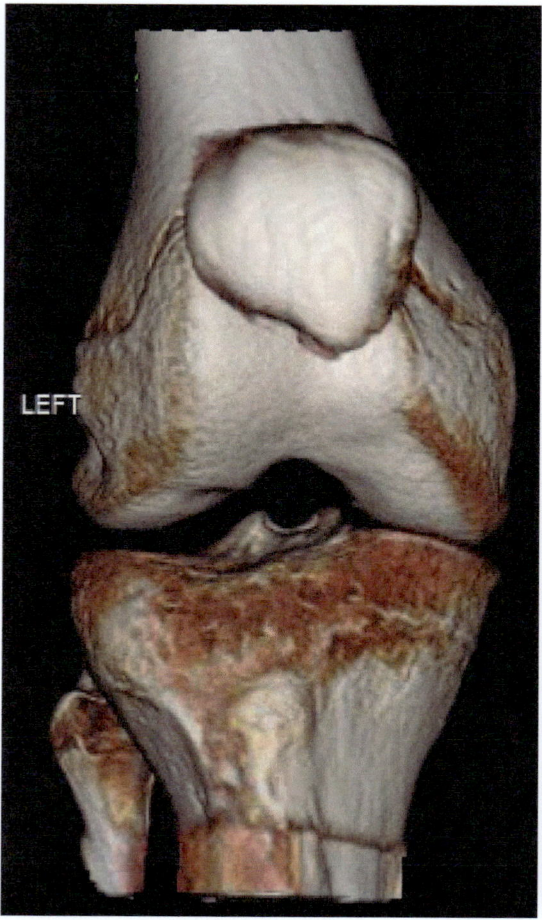

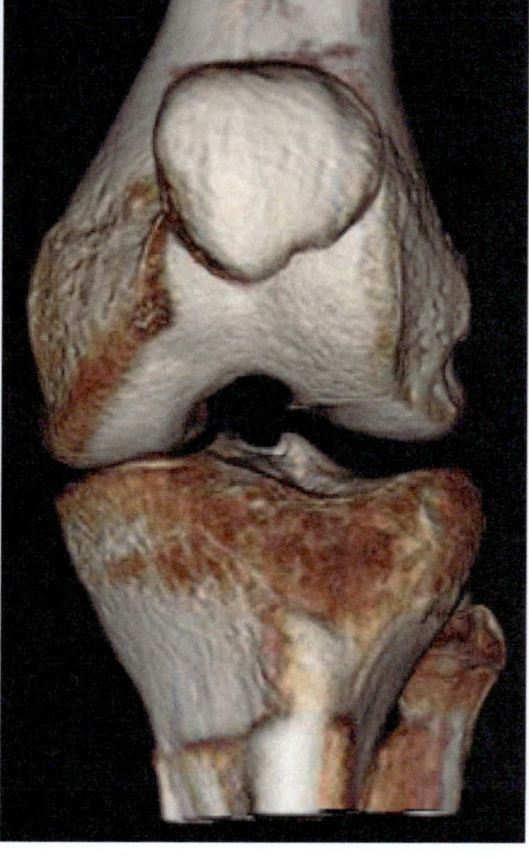

Fig. 28.3 3D reconstruction of dynamic CT imaging in bilateral knees demonstrates one image obtained during a sequence of knee flexion and extension. Visualization throughout range of motion allows for qualitative assessment of patellar tracking, while measurements performed in corresponding 2D axial or sagittal cuts allow for quantitative measurements of patellar position and morphology

milder lateralization (<2 quadrants) are not [18]. That suggests that the degree of patellar lateralization during ROM is related to symptoms, which may have utility during physical examination assessment such as the J-sign. Dynamic CT can improve patellar tracking assessments since intra- and inter-observer reliability of the visual assessment of the J-sign is inadequate ($k < 0.60$) and the agreement between visual and dynamic CT is between 53 and 68% [19]. The causes of a J-sign can also be better understood with the use of dynamic CT. At low flexion angles, both trochlear dysplasia (represented by the lateral trochlear inclination (LTI)) and lateral quadriceps vector (represented by the tibial tuberosity to posterior cruciate ligament (TT–PCL) distance) are correlated with the bisect offset index, a sign of patellar lateralization. However, only lateral trochlear inclination has been shown to be correlated with lateral tilt, another sign of maltracking. At high flexion angles, bisect offset index and lateral tilt are correlated with only lateral TT–PCL distance [20]. Such findings help us to better understanding and evaluate the validity of the current definition of abnormal tracking; dynamic CT studies of larger populations of asymptomatic patients may better distinguish abnormal tracking from normal tracking evaluation with dynamic CT imaging modalities.

Dynamic CT can also be applied postoperatively to assess and evaluate the alterations in anatomic parameters of various PF instability correction techniques (isolated MPFL reconstruction [21] vs MPFL reconstruction with tibial tuberosity osteotomy [22]) to determine if the underlying anatomic abnormalities had been correctly addressed. Gobbi et al. demonstrated a lack of correction of patellar tracking parameters in patients that underwent indicated isolated MPFL reconstruction [21]. Though an interesting finding, clinically none of the patients had recurrence of dislocations. On the other hand, Elias et al. reported that MPFL reconstruction with tibial tuberosity realignment reduces patellar lateral shift and tilt at low flexion angles [22], suggesting that further investigation into the roles of each procedure at different flexion angles will continue to improve our understanding of maltracking and its (potential) role in patellar instability, patellofemoral pain, and patellofemoral load/chondrosis.

28.3 Finite Element Analysis

Recent investigations applying finite element analyses have aimed to address factors that contribute to PF disorders and treatments. Utilizing finite element modeling (FEM) [23–25], researchers have been able to evaluate the kinematic behavior of PF articulation in various disease settings and simulate morphological changes using patient-specific models.

Because of the complexity of patellofemoral joint kinematics, which include the static soft tissue, static osseous, dynamic and alignment-related factors that contribute to stability, the application of FEM has allowed for a greater understanding of the individual factors, as well as the interaction between those factors and their roles in PF mechanics. Studies of articular geometry [26, 27], orientation of the patellar tendon [28], rotational alignment of the femur/tibia [27, 29], vastus medialis obliquus (VMO) functionality [30] have increased our understanding of PF reaction forces, contact mechanics, and kinematics (including patellar tracking).

Using a geometric statistical model, Fitzpatrick et al. demonstrated that the shape of the articular surface in the patellofemoral joint had the greatest influence in PF contact variations with larger PF size having increased contact and lower contact pressure. This was followed by patellar height (5 mm of patellar alta results in a 25% increase in contact pressure in midflexion) and then the contributions of trochlear morphology (more conformity confers lower peak contact pressures) [26]. Elias et al. demonstrated that an increase in PF contact pressures occurs with a lateralized patellar tendon through a computational analysis of external rotation of the tibia [28]. A similar computational analysis by Besier et al. revealed that a 15° increase in external rotation of the femur resulted in a 10% increase in PF contract pressures (shifting the pressure from the lateral patellar facet to the medial facet). In this study, patellar cartilage was shown to be more sensitive to these changes in femoral rotation with a greater increase in shear stresses in the patellar cartilage than in the femoral cartilage [29]. A subsequent computational analysis performed by Elias et al., looking PF contact force variations with VMO functionality revealed that with decreasing VMO force there is an analogous increase in lateral patellar contact forces [30]. Rezvanifar et al. evaluated the influence of trochlear dysplasia (represented by the lateral trochlear inclination), patella alta (represented by the Caton-Deschamps index; CD), and lateral tuberosity position (represented by the TT-PCL) on tracking (represented by the bisect offset Index and lateral patellar tilt) during knee squatting. Modifying the LTI, CD, and TT-PCL to represent mild to severe abnormalities, the authors demonstrated that a shallow trochlear groove increases lateral patellar maltracking. They also found that a lateralized tibial tuberosity in combination with trochlear dysplasia increases lateral patellar tracking and the risk of patellar instability. In this study, patella alta had relatively little influence on patellar tracking when in combination with trochlear dysplasia due to the limited articular constraint provided by the trochlear groove [27].

Patient-specific models can be used to perform simulated TTO [22, 31] and MPFL reconstruction [32–34] with analysis of the resultant effects on PF kinematics, contact pressures, and reaction forces. Application of this technique has also improved our understanding of the influence of tuberosity lateralization on the MPFL graft function and subsequent maltracking patterns [35, 36]. Through this method, simulated anteriorization TTO of 1.25 cm and 2.5 cm has shown to be effective in reducing patellofemoral contact forces, especially at smaller knee flexion angles. The total resultant PF contact force substantially increased with flexion but decreased as the tibial tubercle was moved anteriorly by 78% at 0° and 12% at 90° of flexion. In accordance, the maximum compressive stress substantially decreased at full extension; however, it increased at 90° of flexion. Substantial effects of tuberosity elevation on tibial kinematics, cruciate ligament forces, tibiofemoral contact forces, and extensor lever arm were found. As TTO anteriorization increased posterior translation of the tibia, the posterior cruciate ligament and tibiofemoral contact forces at larger flexion angles considerably increased, whereas the anterior cruciate ligament and tibiofemoral contact forces at near full extension angles decreased. Overall, the extent of changes depends on the magnitude of anteriorization, joint flexion angle, and loading.

Similar modeling studies have advanced our understanding of MPFL reconstruction by reinforcing anatomic placement of the femoral tunnel, as small deviations have been shown to result in increased PF contact pressures [32–34]. In a study performed by Oka et al., they sought to determine the optimal femoral insertion site based on three criteria for the MPFL reconstruction: the graft should remain isometric from 0 to 60° of knee flexion, be taut in full extension, and slacken at >60° of knee flexion. They showed that using simulated models their "optimal insertion sites" were analogous to that of the anatomic insertion site, which was just distal to the adductor tubercle [32]. Such a model to determine femoral insertion site was further reinforced by Sanchis et al. comparing parametric models of anatomic, non-anatomic/physiometric, and non-anatomic/non-physiometric MPFL reconstructions. In reconstructions that were anatomic/physiometric, the contact pressures in the PF articulation were increased from 0 to 30° but then decreased from 60 to 120° of knee flexion as the MPFL reconstruction slackened. They showed that if the insertion site was moved anteriorly (non-anatomic) it would be non-physiometric in behavior by having no tension from 0–30° but with increased tension and PF contact from 60 to 120° [33]. This is similar to previous findings based on FEM studies that showing increased graft tension/restraint with anteriorization of the femoral insertion site; however, these findings were performed at a static 30° knee flexion [34]. Anatomic reconstruction is of utmost importance as it can have a dramatic influence on the tensioning of the graft throughout ROM and the resultant PF contact pressures. The goal is to create a reconstruction that remains functional during the first 30° of knee flexion until the trochlear groove captures the patella and then subsequently slackens as the two attachment points converge towards each other.

Collectively, these FEM models hold great potential in uncovering important factors that affect our ability to diagnose and treat for PF instability, and to tailor treatments based on individual pathoanatomy. With advances in technology and the validation of these models, there will be continued insight into each PF disease process and their respective ideal treatments.

28.4 Suture Tape Augmentation of MPFL Stabilization Surgery

The primary surgical treatment of patellofemoral instability consists of an MPFL reconstruction. Most commonly, tendon autograft/allograft tissue is utilized to reconstruct the MPFL. Harvesting hamstring autograft can lead to deleterious changes in joint mechanics and gait patterns [37, 38]. Similarly, due to cost and availability of allograft, surgeons may be limited in their options. MPFL repair has been shown to have inferior results when compared to MPFL reconstruction [39]. As a result, suture tape augmentation of an MPFL repair has recently been explored to determine if it may serve as an equivalent treatment option for graft-based reconstruction [40]. Mehl et al. performed a biomechanical study comparing suture tape-augmented MPFL repair to MPFL reconstruction with allograft in ten fresh frozen cadaveric knees. They determined that suture tape-augmented repair displayed equivalent PF contact pressures and joint kinematics throughout all knee ROM at a preload of 2 N. While there are known higher failure rate with isolated MPFL repair, it is not yet known whether the suture tape augmentation may negate this risk [40]. A recent cadaveric study performed by Skamoto et al. demonstrated equivalent maximal patellofemoral contact pressures when comparing knees with suture tape MPFL reconstruction fixed at 60–90° of flexion with native knees. Fixation of the suture tape at lower degrees of flexion was found to result in abnormally increased PF maximal contact pressures [41]. At this time, no clinical studies have been conducted to investigate this novel technique. In addition, similarly to the concept of its use along with anterior cruciate ligament reconstruction [42]; suture tape augmentation may be used along with MPFL reconstruction to increase load to failure and decrease elongation of the construct in the early post-operative period. While further studies are needed to better understand the role of such a technique, this serves as preliminary evidence that MPFL repair with suture tape augmentation may be a future alternative for reconstruction techniques with the benefit of not requiring a soft tissue graft.

28.5 Conclusion

In summary, application of these advances to growing areas of inquiry studying PF disease have led to avenues of tremendous potential to improve our ability to accurately diagnose and treat patellofemoral disorders. From dynamic/3D-CT to PPTD testing, individualized diagnoses and quantitative assessments may be made as to the reason for a patient's PF symptoms. FEM analyses may then be applied to understanding these diagnoses, identifying individual alterations of pathoanatomy and potentially resultant changes with patient-specific treatments. And lastly, we are breaking into a new era of biologic treatments and implantable materials that will undoubtedly have a significant impact on future surgical techniques in the management of PF instability.

References

1. Yamada Y, Toritsuka Y, Horibe S, et al. Patellar instability can be classified into four types based on patellar movement with knee flexion: a three-dimensional computer model analysis. J ISAKOS Jt Disord Orthop Sport Med. 2018;3:328–35. https://doi.org/10.1136/jisakos-2018-000220.
2. Smith TO, Clark A, Neda S, et al. The intra- and inter-observer reliability of the physical examination methods used to assess patients with patellofemoral joint instability. Knee. 2012;19:404–10. https://doi.org/10.1016/j.knee.2011.06.002.

3. Smith TO, Davies L, Donell ST. The reliability and validity of assessing medio-lateral patellar position: a systematic review. Man Ther. 2009;14:355–62. https://doi.org/10.1016/j.math.2008.08.001.
4. Tompkins MA, Rohr SR, Agel J, Arendt EA. Anatomic patellar instability risk factors in primary lateral patellar dislocations do not predict injury patterns: an MRI-based study. Knee Surg Sports Traumatol Arthrosc. 2018;26:677–84. https://doi.org/10.1007/s00167-017-4464-3.
5. Egusa N, Mori R, Uchio Y. Measurement characteristics of a force-displacement curve for chronic patellar instability. Clin J Sport Med. 2010;20:458–63. https://doi.org/10.1097/JSM.0b013e3181fb5350.
6. Fithian DC, Mishra DK, Balen PF, et al. Instrumented measurement of patellar mobility. Am J Sports Med. 1995;23:607–15. https://doi.org/10.1177/036354659502300516.
7. Joshi RP, Heatley FW. Measurement of coronal plane patellar mobility in normal subjects. Knee Surg Sports Traumatol Arthrosc. 2000;8:40–5. https://doi.org/10.1007/s001670050009.
8. Kujala UM, Kvist M, Osterman K, et al. Factors predisposing Army conscripts to knee exertion injuries incurred in a physical training program. Clin Orthop Relat Res. 1986:203–212.
9. Ota S, Nakashima T, Morisaka A, et al. Comparison of patellar mobility in female adults with and without patellofemoral pain. J Orthop Sports Phys Ther. 2008;38:396–402. https://doi.org/10.2519/jospt.2008.2585.
10. Reider B, Marshall JL, Warren RF. Clinical characteristics of patellar disorders in young athletes. Am J Sports Med. 1981;9:270–4. https://doi.org/10.1177/036354658100900419.
11. Wong Y, Ng GYF. The relationships between the geometrical features of the patellofemoral joint and patellar mobility in able-bodied subjects. Am J Phys Med Rehabil. 2008;87:134–8. https://doi.org/10.1097/PHM.0b013e31815b62b9.
12. Teitge RA, Faerber WW, Des Madryl P, Matelic TM. Stress radiographs of the patellofemoral joint. J Bone Joint Surg Am. 1996;78:193–203. https://doi.org/10.2106/00004623-199602000-00005.
13. Leal A, Andrade R, Flores P, et al. High heterogeneity in in vivo instrumented-assisted patellofemoral joint stress testing: a systematic review. Knee Surg Sport Traumatol Arthrosc. 2019;27:745–57. https://doi.org/10.1007/s00167-018-5043-y.
14. Leal A, Andrade R, Hinckel BB, et al. A new device for patellofemoral instrumented stress-testing provides good reliability and validity. Knee Surg Sport Traumatol Arthrosc. 2020;28(2):389–97. https://doi.org/10.1007/s00167-019-05601-4.
15. Leal A, Andrade R, Flores P, et al. Unilateral anterior knee pain is associated with increased patellar lateral position after stressed lateral translation. Knee Surg Sport Traumatol Arthrosc. 2020;28(2):454–62. https://doi.org/10.1007/s00167-019-05652-7.
16. Leal A, Andrade R, Hinckel B, et al. Patients with different patellofemoral disorders display a distinct ligament stiffness pattern under instrumented stress testing. J ISAKOS Jt Disord Orthop Sport Med. 2020;5:74–9. https://doi.org/10.1136/jisakos-2019-000409.
17. Tanaka MJ, Elias JJ, Williams AA, et al. Correlation between changes in tibial tuberosity-trochlear groove distance and patellar position during active knee extension on dynamic kinematic computed tomographic imaging. Arthroscopy. 2015;31:1748–55. https://doi.org/10.1016/j.arthro.2015.03.015.
18. Tanaka MJ, Elias JJ, Williams AA, et al. Characterization of patellar maltracking using dynamic kinematic CT imaging in patients with patellar instability. Knee Surg Sports Traumatol Arthrosc. 2016;24:3634–41. https://doi.org/10.1007/s00167-016-4216-9.
19. Best MJ, Tanaka MJ, Demehri S, Cosgarea AJ. Accuracy and reliability of the visual assessment of patellar tracking. Am J Sports Med. 2020;48(2):370–5. https://doi.org/10.1177/0363546519895246.
20. Elias JJ, Soehnlen NT, Guseila LM, Cosgarea AJ. Dynamic tracking influenced by anatomy in patellar instability. Knee. 2016;23:450–5. https://doi.org/10.1016/j.knee.2016.01.021.
21. Gobbi RG, Demange MK, de Ávila LFR, et al. Patellar tracking after isolated medial patellofemoral ligament reconstruction: dynamic evaluation using computed tomography. Knee Surg Sport Traumatol Arthrosc. 2017;25:3197–205. https://doi.org/10.1007/s00167-016-4284-x.
22. Elias JJ, Carrino JA, Saranathan A, et al. Variations in kinematics and function following patellar stabilization including tibial tuberosity realignment. Knee Surg Sport Traumatol Arthrosc. 2014;22:2350–6. https://doi.org/10.1007/s00167-014-2905-9.
23. Fernandez JW, Hunter PJ. An anatomically based patient-specific finite element model of patella articulation: towards a diagnostic tool. Biomech Model Mechanobiol. 2005;4:20–38. https://doi.org/10.1007/s10237-005-0072-0.
24. Baldwin MA, Clary C, Maletsky LP, Rullkoetter PJ. Verification of predicted specimen-specific natural and implanted patellofemoral kinematics during simulated deep knee bend. J Biomech. 2009;42:2341–8. https://doi.org/10.1016/j.jbiomech.2009.06.028.
25. Hinckel BB, Demange MK, Gobbi RG, et al. The effect of mechanical varus on anterior cruciate ligament and lateral collateral ligament stress: finite element analyses. Orthopedics. 2016;39:e729–36. https://doi.org/10.3928/01477447-20160421-02.
26. Fitzpatrick CK, Baldwin MA, Laz PJ, et al. Development of a statistical shape model of the patellofemoral joint for investigating relationships between shape and function. J Biomech. 2011;44:2446–52. https://doi.org/10.1016/j.jbiomech.2011.06.025.
27. Rezvanifar SC, Flesher BL, Jones KC, Elias JJ. Lateral patellar maltracking due to trochlear dysplasia: a

computational study. Knee. 2019;26:1234–42. https://doi.org/10.1016/j.knee.2019.11.006.
28. Elias JJ, Saranathan A. Discrete element analysis for characterizing the patellofemoral pressure distribution: model evaluation. J Biomech Eng. 2013;135:1–6. https://doi.org/10.1115/1.4024287.
29. Besier TF, Gold GE, Delp SL, et al. The influence of femoral internal and external rotation on cartilage stresses within the patellofemoral joint. J Orthop Res. 2008;26:1627–35. https://doi.org/10.1002/jor.20663.
30. Elias JJ, Kilambi S, Cosgarea AJ. Computational assessment of the influence of vastus medialis obliquus function on patellofemoral pressures: model evaluation. J Biomech. 2010;43:612–7. https://doi.org/10.1016/j.jbiomech.2009.10.039.
31. Shirazi-Adl A, Mesfar W. Effect of tibial tubercle elevation on biomechanics of the entire knee joint under muscle loads. Clin Biomech. 2007;22:344–51. https://doi.org/10.1016/j.clinbiomech.2006.11.003.
32. Oka S, Matsushita T, Kubo S, et al. Simulation of the optimal femoral insertion site in medial patellofemoral ligament reconstruction. Knee Surg Sport Traumatol Arthrosc. 2014;22:2364–71. https://doi.org/10.1007/s00167-014-3192-1.
33. Sanchis-Alfonso V, Alastruey-López D, Ginovart G, et al. Parametric finite element model of medial patellofemoral ligament reconstruction model development and clinical validation. J Exp Orthop. 2019;6:32. https://doi.org/10.1186/s40634-019-0200-x.
34. DeVries Watson NA, Duchman KR, Bollier MJ, Grosland NM. A finite element analysis of medial patellofemoral ligament reconstruction. Iowa Orthop J. 2015;35:13–9.
35. Tanaka MJ, Cosgarea AJ, Forman JM, Elias JJ. Factors influencing graft function following MPFL reconstruction: a dynamic simulation study. J Knee Surg. 2021;34(11):1162–9. https://doi.org/10.1055/s-0040-1702185.
36. Elias JJ, Tanaka MJ, Jones KC, Cosgarea AJ. Tibial tuberosity anteriomedialization vs. medial patellofemoral ligament reconstruction for treatment of patellar instability related to malalignment: computational simulation. Clin Biomech (Bristol, Avon). 2020;74:111–7. https://doi.org/10.1016/j.clinbiomech.2020.01.019.
37. Hardy A, Casabianca L, Andrieu K, et al. Complications following harvesting of patellar tendon or hamstring tendon grafts for anterior cruciate ligament reconstruction: systematic review of literature. Orthop Traumatol Surg Res. 2017;103:S245–8. https://doi.org/10.1016/j.otsr.2017.09.002.
38. Webster KE, Wittwer JE, O'Brien J, Feller JA. Gait patterns after anterior cruciate ligament reconstruction are related to graft type. Am J Sports Med. 2005;33:247–54. https://doi.org/10.1177/0363546504266483.
39. Puzzitiello RN, Waterman B, Agarwalla A, et al. Primary medial patellofemoral ligament repair versus reconstruction: rates and risk factors for instability recurrence in a young, active patient population. Arthroscopy. 2019;35:2909–15. https://doi.org/10.1016/j.arthro.2019.05.007.
40. Mehl J, Otto A, Comer B, et al. Repair of the medial patellofemoral ligament with suture tape augmentation leads to similar primary contact pressures and joint kinematics like reconstruction with a tendon graft: a biomechanical comparison. Knee Surg Sport Traumatol Arthrosc. 2020;28:478–88. https://doi.org/10.1007/s00167-019-05668-z.
41. Sakamoto Y, Sasaki S, Kimura Y, et al. Patellofemoral contact pressure for medial patellofemoral ligament reconstruction using suture tape varies with the knee flexion angle: a biomechanical evaluation. Arthroscopy. 2020;36:1390–5. https://doi.org/10.1016/j.arthro.2019.12.027.
42. Smith PA, Bradley JP, Konicek J, et al. Independent suture tape internal brace reinforcement of bone-patellar tendon-bone allografts: biomechanical assessment in a full-ACL reconstruction laboratory model. J Knee Surg. 2019. https://doi.org/10.1055/s-0039-1692649.

MIX
Papier aus verantwortungsvollen Quellen
Paper from responsible sources
FSC® C105338

If you have any concerns about our products,
you can contact us on
ProductSafety@springernature.com

In case Publisher is established outside the EU,
the EU authorized representative is:
Springer Nature Customer Service Center GmbH
Europaplatz 3, 69115 Heidelberg, Germany

Printed by Libri Plureos GmbH
in Hamburg, Germany